Diagram techniques in group theory

Diagram techniques in group theory

GEOFFREY E. STEDMAN

Reader, Department of Physics, University of Canterbury
Christchurch, New Zealand

CAMBRIDGE UNIVERSITY PRESS

CAMBRIDGE
NEW YORK PORT CHESTER
MELBOURNE SYDNEY

CAMBRIDGE UNIVERSITY PRESS
Cambridge, New York, Melbourne, Madrid, Cape Town, Singapore, São Paulo, Delhi

Cambridge University Press
The Edinburgh Building, Cambridge CB2 8RU, UK

Published in the United States of America by Cambridge University Press, New York

www.cambridge.org
Information on this title: www.cambridge.org/9780521119702

First published 1990
This digitally printed version 2009

A catalogue record for this publication is available from the British Library

Library of Congress Cataloguing in Publication data

Stedman, Geoffrey E. (Geoffrey Ernest), 1943–
Diagram techniques in group theory.

Includes index.
1. Groups, Theory of. 2. Field theory (Physics)
3. Graph theory. I. Title.
QC20.7.G76S84 1989 530.1′5222 88-25830

ISBN 978-0-521-32787-9 hardback
ISBN 978-0-521-11970-2 paperback

Contents

To Rachel and Tim

Preface

Physicists have a love–hate relationship with diagrams. Feynman diagrams, and more generally field theoretic diagrams in all areas of physics, are universally favoured for their handwaving insight into the interaction picture of quantum mechanics as much as for their calculational advantages. Equally, everyone but the inventor hates learning idiosyncratic hieroglyphics for a special task. The developer of diagram techniques risks being labelled as an unfortunate addict.

I hope to persuade the reader that, when diagrams are used in such areas as group theory, the method is essentially unique, and as ubiquitous as the more well-accepted field theoretic diagram. The reputation of idiosyncrasy for the technique (whatever may be said of the practitioners) is a gross libel. More than that, the two methods are compatible, and may be combined into a technique with a power superior to either field theory or group theory on its own. We show this explicitly by developing both the group diagram technique and the field theoretic technique from scratch, and by applying them jointly in a variety of condensed matter problems.

This compatibility is founded on the Wigner–Eckart theorem of compact group theory: physical matrix elements are proportional to group coupling coefficients and reduced matrix elements. The coupling coefficients enshrine the geometrical (or group) properties of the matrix element, and the reduced matrix element its physical (or dynamical) magnitude. Here we use 'geometrical' to denote the effect of group transformations in abstract group space, e.g. in mixing degenerate eigenstates of a group-symmetric Hamiltonian, and 'physical' or 'dynamical' to denote the effect of the choice of the general type of interaction and the parentage of the wavefunction in such a matrix element. In this way we may separate the geometrical from the physical parts of the calculation. Group theory permits the calculation of the required property once and for all, for each orientation of the system with respect to the interaction probes, given the results for a sufficient number and variety of (primitive or irreducible) representative cases. In particular, one knows in advance whether or not the orientational average permitted by any symmetry will vanish.

If there were no other reason for considering diagram techniques in group theory, consistency and coherence alone would determine their value in any field theoretic context, or any area of theoretical physics where perturbation theory and symmetry were relevant. In fact there are many other such reasons. The bookkeeping (database) advantages and the mnemonic (icon) value of diagram techniques are readily imagined, and indeed are very powerful advantages. On the other hand, the most basic of all, yet often misunderstood, point deserves emphasis: diagram techniques are perfectly adequate as a stand-alone calculational method in which all phases and factors, customarily the prerogative of the algebraic approach, may be included and manipulated graphically as desired.

In every application, the name of the game is to invent elegant and unambiguous rules for the interconversion between algebraic formulae and diagram. We need elegance for the method to be useful; the structural clarity of a diagram can then be a decisive advantage. In Feynman's hands, a diagram technique associated with the interaction picture of quantum mechanics and incorporating manifest Lorentz covariance not only bypassed the need for a prodigious facility for abstract thinking, but was wedded to a new calculational method which continues to outclass its rivals. Nor can we tolerate the ambiguity of a construction drawing.

Group theory is one obvious case of a theoretical formalism where the use of diagrams introduces the above advantages, and we focus on that for the sake of homogeneity. While the earlier chapters of the book will indicate some ground rules for more extended applications as well, we certainly do not cover a gamut of diagram applications in physics. Many topics, such as Young tableaux, Dynkin diagrams, simplex theory and lattice graphs, as used for example in the statistical mechanics of phase transitions, have little resemblance to the style of diagrams discussed under our general rules. Our discussion is oriented rather towards those techniques associated with a 'network' approach: nodes connected by lines, as in circuit theory or as in the mathematician's graph, where the length or orientation of the line is not significant. In style, as well as content, I am writing more for the electronics buff than for the topologist. Young tableaux are certainly very relevant to the work on symmetric and unitary groups and are used in §4, but without detailed discussion.

Diagram techniques are not commonly used in so general a manner. One deterrent has been the fearsome complexity of the proofs and even of the applications of the Feynman or field theoretic diagram techniques. In §7 we give a gentle introduction, omitting no details, to field theoretic diagrams in one area of physics that is poorly served by the monograph and review literature, and that is particularly relevant to applications of group theory. Another deterrent is that diagrams tend to have most value in problems of a particularly tedious type. Finally, the true generality of the method has been obscured by the somewhat individualistic notations of individual workers in the field. An important exception to this statement is the use of diagrams in coupling calculations for the rotation group. This technique is also now well accepted. An elementary review of this is included in §2. Exactly the same problems of notation arise in principle in the algebraic formalism; it is only the inertia arising from customary usage that makes the choice of convention less conspicuous in the algebraic application.

Learning to 'think diagrams' can be the major hurdle for the novice. For that reason I spend §1 on elementary applications in vector analysis. These illustrate all the basic manoeuvres. Mastering this undergraduate level section – which should be easy – will place the inexperienced reader in a good position for tackling all manner of geometric manipulations. Similarly, §2 is an obvious entry point for those with some familiarity with the diagram techniques for the theory of angular momentum. A more expert reader will merely wish to skim such sections. Indeed this whole work is pitched not at the 'front bench', who know it all or can work it out independently in any case, but at the relative novice. For this reason also I do not attempt to go into the more fashionable (and certainly important) areas such as exceptional groups, superstrings etc., though we get as far as displaying the group theoretic content of the triangle diagram in the chiral anomaly. Since the relativistic field theory end of the spectrum is in any case well served by the various publications of Cvitanovic (1976, 1984), our applications are biassed towards condensed matter theory. We concentrate on clarifying the foundations, hopefully to the point at which their use becomes second nature and at which their application in growth areas become obvious.

The material is presented in interactive mode. Many simple – and some not so simple – exercises are worked through. They are called problems, because on the whole they should be regarded as compulsory exercises for any reader, especially near the start of a chapter. Looking at diagrams is no substitute for doodling with them.

I am grateful particularly to Phil Butler, Mike Reid, Brian Wybourne, Brian Judd, Alistair McLellan, Doug Newman and Bill Moreau for their comments over the years, and especially to those students who gallantly trained in these techniques, particularly Stephen Payne, Richard Black, Clare Churcher and Bruce McKenzie. My special thanks go to Mrs Koos van der Borch for draughting the many diagrams.

1

Elementary examples

The origins of graph theory are humble, even frivolous.

Biggs, Lloyd and Wilson (1976)

A moment occurred – I remember distinctly – when I looked at them . . . and I thought consciously: Wouldn't it be funny if this turns out to be useful, and the *Physical Review* would be all full of these funny-looking pictures?

R.P. Feynman, quoted in Schweber (1986)

To the student who wishes to use graphical methods as a tool, it cannot be emphasised too strongly that practice in the use of that tool is as essential as a knowledge of how to use it. The oft-repeated pedagogical phrase 'we learn by doing' is applicable here.

Running (1927)

1.1 Introduction

This chapter is introductory, and aims to help the reader to become bilingual, that is to translate the process of mathematical deduction from algebraic language to diagram language. The sample text used – vector analysis – affords a particularly simple and convenient illustration for the translation process; we have tried to strike a reasonable balance between precision and readability.

The material is presented in interactive mode. Looking at diagrams is no substitute for doodling with them. Especially in the early sections, our problems may seem trivial; nevertheless they have been graded to introduce and illustrate an extraordinary number of principles whose generality will emerge only much later in the book, and which are itemised by way of comments.

We shall not attempt to relate this material to formal developments of graph theory, but a few concepts will have exact or close counterparts (Harary 1969).

1.2 Basic rules

Rule 1

A node corresponds to a mathematical expression (a matrix element, a tensor, or a function for example).

$$T \leftrightarrow (T).$$

Comments

(a) Nodes for different expressions may be distinguished by virtue of different algebraic labels, by different geometric design, or by different types of lines connected to the node (see Comment (c)(i) of Rule 2).

(b) The complex conjugate of the mathematical expression will be indicated explicitly by adding an asterisk to the diagram near the node.

(c) Alternatively, a node may be called a vertex.

Rule 2

A terminal at a node corresponds to one or more indices (or arguments).

Comments

(a) The value of each index may be assigned by adding the algebraic label(s) to the terminal. For example:

$$T_\alpha \leftrightarrow (T)\!-\!_\alpha, \quad g_{\mu\nu} \leftrightarrow {}_\mu\!-\!(g)\!-\!_\nu.$$

(b) The combination of the (graphical) terminals and (explicit algebraic) indices so introduced should suffice to define the expression uniquely for the intended purpose. Any index whose graphical representation is irrelevant may be suppressed by subsuming these indices into the definition of a node. This might be done either by using a separate geometrical shape for each index, or by attaching an algebraic index explicitly to each node:

$$\gamma^\mu_{ij} \leftrightarrow {}_i\!-\!(\gamma^\mu)\!-\!_j = {}_i\!-\!(\gamma)\!-\!_j \text{ (with vertical line } \mu\text{)}.$$

(c) For several reasons it is often useful to attach sizeable *lines* to these terminals.

(i) One reason, and the justification for raising the matter here, is that different types of index (cartesian, spacetime, group theoretic), are conveniently distinguished by using different types of lines – solid, broken, wavy, etc.

(ii) Two more important reasons are given in Rule 6; the line allows one to indicate graphically the equation and summation of two suffices on different nodes, and the terminal, which then denotes the line–node junction, is the proper home of an index parity.

(iii) An alternative name for a line is a leg.

(iv) *Internal lines* connect any two nodes within the complete figure, and include those lines connecting two terminals on the same node. *External lines* correspond to indices characterising the expression as a whole; or (diagrammatically speaking) to loose ends.

(d) One terminal (and its attached line) may conveniently act for two indices if one index represents a set of which the other is a member. For example, a spherical harmonic is labelled by the angular momentum l and projection m of a spherical tensor:

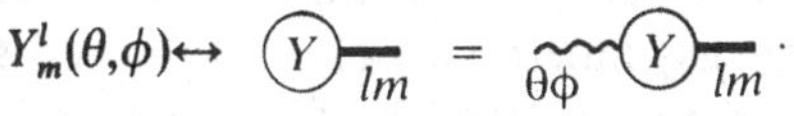

$$Y^l_m(\theta,\phi) \leftrightarrow \quad (Y)\!\!-_{lm} \quad = \quad {}_{\theta\phi}\!\sim\!(Y)\!\!-_{lm} .$$

The irreducible representation label and the partner (or component) label, which distinguishes basis functions within that irreducible representation, for any group are natural companions for such joint treatment.

(e) A node with n terminals is said to have *degree* (or order) n. If N is the number of indices in the corresponding mathematical expression, n is not greater than N.

Rule 3

Translation or rotation of a diagram or of any part of it does not alter its significance. For example:

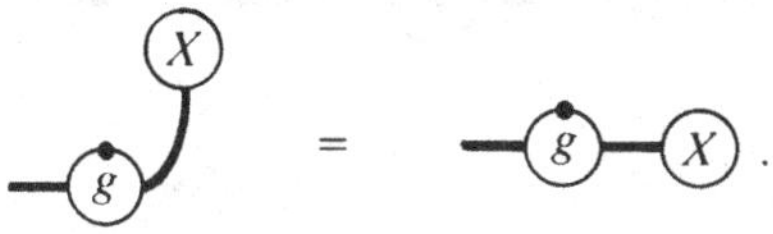

$$(g) \;\; (X) \quad = \quad (g)\!-\!(X) .$$

Comments

(a) Reflections are not allowed.

(b) No significance is attached to the crossing of two lines. Think of them as insulated wires, free to move on a plane.

(c) The definition of a node must be sharpened to permit this, since the order of similar indices affects its value in general. We adopt the following:

Rule 4

Indices are read in anticlockwise order, starting from some conventional point:

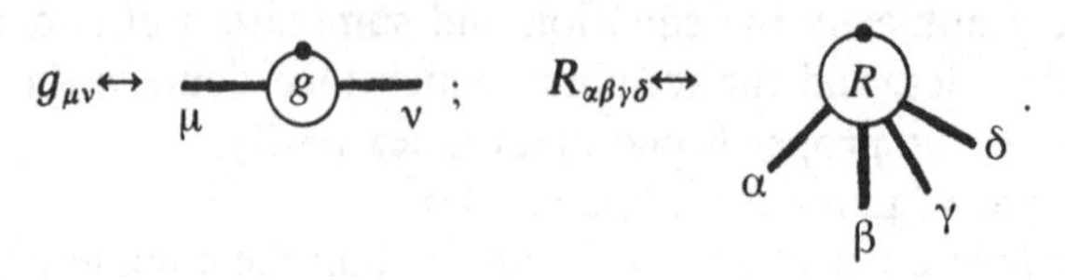

The point is indicated explicitly where necessary as a dot on the node.

Comments

(a) The order is clearly preserved under translations and rotations but not reflections.

(b) Expressions with cyclic symmetry need no conventional point to specify the first index. Otherwise it is vital to have some geometric device, such as a dot or an asymmetrical polygonal shape for the node. For example, in the first figure above, the dot would be unnecessary for a symmetric tensor.

(c) If at least one line is of a distinct type, it may be taken as the reference point, and again there is no need of a geometric device. (The 3jm symbol furnishes an example of (b) for simply reducible simple phase groups (§2.2), and of (c) for groups with multiplicity in the reduction of the Kronecker products of representations (§3.3.2), respectively.)

Rule 5

Corresponding labels on external lines on each side of a diagram equation are identical. For example:

$$g_{\mu\nu}=\eta_{\mu\nu}+h_{\mu\nu} \leftrightarrow$$

Comments

(a) This admittedly suppresses the suggestion in Comment (a) on Rule 2, with the goal of reducing the algebraic notation to a minimum.

(b) A minor breakdown of Rule 3 occurs in the context of this rule; the 'free' ends of the external lines of the diagram must be 'tied down' in corresponding positions on the two sides of an equation, so that it is clear which label corresponds to which.

(c) The diagram will be called *open* or *closed* according as it has some or no external lines.

Rule 6

A rule for pairing two indices (or sets of indices) *is denoted by linking the free ends of the corresponding lines in the diagram.* For example:

$$\Sigma_\nu g_{\mu\nu} X_\nu \leftrightarrow$$

Comments

(a) In the majority of cases the rule will simply represent summation over the label in question. In all cases, see (c), some summation of labels will be intended.

(b) The summed label is redundant (or a dummy), and may be omitted from the internal line in the diagram, as in the example above.

(c) If the line represents two labels (see Comment (d) on Rule 2) the retention of the set label would signify that only the member label (m, in that case) should be summed over. For example, if as mentioned above the node representing a spherical harmonic Y^l_m might be labelled using a single line carrying both l and m labels, a sum over m in an expression involving two spherical harmonics would be implied by an internal line carrying only the l label.

(d) In view of Rule 5, the implication of summation through omission only holds for internal lines.

(e) We mention in passing that Feynman diagrams afford an example where the linking of two nodes by a line depicts a rule for pairing indices at terminals which is more complicated than simple equivalence. In quantum field theory, a line depicts the quantum amplitude for a spacetime propagation:

$$D_{\mu\nu}(x,y) = -\langle T\{A_\mu(x)A_\nu(y)\}\rangle \leftrightarrow$$

It is highly relevant to the power of Feynman's method that this particular (more complicated) interpretation of a line corresponds to such a nontrivial expression! The condensed matter analogue will be discussed in detail in §7.

(f) The pairing rule may be *intransitive*, i.e. nonreciprocal in the two nodes; exchanging the nodes might alter the rule:

(I) For example, one may want to depict which end of the line corresponds to a contravariant index, and which the covariant counterpart.

(II) Or, as in quantum field theory, we may wish to couple a creation with an annihilation operator, on account of Wick's theorem (§7.2.6). This is the significance of the arrow on the above Feynman propagator; an excitation of the system is thus traced through its life history.

(III) Or again, we may wish to indicate that the labels which are related through the line are of conjugate character under the transformation of some symmetry group.

In all these cases, some geometrical device is required to break the symmetry implied by a simple link:

(i) Any symbol lacking symmetry under a rotation by 180° may be added to the line. One might choose either of the notations:

$\eta_{\mu\nu}\leftrightarrow$ μ (η) ν = μ ν = μ ν .

We shall use both, but not interchangeably:

(A) A common choice is an *arrow*, as in a Feynman diagram (see the above example of $D_{\mu\nu}(x,y)$). We shall use this in case (II) above.

(B) A sideways pointing *stub* is also used when it is helpful to emphasise when a node is meant. The node so depicted may include a phase factor, or labels on each side may be different. We shall use this in case (III) above. In particular, we use the stub notation in place of the arrow used by some (but not all, see Sandars 1969a) workers in the quantum theory of angular momentum. We do this to avoid confusion and inconsistencies of notation (see below), and to emphasise the distinction between the group labels on each side of the stub, particularly for subgroups of O(3) or for U(N) (§§3, 5).

(ii) Conventional signs or *parities* may be assigned to the terminals.

Take as an example case (I) above: the distinguishing of contravariant and covariant indices in a relativistic calculation. In following suggestion (ii), we may assign a parity + to all terminals associated with contravariant indices and the opposite parity − to all terminals associated with covariant indices, and then require that every internal line connects terminals of opposite parity. This will preserve Lorentz invariance in the internal summations in this case. We may now make the earlier notation more precise by adding arrows say from contravariant to covariant labels:

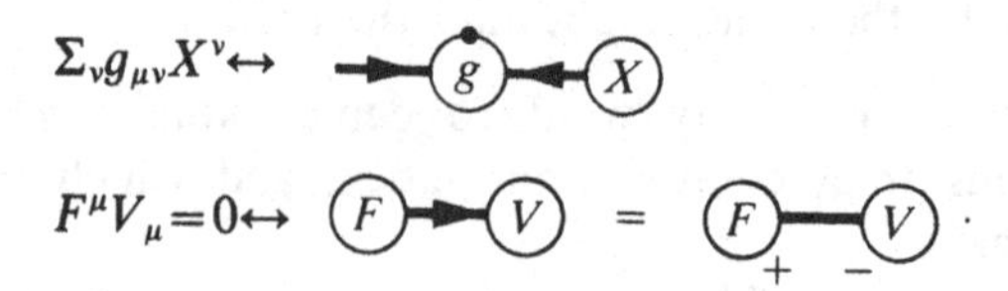

The full justification of these notational choices and assignments is quite complicated, and will be introduced progressively in §§2.5, 3.6, 5.3. We warn the reader that in the diagram literature two uses of the arrow are often confused and conflated to the point of inconsistency. The fundamental problem, which requires us to use two distinct symbols (arrow and stub) for distinct purposes, is that the action of complex conjugation (with which conjugacy, amongst others, we associate the arrow) and that of transformation by a metric tensor (for which we use a stub) are not equivalent.

1.3 Vector analysis: a diagram primer

Following these rules, we shall write a vector as a node of degree 1:

$$\mathbf{a}_\alpha \leftrightarrow \text{(a)—}\alpha \; .$$

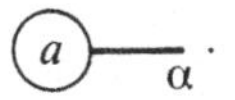

For a 3-vector, the label α may be written as x, y, z in a cartesian basis. By Rule 6, the scalar product of two vectors is the combination

$$\mathbf{a}\cdot\mathbf{b} \leftrightarrow \text{(a)—(b)} \; ;$$

A line joining two nodes has the significance of a Kronecker delta symbol $\delta_{\alpha\beta}$ in the basis labels α, β.

What is the vector product? It must be a network with one external line of similar type. Clearly this calls for a node of degree 3, which will make the topology nontrivial:

The *antisymmetric symbol* $\varepsilon_{\alpha\beta\gamma}$ (or Levi–Citiva tensor density) is denoted by:

$$\varepsilon_{\alpha\beta\gamma} \leftrightarrow$$

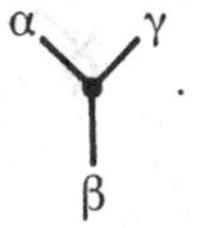

α γ
β

Comments

(a) Note the anticlockwise ordering convention in these figures, in conformity with Rule 4.

(b) The cyclic symmetry of the antisymmetric symbol makes it unnecessary to have any geometrical shape for the corresponding node to render the expression unambiguous.

(c) This allows us to write the vector product as:

$$(\mathbf{a} \times \mathbf{b})_\alpha \leftrightarrow$$

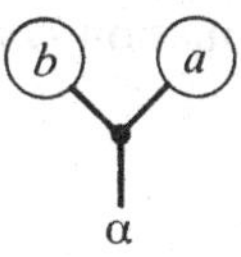

b a
α

(d) The triple scalar product $[\mathbf{a}\,\mathbf{b}\,\mathbf{c}] \equiv \mathbf{a}\cdot\mathbf{b} \times \mathbf{c}$ is then formed by contracting the antisymmetric node with three vectors:

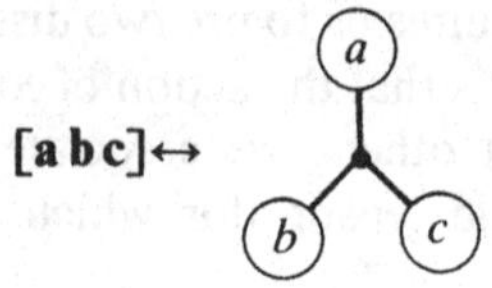

$[\mathbf{a}\,\mathbf{b}\,\mathbf{c}] \leftrightarrow$

The diagram shows that this is clearly a symmetrical definition under cyclic permutations of the component vectors.

(e) However, reflection symmetry is certainly violated:

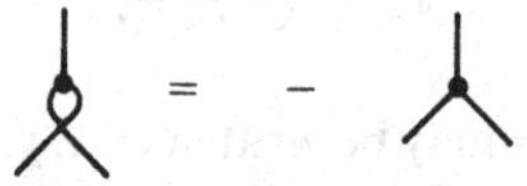

(we omit labels in conformity with Rule 5). It is essential to treat the untangling of lines at a node as a nontrivial problem, since the anticlockwise order may be altered.

Problem 1.1

Prove that $\mathbf{a} \times \mathbf{a} = 0$.

Solution

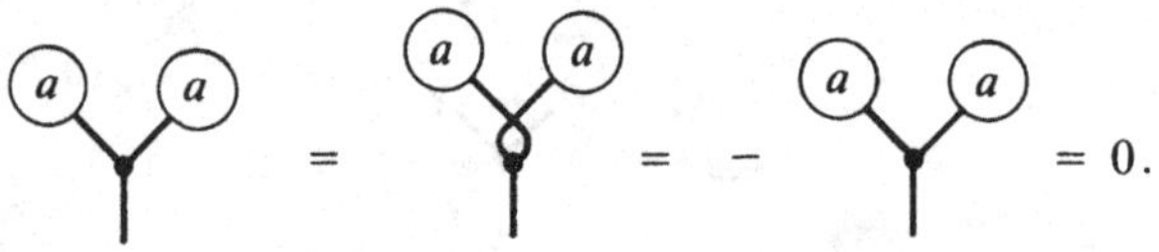

Comment

In the second step, the nodes have been moved past each other. The crossing of lines is unimportant; only the change of rotational order at the node is significant.

Problem 1.2

Show that $\varepsilon_{\alpha\gamma\gamma} = 0$. (A summation convention is implied.)

Solution

Comments

(a) This illustrates a line 'untangling' operation. In the second step the twist is pushed down so as to flip the line harmlessly at its lowest part.

(b) Obviously other methods could be used, as in the previous problem; each term in the sum vanishes.

(c) A generalisation of this result is presented as Problem 1.7.

Problem 1.3

Show that $\mathbf{a}\times(\mathbf{b}\times\mathbf{c})$ is different to $(\mathbf{a}\times\mathbf{b})\times\mathbf{c}$.

Solution

These expressions equal

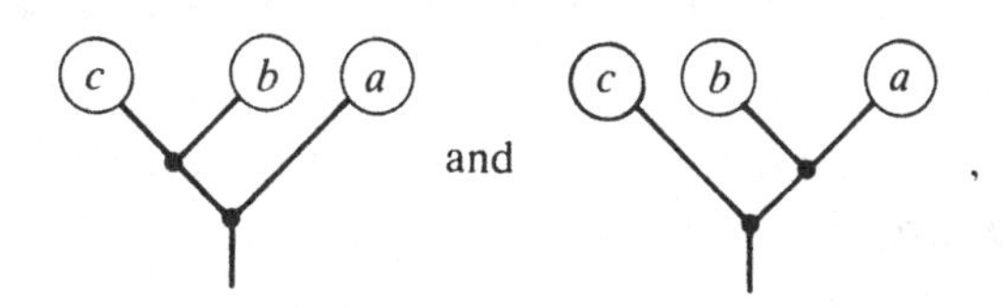

respectively. These obviously have different topologies. A full proof requires a new result, which follows.

Problem 1.4

Write the vector identity $\mathbf{a}\times(\mathbf{b}\times\mathbf{c})=\mathbf{b}(\mathbf{a}\cdot\mathbf{c})-\mathbf{c}(\mathbf{a}\cdot\mathbf{b})$ in diagram form.

Solution

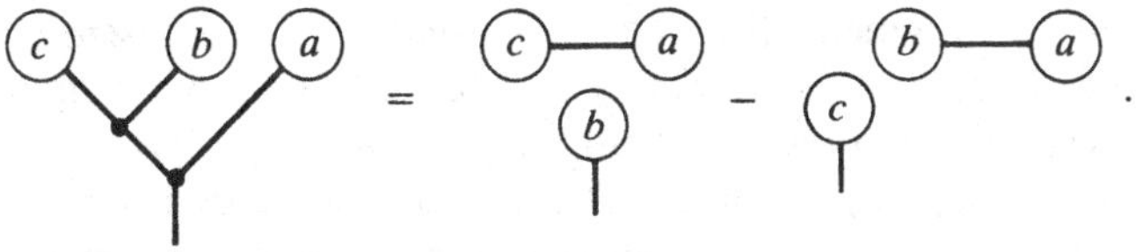

Comments

(a) This incidentally is our first example of diagram multiplication, whose form follows from our rules: disconnected diagrams correspond to multiplicatively related algebraic expressions.

(b) Since the vectors are general, we can also write this as

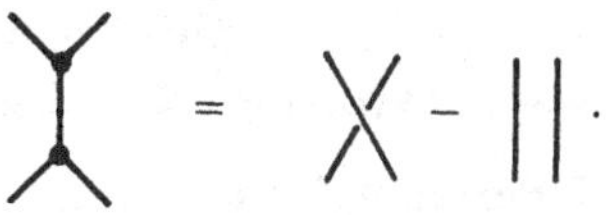

(c) This is our prototype of a reduction theorem or depolymerisation rule. A complicated diagram may be simplified by the above replacement, which is then a general rule for the simplification of networks:

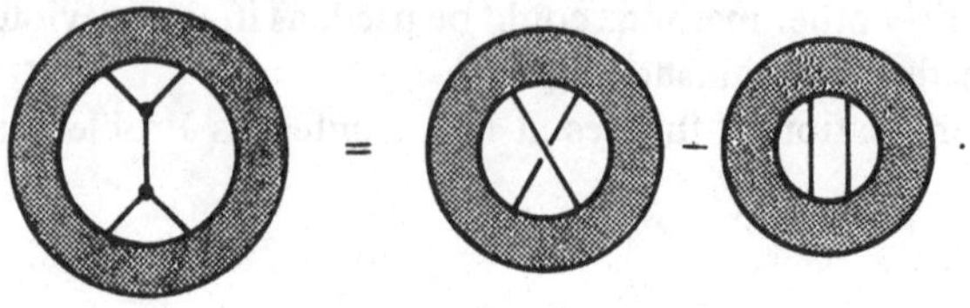

(d) The corresponding algebraic expression may be read off as

$$\varepsilon_{\alpha\beta\gamma}\varepsilon_{\mu\nu\gamma}=\delta_{\alpha\mu}\delta_{\beta\nu}-\delta_{\alpha\nu}\delta_{\beta\mu}.$$

Problem 1.5

Prove that $\varepsilon_{\alpha\beta\gamma}\varepsilon_{\alpha\beta\gamma}=6$.

Solution

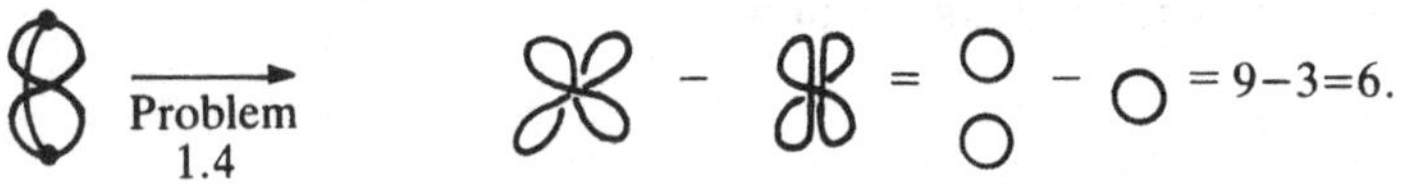

Comments

(a) Note that the application of Problem 1.4 centres on identifying the backbone of the original diagram as the 'double-Y' of the left side of Problem 1.4, Comment (b). The first step in the present solution corresponds to linking the four external lines of this double-Y figure. Also note that a simple loop within a diagram corresponds to the contraction of a line, and so to $\Sigma_{\alpha=1}^{3}\delta_{\alpha\alpha}=3$, and that two such loops correspond to $3^2=9$.

(b) Note that the natural ordering with our anticlockwise convention for the ordering of indices at each node requires a crossing of the lines linking the two nodes. It is significant that the leading term on the right side of Problem 1.4 is crossed. While the expression

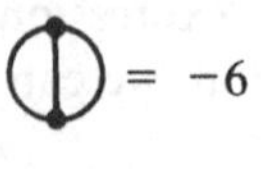

is less tangled, it inevitably carries a negative sign.

(c) The following are ways of avoiding the crossing and the overall sign:

(i) Define a new node, for example by adding a + for, say, anticlockwise, and − for clockwise, ordering at each node, so as to distinguish cyclic and interchange permutations of the antisymmetric symbol. (This has nothing to do with the parity discussed in Comment (f)(ii) on Rule 6, and so is incompatible with it.) For example:

This convention is commonly used in the quantum theory of angular momentum by Jucys and coworkers, also ElBaz, Brink and Satchler, etc.

(ii) If possible, divide the nodes into two classes so that no lines connect nodes in the same class; the diagram is said to have a *bipartite* graph (Harary 1969). Such situations are common (see for example §2.5). We then use a clockwise ordering convention for nodes in one class, and an anticlockwise ordering convention for the other nodes. In effect, this is an implicit form of (i).

(iii) Insert permutation matrices explicitly into the connecting lines. In the case of the antisymmetric symbol, and more generally for 3jm symbols in simple phase groups, the permutation matrices are simple phase factors. In any case they may readily be given diagram form (see §3.3.2). Where necessary we use this method in view of its explicitness.

(d) None of the three solutions in (c) is particularly natural, nor is it necessarily easy to use. If a diagram is not too complicated, it is better to accept the purely topological complication, namely the crossing of lines between nodes, as an inevitable and standard feature. With a little experience the consequences are surprisingly easy to live with.

Problem 1.6

Prove that

$$\Sigma_{\beta\gamma}\varepsilon_{\alpha\beta\gamma}\varepsilon_{\tau\beta\gamma}=2\delta_{\alpha\tau}.$$

Solution

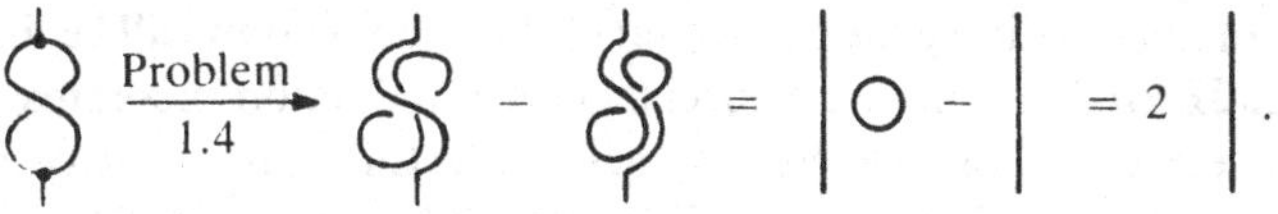

Problem 1.4

Comments

(a) Again the natural diagram representation requires the crossing of lines.

(b) We now use symmetry principles (applied to the three dimensional rotation group) to indicate that the result must be proportional to $\delta_{\alpha\beta}$. Just as **a**·**b** is the only invariant linear in each of the vectors **a**, **b**, the only second rank cartesian tensor which is invariant under rotation is $\delta_{\alpha\beta}$. Likewise the only invariant third rank tensor is $\varepsilon_{\alpha\beta\gamma}$, corresponding to the invariance of the triple product [**a b c**]. If combinations of such tensors are made in a rotationally invariant way, e.g. by contraction, the only possible result is another invariant tensor, or ultimately a rotational invariant. More complicated examples of invariant second rank tensors that reduce to the Kronecker delta symbol are given in Problem 1.8. A proof of the central importance of scalar and vector products in forming SO(3) tensor invariants using integrity basis theory is given in §8.6.1.

Problem 1.7
Show that the following networks, which are constructed from the node $\varepsilon_{\alpha\beta\gamma}$ and which have just *one* external line, vanish:

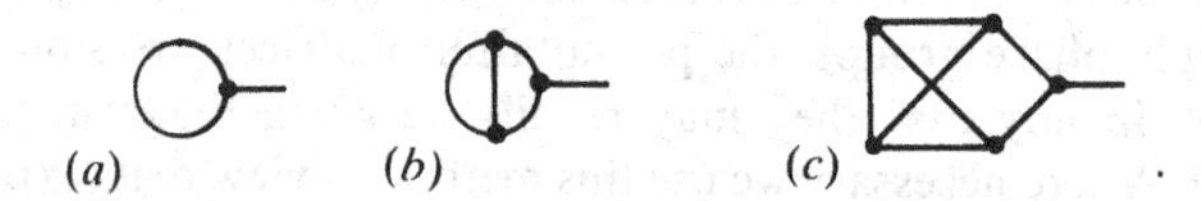

Solution
For (a), see Problem 1.2. (b) follows directly from Problems 1.1, 1.2 and 1.6; use Problem 1.1 to recast Problem 1.6 without the internal crossing, then use the result to show that (b) is proportional to (a). (c) also requires Problem 1.4.

Comments
(a) This also is a straightforward consequence of rotational symmetry. No vector can be rotationally invariant.

(b) *This is our prototype of the reduction theorem JLV1.* The labelling is drawn from the first workers (Jucys, Levinson and Vanagas 1960) who enunciated these theorems for the quantum theory of angular momentum (the case of full rotation group symmetry). The connection will be discussed in §2. The JLV*n* theorems are a special class of reduction theorems of basic importance in group theory. We shall generalise this example progressively. The case of the rotation group will be studied in detail in §2, and more general cases in §3.

Problem 1.8
Show that the following networks, which are constructed from the node $\varepsilon_{\alpha\beta\gamma}$

and which have just *two* external lines, are each proportional to the Kronecker symbol $\delta_{\alpha\beta}$.

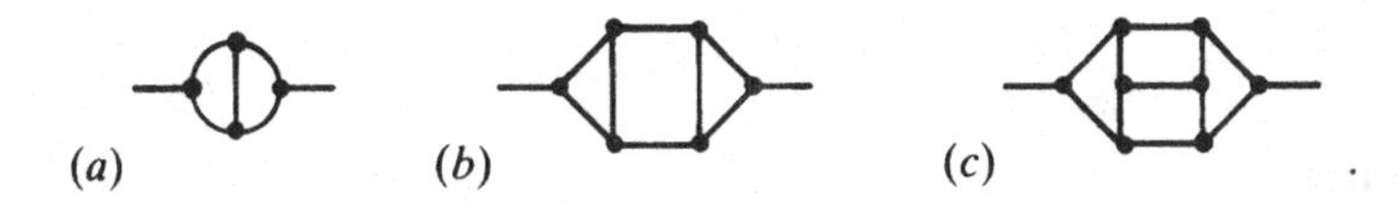

Solution

Use Problem 1.4. The proportionality factors are 2, -2 and 8, respectively, in the final answers.

Comments

(a) As anticipated in Comment (b) of Problem 1.6, this is a consequence of rotational invariance.

(b) This is our prototype of the reduction theorem JLV2. It is related to *Schur's lemma* for the rotation group: an invariant matrix is proportional to the unit matrix.

(c) The constant of proportionality in this relation may be determined by taking traces:

$= c$

$= c \quad = 3c.$

One may think of the external lines as having been pinched together:

$= \frac{1}{3}$.

Problem 1.9

Show that the following networks, which are constructed from the node $\varepsilon_{\alpha\beta\gamma}$ and which have just *three* external lines, are each proportional to $\varepsilon_{\alpha\beta\gamma}$.

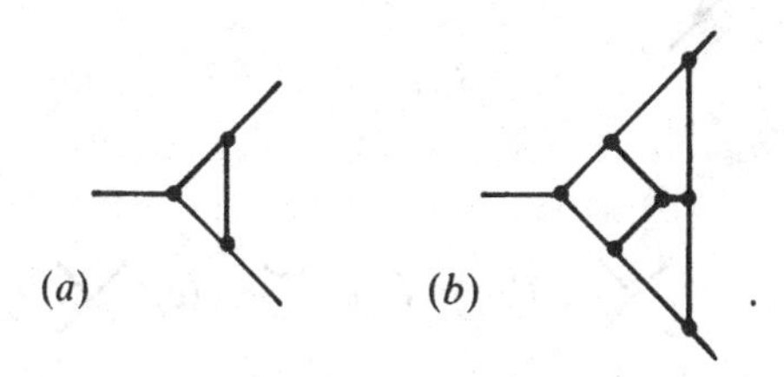

Solution

Use Problem 1.4, etc. The proportionality constants are −1, −4, respectively.

Comments

(a) Naturally, this also is a consequence of rotational invariance, since $\varepsilon_{\alpha\beta\gamma}$ is the only invariant tensor with rank 3.

(b) This is our prototype of the reduction theorem JLV3, and is related to the *Wigner–Eckart theorem.*

(c) The numerical coefficient may be determined by contracting the expression against another antisymmetric symbol:

= c ; $\xrightarrow[1.5]{\text{Problem}}$ $-6c$.

Again, the result suggests the mnemonic: one may think of the external lines as having been pinched together.

(d) Phases are avoided if the lines are crossed before the pinching operation (see Comment (b) on Problem 1.5):

$= \frac{1}{6}$.

Problem 1.10

Find an analogous result for each of the following networks, which are constructed from the node $\varepsilon_{\alpha\beta\gamma}$ and which have just *four* external lines. In (a) for example, we show that:

$$\Sigma_{\gamma\delta\lambda\tau}\varepsilon_{\alpha\beta\gamma}\varepsilon_{\gamma\delta\tau}\varepsilon_{\delta\lambda\mu}\varepsilon_{\lambda\nu\tau} = \delta_{\alpha\mu}\delta_{\beta\nu} - \delta_{\alpha\nu}\delta_{\beta\mu}.$$

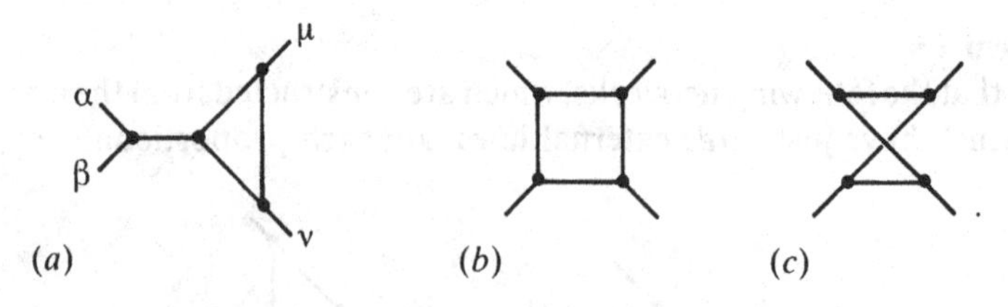

Solution

(*a*) − , (*b*)) (+ , (*c*) + .

Comments

(a) The value of diagram techniques even at this rudimentary level should be clear by now: it is easier to visualise where simplifications may be found in a complicated network by searching for a reducible linkage than by examining a complicated algebraic expression.

(b) The right side of example (a) of this problem and Problem 1.4 are proportional. This, however, does not indicate a simple generalisation of the earlier theorems, as examples (b) and (c) show. Clearly, we obtain different right sides, even allowing for different independent overall numerical factors. This is because there is more than one rotationally invariant tensor $T_{\alpha\beta\gamma\delta}$ of fourth rank. In fact there are three, which may be taken as the three products of Kronecker symbols in the expansion (for any invariant fourth rank tensor)

$$T_{\alpha\beta\gamma\delta} = a_1\delta_{\alpha\beta}\delta_{\gamma\delta} + a_2\delta_{\alpha\delta}\delta_{\beta\gamma} + a_3\delta_{\alpha\gamma}\delta_{\beta\delta} \leftrightarrow$$

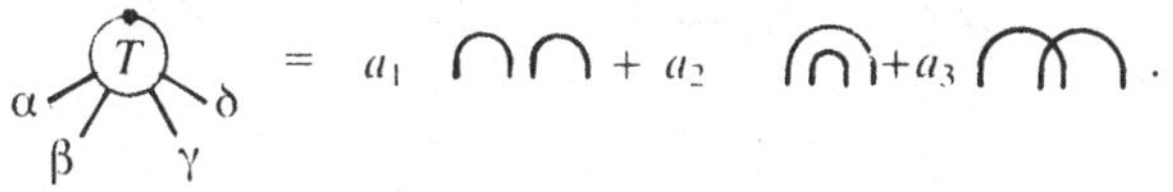

The coefficients a_i are arbitrary (in the simple examples above they are integral). This corresponds to the fact that by Problem 1.4, for example, any invariant in four vectors (such as $(\mathbf{a}\times\mathbf{b})\cdot(\mathbf{c}\times\mathbf{d})$) may be written as a linear combination of the three scalar products $\mathbf{a}\cdot\mathbf{b}\,\mathbf{c}\cdot\mathbf{d}$, $\mathbf{a}\cdot\mathbf{c}\,\mathbf{b}\cdot\mathbf{d}$, $\mathbf{a}\cdot\mathbf{d}\,\mathbf{b}\cdot\mathbf{c}$. A full analysis of the invariants of the rotation group is given in §8.6.1.

(c) Note that the final diagrams in examples (b), (c) satisfy the obvious symmetries of the initial diagrams under 90° or 180° rotations etc.

Problem 1.11

Prove that:

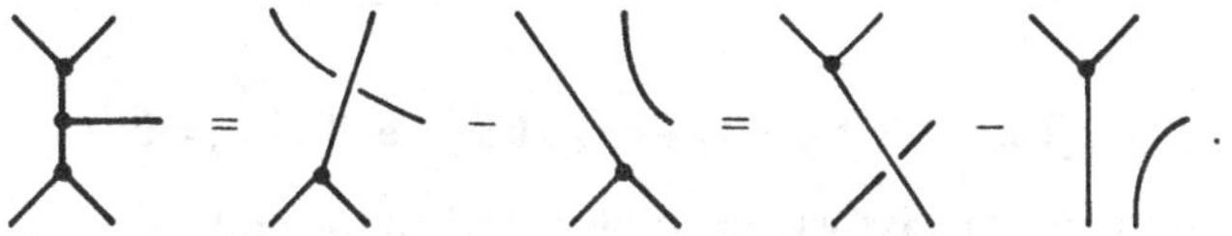

Solution

Use Problem 1.4 repeatedly in different ways.

Comments

(a) This problem leads on to the next, and also illustrates the possibility of alternative final forms in a diagrammatic (or algebraic) reduction.

(b) This ambiguity is unavoidable for invariant diagrams with five or more external lines. The origin of the choice is that any 3-vector may be written in terms of any three noncoplanar 3-vectors as basis, as demonstrated in the following problem.

Problem 1.12
Prove:

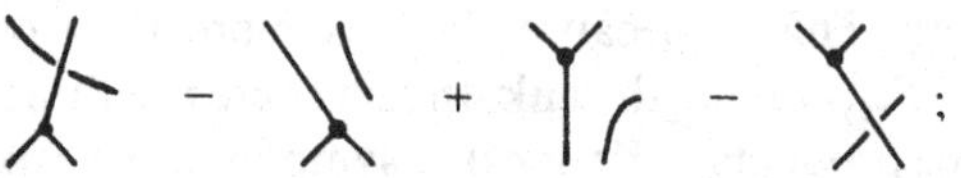

(*a*) − + − ;

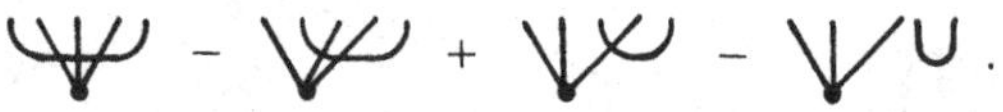

(*b*) − + − .

Solution
Rearrange the right sides of the answer to Problem 1.11. In converting form (a) to form (b), lay out the diagram as though it consisted of pieces of string, and move the ends around. In doing so, note that one external line (in the right centre of (a), and moving to the right top of (b)) is never connected to the antisymmetric vertex.

Comments
(a) This corresponds to the vector identity

$$\mathbf{d} = (\mathbf{d}\cdot\mathbf{A})\mathbf{a} + (\mathbf{d}\cdot\mathbf{B})\mathbf{b} + (\mathbf{d}\cdot\mathbf{C})\mathbf{c},$$

where $\mathbf{A}$, $\mathbf{B}$, $\mathbf{C}$ are the reciprocal vectors $\mathbf{A} = \mathbf{b}\times\mathbf{c}/[\mathbf{a}\,\mathbf{b}\,\mathbf{c}]$, etc. This connection may be seen, for example, by rewriting the vector identity in the form

$$\mathbf{e}\cdot\mathbf{a}[\mathbf{b}\,\mathbf{c}\,\mathbf{d}] - \mathbf{e}\cdot\mathbf{b}[\mathbf{a}\,\mathbf{c}\,\mathbf{d}] + \mathbf{e}\cdot\mathbf{c}[\mathbf{a}\,\mathbf{b}\,\mathbf{d}] - \mathbf{e}\cdot\mathbf{d}[\mathbf{a}\,\mathbf{b}\,\mathbf{c}] = 0,$$

which corresponds precisely to example (b) when the vectors **a**, **b**, **c**, **d**, **e** are attached from left to right to the external lines in each term.

(b) The different forms of examples (a), (b) show that it is not always easy to recognise this cancellation in a complicated network. Note that one external line is never connected to the antisymmetric node, and that there is an alternating symmetry in the other four labels.

A similar development of vector algebra is given by ElBaz and Nahabetian (1977).

1.4 Generalised Kronecker deltas and symmetrisers

Often label-exchange symmetries are relevant to some expression; we wish to consider a sum over permutations, possibly including the sign of each permutation, in forming a final answer. We need an abbreviated diagram notation.

Following Penrose (1971), we define *generalised Kronecker delta* symbols of symmetric and antisymmetric type, depicting them by open and solid bars, respectively:

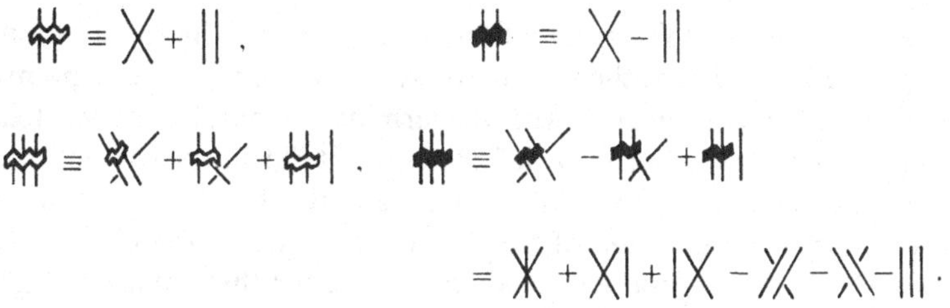

Comments

(a) These generalised Kronecker delta symbols correspond to $+1$ if the lower labels are an even permutation of the upper labels (reading the order in anticlockwise convention about the bar), and to $+1$ (-1) for the symmetric (antisymmetric, respectively) generalised Kronecker delta symbol when the lower labels are an odd permutation of the upper labels; and to zero otherwise.

(b) Note that the crossed diagram naturally has the positive sign in the antisymmetric case. In this we follow Penrose (1971) among others. Different workers have different conventions here.

(c) The usefulness of these definitions may be demonstrated in previous exercises. For example, the right side of the diagram in Comment (b) of Problem 1.4 is just the antisymmetric generalised Kronecker symbol on two labels.

(d) The zigzag lines denote that these are not normalised, in the sense of not being idempotent. *Symmetrisers* and *antisymmetrisers*, i.e. idempotent projection operators, may be derived by adding a normalisation factor $1/n!$, where n is the number of labels. We use a straight line to denote the inclusion of this normalisation:

It is convenient to retain both notations, since both forms have their natural uses.

(e) A fuller account of the properties of these symbols is given in §4.3, and illustrations are spread through §§3–5.

Problem 1.13

Prove that

= .

Solution

Square the result of Problem 1.12. In doing so, first rewrite the solution to Problem 1.12 in yet another form where three terminals are pointing upwards and two downwards, with the terminal uniquely unconnected to the antisymmetric symbol now at bottom right. Then mirror reflect the resulting expression, as in a lake, to obtain the first of the following diagrams. On the right side of the following diagram, the circled cross denotes a 'Kronecker product', in which each of the three diagrams in the top half is paired with each of the three diagrams in the bottom half in turn (the adjacent terminals in the centre then being connected), and the resulting diagrams summed.

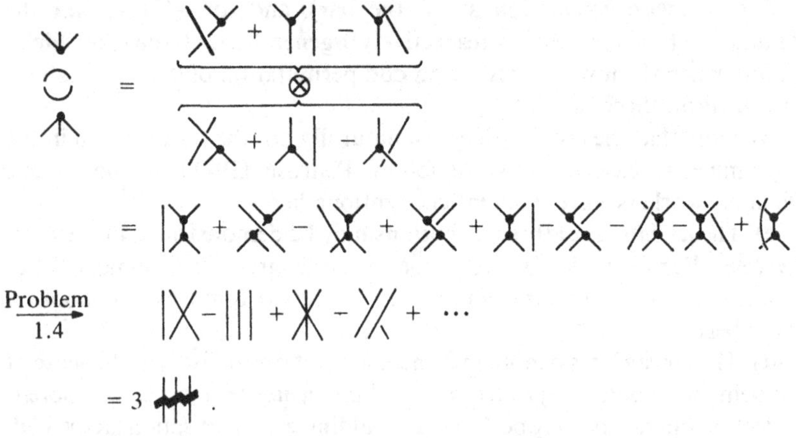

= + − ⊗ + −

= + − + + − − − +

Problem 1.4 → − + − + ...

= 3 .

Comments

(a) Again this result may be regarded as a reduction theorem in which lines are pinched together, in this case irrespective of the nature of the network, if any, to which they are joined.

(b) In tensor notation, this corresponds to the identity

$$\varepsilon_{\alpha\beta\gamma}\varepsilon_{\lambda\mu\nu}=\delta_{\alpha\lambda}\delta_{\beta\mu}\delta_{\gamma\nu}+\delta_{\alpha\nu}\delta_{\beta\lambda}\delta_{\gamma\mu}+\delta_{\alpha\mu}\delta_{\beta\nu}\delta_{\gamma\lambda}$$
$$-\delta_{\alpha\lambda}\delta_{\beta\nu}\delta_{\gamma\mu}-\delta_{\alpha\mu}\delta_{\beta\lambda}\delta_{\gamma\nu}-\delta_{\alpha\nu}\delta_{\beta\mu}\delta_{\gamma\lambda}.$$

In short, the antisymmetric symbol is an independent rotational invariant only in first order (see §8.6.1).

(c) The corresponding vector identity is

$$[\mathbf{a}\,\mathbf{b}\,\mathbf{c}][\mathbf{d}\,\mathbf{e}\,\mathbf{f}] = \begin{vmatrix} \mathbf{a}\cdot\mathbf{d} & \mathbf{a}\cdot\mathbf{e} & \mathbf{a}\cdot\mathbf{f} \\ \mathbf{b}\cdot\mathbf{d} & \mathbf{b}\cdot\mathbf{e} & \mathbf{b}\cdot\mathbf{f} \\ \mathbf{c}\cdot\mathbf{d} & \mathbf{c}\cdot\mathbf{e} & \mathbf{c}\cdot\mathbf{f} \end{vmatrix}.$$

Problem 1.14

Prove that:

$$= 0.$$

Solution

Expand the left side:

$$= \frac{1}{4}\left(\quad - \quad + \quad - \quad \right);$$

then use Problem 1.13 and Problem 1.12.

Comments

(a) This relation indicates a new source of alternative forms for the reduction of invariant diagrams with eight external lines, similar to that discussed in Problem 1.11 for invariant diagrams with five external lines.

(b) Generalisations of this are discussed by Penrose (1971).

(c) The analogous vector identity is (e.g. Minard, Stedman and McLellan, 1983)

$$\begin{vmatrix} \mathbf{a}\cdot\mathbf{e} & \mathbf{a}\cdot\mathbf{f} & \mathbf{a}\cdot\mathbf{g} & \mathbf{a}\cdot\mathbf{h} \\ \mathbf{b}\cdot\mathbf{e} & \mathbf{b}\cdot\mathbf{f} & \mathbf{b}\cdot\mathbf{g} & \mathbf{b}\cdot\mathbf{h} \\ \mathbf{c}\cdot\mathbf{e} & \mathbf{c}\cdot\mathbf{f} & \mathbf{c}\cdot\mathbf{g} & \mathbf{c}\cdot\mathbf{h} \\ \mathbf{d}\cdot\mathbf{e} & \mathbf{d}\cdot\mathbf{f} & \mathbf{d}\cdot\mathbf{g} & \mathbf{d}\cdot\mathbf{h} \end{vmatrix} = \mathbf{0}.$$

(d) Problem 1.12 can be rewritten as

$$= 0.$$

1.5 Matrix analysis

We have, directly from the rules of §1.2, that a matrix element M_{ab} is represented by

$$M_{ab} \leftrightarrow \ {}_{a}\!\!-\!(M)\!-\!{}_{b}\ .$$

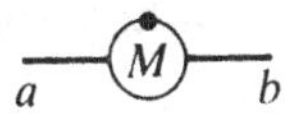

(Note the necessity of a dot – or some such device – for an asymmetric matrix.) Symmetric (S), hermitian (H), orthogonal (O) and unitary (U) matrices therefore satisfy the relations:

respectively. These will be of vital importance in later work.

Problem 1.15
Express the determinant and permanent of a matrix in diagram terms.

Solution

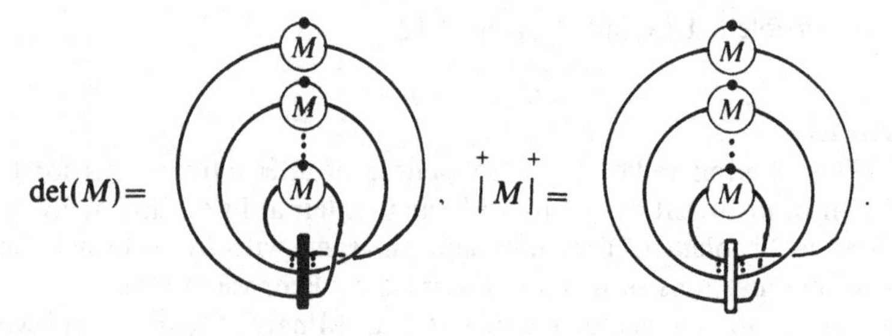

Comments

(a) The symmetriser allows for the $n!$ permutations of the row labels, each of which guarantees the same contribution.

(b) Note the crossings in such diagrams, including those intrinsic in the symmetrisers. By contrast, the diagram corresponding to the *trace* of a product of matrices is characteristically planar:

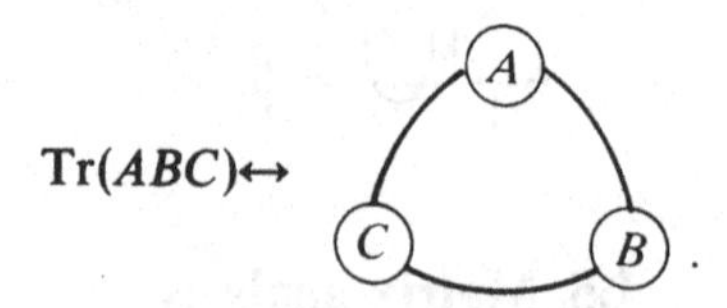

Problem 1.16
Prove that the eigenvalues of a hermitian matrix are real, and that the eigenvectors are orthogonal.

Solution

Without loss of generality, we normalise the sum of squares in $\mathbf{V}^i \cdot \mathbf{V}^{i*}$ to unity. We depict the eigenvector $\mathbf{V}^i$ as a node of degree 2, one line (solid) carrying the component labels $\alpha, \beta, \ldots$, and the other (broken) carrying the eigenvector label i:

$$(V^i)_\alpha \leftrightarrow$$

The eigenvalue λ_i is depicted as a cross (a diagonal matrix) on the broken line:

$$\lambda_i \leftrightarrow$$

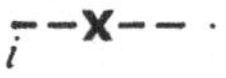

The proof may then be written out in parallel to the usual formalism, starting by

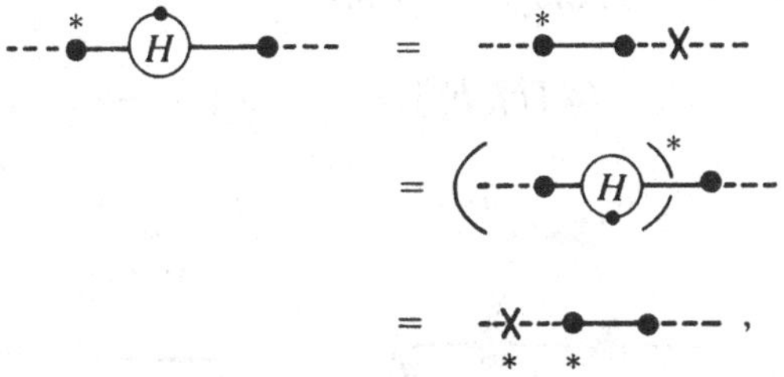

or, in algebra, the relation $(\lambda_i - \lambda_j^*)(\mathbf{V}^{j*} \cdot \mathbf{V}^i) = 0$. This relation shows two things. First, for $i = j$, when the norm of $\mathbf{V}^i$ cannot vanish, the eigenvalue λ_i must be real. Second, and using this first result, when $\lambda_i \neq \lambda_j$, the inner product $(\mathbf{V}^{j*} \cdot \mathbf{V}^i)$ must therefore vanish.

Comments

(a) This rather complicated example indicates how it is possible for all standard matrix theorems to be reproduced in diagram form. Other notations for basis states will be introduced in §§2.7.1, 2.7.2.

(b) For a degenerate set of eigenvalues, when $i \neq j$ but $\lambda_i = \lambda_j$, the inner products $(\mathbf{V}^{j*} \cdot \mathbf{V}^i)$ may still be made to vanish by an appropriate change of basis.

1.6 Second quantisation algebra

Another example is that of the algebra of boson and fermion second quantisation operators. It illustrates some of our rules as well as some earlier notational definitions. However, in contrast to the earlier parts of this chapter, this is not essential reading for a novice, and indeed is a rather advanced illustration, to be skipped on a first reading. We shall use a notation that is more complicated than is strictly necessary: we shall differentiate creation from annihilation operators by a special label (leg). This will enhance the elegance of the method.

A superscripted sign distinguishes creation and annihilation operators: $b_i^- \equiv b_i$, $b_i^+ \equiv b_i^\dagger$. We denote the matrix elements $\langle a|b_i^\eta|a'\rangle$ of each type of boson operator jointly by the diagram:

$$\langle a|b_i^\eta|a'\rangle \leftrightarrow$$

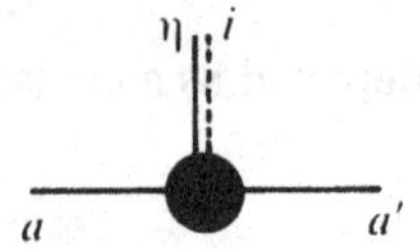

For fermions $f_i^\dagger, f_i$, we use the same notation with an open circle for the node.

The matrix elements of the appropriate commutation or anticommutation relations take the diagram forms

$$\langle a|[b_i^\eta, b_j^{\eta'}]|a'\rangle = \delta_{ij}\delta_{aa'}\varepsilon_{\eta\eta'} \leftrightarrow$$

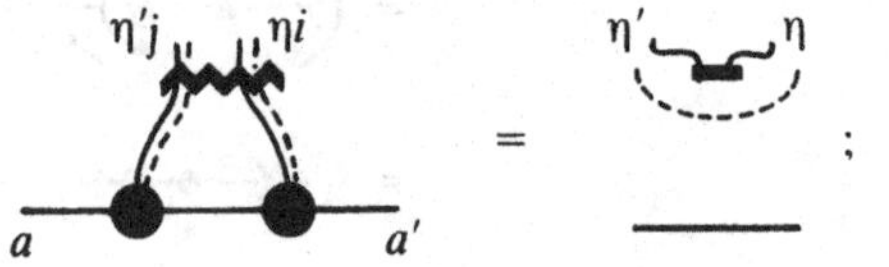

$$\langle a|\{f_i^\eta, f_j^{\eta'}\}|a'\rangle = \delta_{ij}\delta_{aa'}\delta_{\eta,-\eta'} \leftrightarrow$$

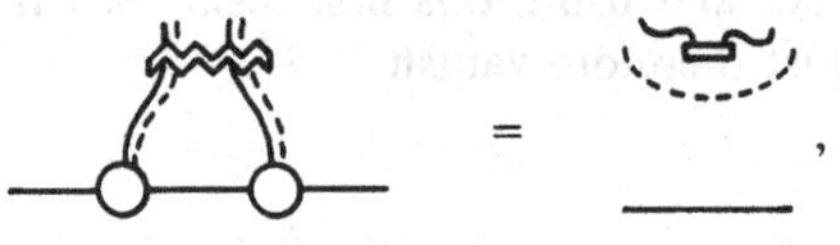

where the antisymmetric symbol has the matrix elements and the diagram form

$$(\varepsilon_{\lambda\mu}) \equiv \begin{array}{c|cc} \lambda \backslash \mu & + & - \\ \hline + & 0 & -1 \\ - & 1 & 0 \end{array}, \quad \varepsilon_{\lambda\mu} \leftrightarrow$$

and its modulus is denoted by an open bar:

$$|\varepsilon_{\lambda\mu}| = \delta_{\lambda,-\mu} \leftrightarrow$$

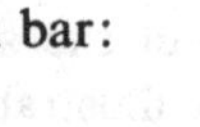

The Kronecker delta symbols act jointly on label pairs; for example,

This notation can be used to prove any algebraic relation for second quantised operators; some examples are given below. The notation has a special elegance when it is associated with the quantum theory of angular momentum (§2.8).

Problem 1.17

Prove that if N_i is the number operator, say for bosons ($N_i \equiv b_i^\dagger b_i$), then

$$[N_i, b_j^\eta] = \eta b_i^\eta \delta_{ij}.$$

Solution

The left side is

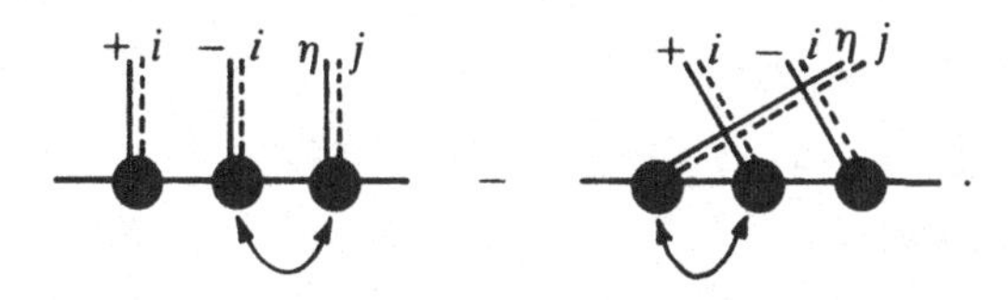

As indicated by the arrows, we now exchange the last two operators in the first term, and the first two in the second term, using the commutation relations to find

as required. We could abbreviate this result by defining a node for the number operator:

$\langle N_i \rangle \leftrightarrow$

so that

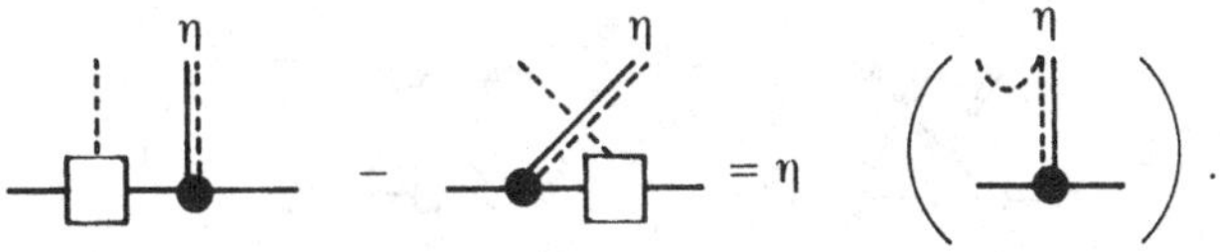

Problem 1.18

Show that for a two-state system ($i=1,2$) the isospin operators

$$\mathrm{T}_+=b_2^\dagger b_1,\ \mathrm{T}_-=b_1^\dagger b_2,\ \mathrm{T}_3\equiv\tfrac{1}{2}(b_2^\dagger b_2-b_1^\dagger b_1)$$

satisfy the commutation relations of angular momentum operators.

Solution

For example, consider $[\mathrm{T}_3, \mathrm{T}_+]$. This becomes

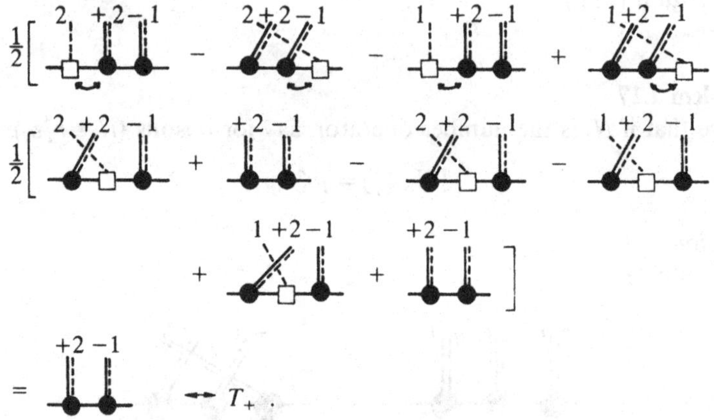

$\frac{1}{2}$[2 +2 −1 − 2 +2 −1 − 1 +2 −1 + 1 +2 −1

$= \frac{1}{2}$[2 +2 −1 + +2 −1 − 2 +2 −1 − 1 +2 −1

+ 1 +2 −1 + +2 −1]

= +2 −1 $\leftrightarrow T_+$.

Problem 1.19

Find the commutation relations of operators of the product form $B_j\equiv f_j^\dagger f_{\bar{j}}$, where $f_i^\dagger$ and $f_{\bar{i}}^\dagger$ are fermion operators for time-reversal conjugate states $i, \bar{i}$. (Since fermionic states have a half-integral spin projection, i and $\bar{i}$ are necessarily distinct.)

Solution

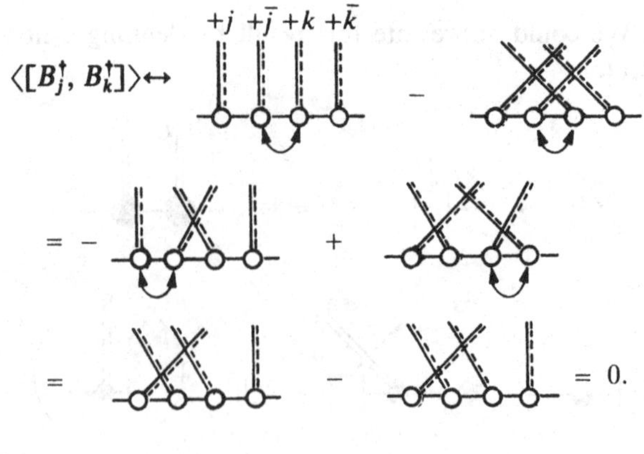

$\langle[B_j^\dagger, B_k^\dagger]\rangle\leftrightarrow$ $+j$ $+\bar{j}$ $+k$ $+\bar{k}$ −

= − +

= − = 0.

For mixed operators, the same sequence of operations throws up the commutators:

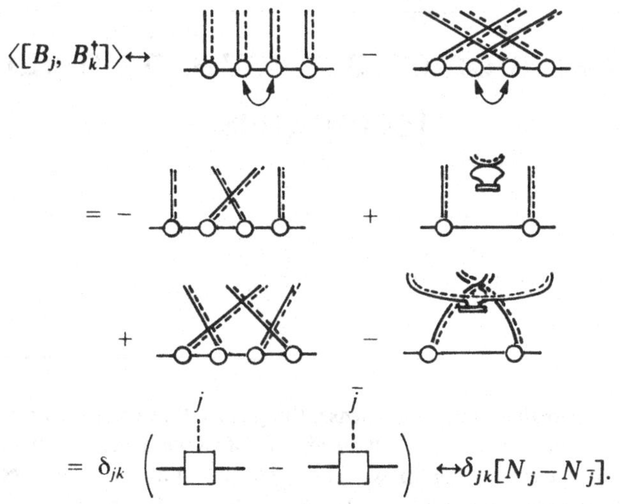

$\langle [B_j, B_k^\dagger] \rangle \leftrightarrow$

$= \delta_{jk}$ $\leftrightarrow \delta_{jk}[N_j - N_{\bar{j}}]$.

Comment

This combination is of interest, e.g., in BCS superconductivity theory, where it describes the creation of a Cooper pair or bound state of electrons with opposite momenta and spin. The fact that these commutation relations are not quite those of a boson operator is crucial to the existence of the energy gap of the superfluid and the superconducting state (Kittel 1963, Schrieffer 1973).

2

Angular momentum coupling diagram techniques

The diagram is really, in a certain sense, the picture that comes from trying to clarify visualization, which is a half-assed kind of vague, mixed-up symbols . . . When I think of electron spin in an atom, I see an atom and I see a vector and a Ψ written somewhere, sort of, or mixed up with it somehow, and an amplitude all mixed up with it . . . Ordinarily I try to get the pictures clearer, but in the end the maths can take over and be more efficient in communicating the idea . . .

R.P. Feynman, quoted in Schweber (1986)

2.1 Introduction

This technique, sometimes known as the diagram technique for the quantum theory of angular momentum, or as the graphical technique of spin algebras, is the best known of the methods described in this book. It is simply the application of a general method to the rotation group in three dimensions. As such, it is an obvious introductory example. Since standard works abound, our presentation does not prove all the relevant group theoretic theorems. We concentrate on demonstrating the logic of the development of the method, spending some time on topics not so well known (in particular the metric role of the 2jm symbol) and to establishing conventions (for nodes, arrows, parities, etc.) consistent with our conventions in other applications.

The method originated in the school of a Lithuanian physicist, A.P. Jucys (Jucys *et al.* 1960, Jucys and Bandzaitas 1964). The most well known English translation of the first of these works is that by the Israel program for scientific translations, under the names (transliterated from Russian forms) Yutsis, Levinson and Vanagas (1962). Further reviews and extensions, not always compatible in detail, are given for example by Brink and Satchler

(1968), Sandars (1969a), ElBaz and Castel (1971, 1972), Lindgren and Morrison (1982), Wilson (1984).

The application to field theoretic diagrams originated in the work of Bolotin, Levinson and Tomalchev (1969) (see Tomalchev 1969 for an English presentation). The adaptation to the true Feynman diagrams of QED (see Biritz 1975a,b, 1979, Kennedy 1982, Stedman and Pooke 1982) is not as direct as for the nonrelativistic case, but has practical value (§8.6.2).

2.2 Rotation matrices, Clebsch–Gordan coefficients and *njm* symbols

We consider the rotation group R(3), or SO(3) – the special orthogonal group in three dimensions, represented by all 3×3 orthogonal matrices with unit determinant. The extension to O(3) merely requires adding the parity π (or signature under spatial inversion), denoted where necessary as a superscript to the representation label j.

True and spin irreps (unitary irreducible representations) of R(3) are labelled by j, an integer or half-integer, respectively. The dimension of the irrep j, $2j+1$, appears in various powers in the following results. We use the notation (Sandars 1969a) $\hat{j}=(2j+1)^{\frac{1}{2}}$, $\check{j}=\hat{j}^{-1}$, $\hat{\hat{j}}=\hat{j}^2$, $\check{\check{j}}=\check{j}^2$, etc.

An element R of SO(3) has a representation matrix in the irrep j which is depicted by

$D^j_{mm'}(R) \leftrightarrow$ [diagram: line from jm to jm' with dashed line to R].

These matrix elements arise from the action of R on basis kets of the form $|jm\rangle$:

$$R|jm\rangle = \Sigma_{m'} D^j_{m'm}(R)|jm'\rangle. \tag{2.1}$$

We postpone giving the basis functions a diagram form until §2.7. We use a spherical basis, in which m is an irrep label of SO(2), the group of rotations in two dimensions.

This representation matrix is unitary:

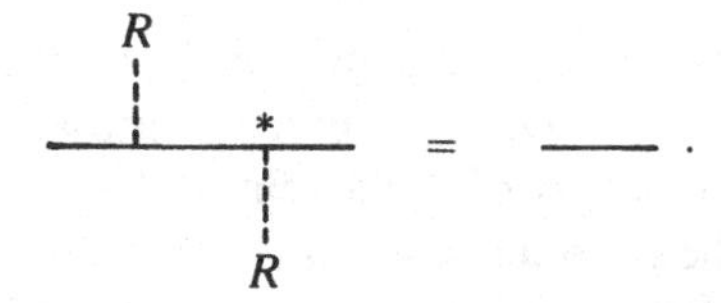

Product basis functions of the form $|jm\rangle|j'm'\rangle$ may be expanded in irreducible parts $\{|JM\rangle\}$ with the Clebsch–Gordan coefficients $\langle JM|jmj'm'\rangle$ as coefficients in the expansion

$$|jm\rangle|j'm'\rangle=\Sigma_{JM}|JM\rangle\langle JM|jmj'm'\rangle. \tag{2.2}$$

Since both bases are orthonormal, the Clebsch–Gordan coefficients form a unitary matrix.

Equivalently, the product basis $|jm\rangle|j'm'\rangle$ defines a matrix representation $D^{j\times j'}$ of SO(3) (from the manner in which rotations mix these kets) which itself is reduced to block-diagonal form by a matrix whose elements are the Clebsch–Gordan coefficients. Hence

$$\begin{aligned} D^{j\times j'}_{mm'nn'}(R)&=D^{j}_{mn}(R)D^{j'}_{m'n'}(R)\\ &=\Sigma_{MM'}\langle jmj'm'|JM\rangle^{*}D^{J}_{MM'}(R)\langle JM'|jnj'n'\rangle^{*} \end{aligned} \tag{2.3}$$

and

$$D^{J}_{MM'}(R)=\Sigma_{mm'nn'}\langle JM|jmj'm'\rangle^{*}D^{j\times j'}_{mm'nn'}(R)\langle jnj'n'|JM'\rangle^{*}. \tag{2.4}$$

Clebsch–Gordan coefficients will be represented diagrammatically by

$\langle JM|jmj'm'\rangle\leftrightarrow$ 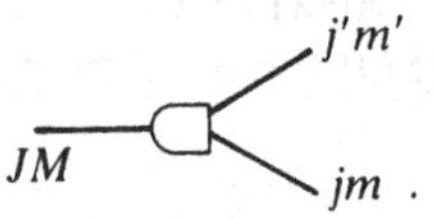

We have chosen the AND gate symbol of digital electronics to suggest the multiplicative connection between irrep labels.

From the properties stated above, we have the diagram relations:

(a) Unitarity:

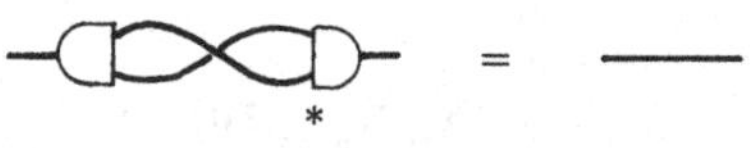

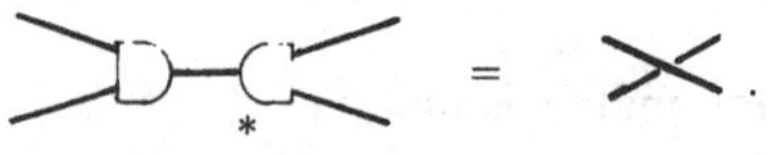

The second relation in particular is known as the *Clebsch–Gordan series*, on account of the summation over the intermediate angular momentum J, itself implied by the omission of the algebraic label.

(b) Reduction of the product representation $j\times j'$; the matrices providing the irreducible decomposition or reduction of the representation matrices for the product basis are the same for all elements R of G. This amounts to the apparently trivial statement that as mere numbers, the

Clebsch–Gordan coeffecients commute with all group operations. Hence, from equations (2.3), (2.4):

$$\begin{aligned}&\Sigma_N D^J_{MN}(R)\langle JN|jmj'm'\rangle\\&=\Sigma_{mm'}\langle JM|jnj'n'\rangle D^j_{nm}(R)D^{j'}_{n'm'}(R),\end{aligned} \tag{2.5}$$

i.e.

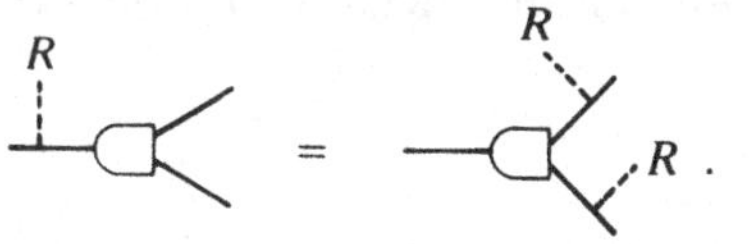

R R $= \quad .R\ .$

This will be our prototype of a *commutativity* property for a diagram node (§2.5).

(c) The Clebsch–Gordan symbol may be symmetrised (forming a 3jm symbol) by applying a conjugacy operation, or 2jm symbol, to the labels J, M. We prepare the way for this. First we obtain the conjugacy operation from the special and renormalised Clebsch–Gordan symbol for SO(3) obtained by setting $J=0$.

2jm symbol

$$\begin{pmatrix} j \\ m \end{pmatrix} = \begin{pmatrix} j & j \\ m & m' \end{pmatrix} = \hat{j}\langle jmjm'|00\rangle \leftrightarrow \quad jm \quad jm'$$

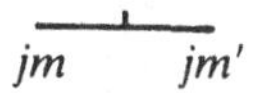

Comments

(a) We have deliberately introduced a variety of notation. The second is less ambiguous, and emphasises that the 2jm symbol may be thought of as a $(2j+1)\times(2j+1)$ matrix $M_{mm'}$ with the component labels m, m' as row and column labels. However, this matrix can always be chosen so that only one element in each row or column is nonzero; since m' is therefore determined from m, only that leading label need be specified.

(b) From the definition, the 2jm symbol couples two kets to the identity irrep, that is, to an invariant ($j=0$):

$$\begin{pmatrix} j & j \\ m & m' \end{pmatrix} |jm\rangle\,|jm'\rangle = |\,(j)00\rangle.$$

The 2jm symbol is analogous to the metric of relativity theory, which as a matrix couples two four-vectors to a Lorentz invariant. A fuller analysis of the metric character of the 2jm symbol for a Lie group is given in §5.3.

(c) The reason for the now-standard 'm' in the name $n$jm symbol is because the value of the symbol (more accurately, of any element of the matrix M) depends on the choice of basis $\{m\}$ for the irrep j.

(d) In the standard spherical (Condon and Shortley 1935) basis, the 2jm symbol has the value

$$\begin{pmatrix} j \\ m \end{pmatrix} = (-1)^{j-m}\delta_{m,-m'}.$$

The Kronecker delta in this expression is a special case of the more general constraint $M = m + m'$ on the Clebsch–Gordan coefficient.

Problem 2.1
Show that the 2jm symbol is invariant under the action of any element of the rotation group on its two lines, acting jointly and in an anti-clockwise sense.

Solution
Because the Clebsch–Gordan node is commutative (see equation 2.5, in Comment (b) after its definition above),

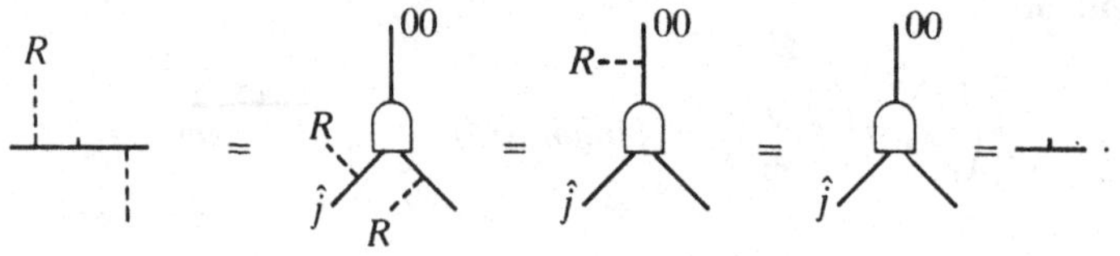

Comments

(a) This is our prototype of an *invariant* node (§2.5). Note that group operators are applied in an anticlockwise sense.

(b) The invariance of the 2jm symbol implies that the 2jm symbol is also the transformation which takes complex conjugate irrep matrices into a standard irreducible form, i.e. it transforms $(\lambda)^*$ to (λ^*):

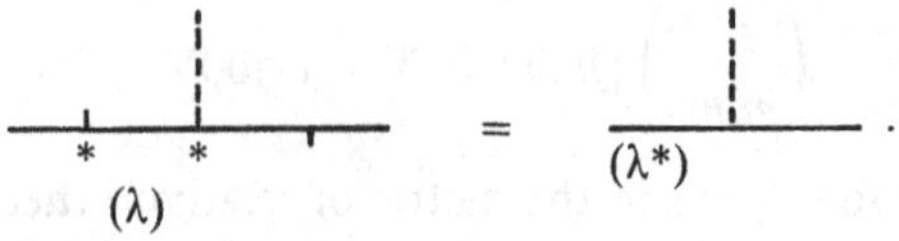

The factor $\hat{j}$ in the definition of the 2jm symbol normalises the definition of the 2jm symbol from the Clebsch–Gordan coefficient so that $M_{mm'}$ is a unitary matrix:

Problem 2.2

Show that

(a) the matrices

$A \equiv MM^{\dagger}$, $B \equiv MM^{*}$ $A \leftrightarrow$, $B \leftrightarrow$

each correspond to a commutative node;

(b) Tr $A = \hat{j}$;

(c) A and $(BB^{\dagger})$ are each equal to the unit matrix:

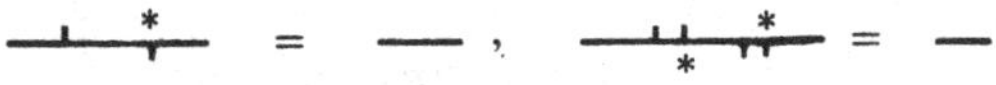

(a stronger result for B is discussed in the comments) so that in particular the 2jm symbol is a unitary matrix in the labels of the two lines;

(d) B is an interchange matrix, reversing the sense of the stub:

Solution

(a) The invariance of the 2jm symbol (Problem 2.1) leads to the result:

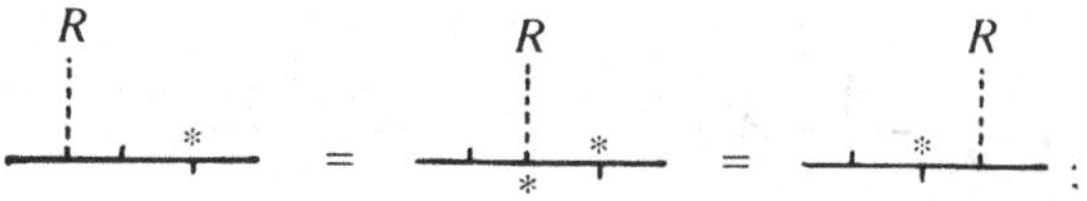

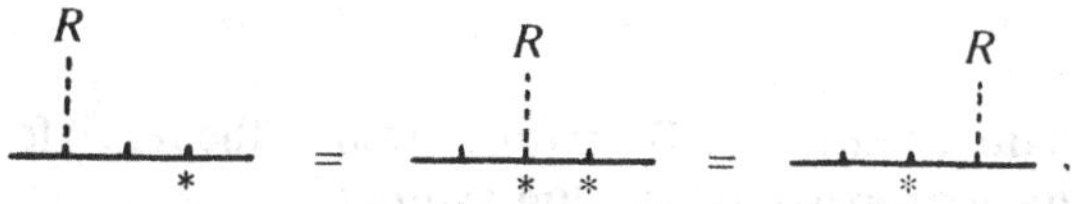

where the last step follows from unitarity. A similar result holds for B.

(b) Taking the trace of A, we obtain, from the definition of the 2jm symbol,

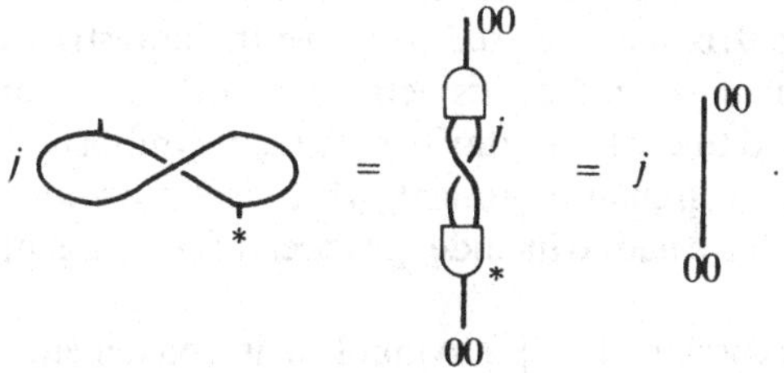

$= j$

(c) A is hermitian; it is the complex conjugate of its π-rotated form. Hence A has real eigenvalues $\{e_m\}$ and orthonormal eigenstates $\{\mathbf{V}^m \leftrightarrow |jm\rangle\}$ (Problem 1.16). The unitary matrix formed from the eigenvectors will transform A into the form diag$(e_j, e_{j-1}, \ldots, e_{-j})$. Since, from part (a) to the problem, A also commutes with the group operators,

$$\Sigma_{m'} e_m D^j_{mm'}(R)|jm'\rangle = \Sigma_{m'} D^j_{mm'}(R) e_{m'} |jm'\rangle.$$

We multiply on the left by $\langle jm|$ to isolate each term in the sum, using the orthogonality of different eigenstates:

$$(e_m - e_{m'}) D^j_{mm'}(R) = 0.$$

In a representation which is irreducible, by definition it is impossible to guarantee that for any choice of m, m' the matrix element $D^j_{mm'}(R)$ is zero for all R. Hence the eigenvalues must themselves all be equal, and A is then proportional to the unit matrix. Part (b) shows then that the proportionality constant is unity, since the trace of A is equal to its dimension. This proof may be transliterated into diagrams as for Problem 1.16:

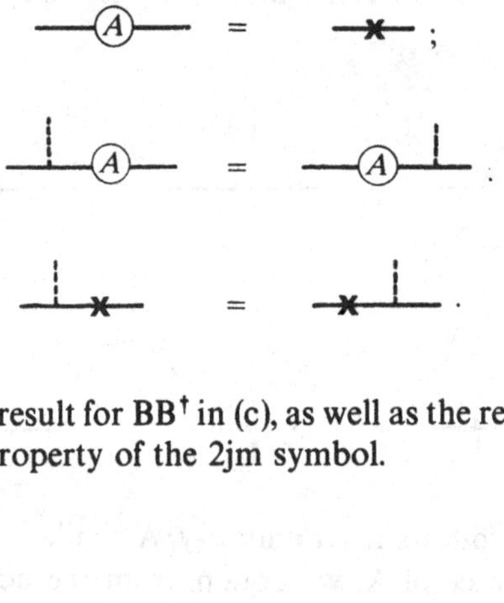

The corresponding result for $\mathrm{BB}^\dagger$ in (c), as well as the result for (d), follows from this unitary property of the 2jm symbol.

Comments

(a) Readers may recognise the argument of part (b) of this solution as the proof of a restricted form of Schur's lemma II; a similar argument for a nonhermitian matrix may be used to prove the unrestricted form (Butler 1981). We shall develop Schur's lemma shortly in Problems 2.6, 2.7. However, we need to use the unitarity of the 2jm symbol in those proofs. To avoid presenting a circular argument, which is not avoided by reordering the material, we have had to include the special derivation of part (b) of this solution.

(b) While discussing the 2jm symbol, it is convenient to introduce a result which may be justified only from the full form of Schur's lemma II, but which is not needed for the proof we later give of the lemma itself: the (possibly nonhermitian) matrix B is also by itself proportional to the unit matrix since it is a commutative matrix. Part (c) of the problem then shows that the proportionality constant η is a phase, which is readily shown (by examining B^2) to be real, and so equal to ± 1. η is called the *2j phase* $\{j\}$. From the explicit form of the 2jm symbol in a spherical basis, we see that $\eta = \pm 1$ as j is even/odd:

$$= \{j\} = (-1)^{2j} .$$

We now combine these definitions to obtain the following.

3jm symbol

$$\begin{pmatrix} J & j & j' \\ M & m & m' \end{pmatrix} = \tilde{J} \begin{pmatrix} J \\ -M \end{pmatrix}^* \langle J-M|jmj'm'\rangle \leftrightarrow$$

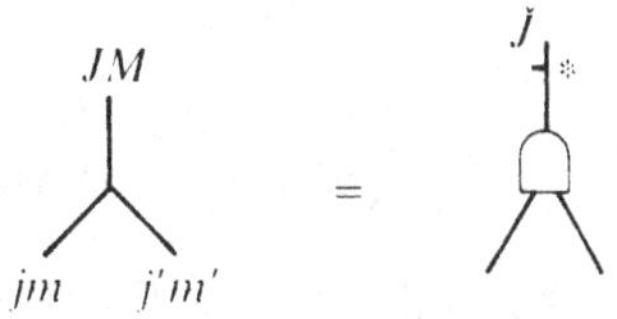

Comments

(a) This has a special importance because it treats the labels more symmetrically. In particular, it couples three kets to an invariant (and could be so defined):

$$\begin{pmatrix} J & j & j' \\ M & m & m' \end{pmatrix}^* |JM\rangle |jm\rangle |j'm'\rangle = |(Jjj')00\rangle.$$

A proof is given in Problem 2.3.

(b) It has the property that it is symmetric under a cyclic permutation of its columns. This allows the above simple diagram form in which no special notation for the vertex is needed.

(c) The Clebsch–Gordan coefficients may be written in terms of the 3jm symbol using the unitary of the 2jm symbol:

(d) The 3jm symbol is symmetric or antisymmetric under an interchange of any two columns as the sum of j values $J+j+j'$ is even or odd. We define the *3j phase* by analogy with the 2j phase as the interchange phase $\{Jjj'\} \equiv (-1)^{J+j+j'}$,

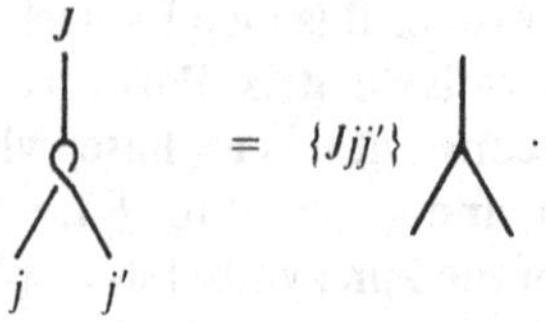

J

$=$ $\{Jjj'\}$

j j'

(e) In a spherical basis, the 3jm symbol satisfies the relation

$$\begin{pmatrix} J & j & j' \\ M & m & m' \end{pmatrix} = \{Jjj'\} \begin{pmatrix} J & j & j' \\ -M & m & m' \end{pmatrix}.$$

This particular form depends on the reality of the 3jm symbol in the Condon and Shortley basis choice. More generally, complex conjugate 3jm symbols are related by a 2jm transformation on each set of labels:

$$\begin{pmatrix} J & j & j' \\ M & m & m' \end{pmatrix}^* = \begin{pmatrix} J \\ M \end{pmatrix} \begin{pmatrix} j \\ m \end{pmatrix} \begin{pmatrix} j' \\ m' \end{pmatrix} \begin{pmatrix} J & j & j' \\ -M & -m & -m' \end{pmatrix}.$$

When generalised to the 3jm symbols of any group, this has been referred to as the *Derome–Sharp lemma* (Butler 1975, 1981). In the light of some results which we shall later prove for other group tensors which have a similar structure, we shall refer to this with its analogues as the *conjugation lemma.* In diagrams:

$*$ $=$ $=$

Throughout this work, successive generalisations of this will be discussed (§§2.5, 2.7, 3.6, 5.3). In brief, any matrix element of a physical operator obeys such a relation, perhaps with a sign included to allow for a time-odd interaction, and may be termed self-conjugate or -anticonjugate, respectively; the same is not true for basis functions, which may usefully be transformed both by the addition of stubs and simultaneously by complex conjugation into a new, conjugate, node.

(f) One step in the proof of the conjugation lemma in the above form is the result that the product of 2j phases in any triad (Jjj') is unity, i.e. $\{J\}\{j\}\{j'\} = 1$. This is clearly consistent with an iteration of the conjugation lemma:

$=$ $*$ $*$ $*$

This condition is certainly fulfilled in SO(3); the number of spin irreps in any triad must be even.

Tables of 3jm values up to $j=8$ are given by Rotenberg *et al.* (1959).

Problem 2.3

Find the consequences of the commutativity of the Clebsch–Gordan coefficient (equation 2.4) for the 3jm symbol.

Solution

R R R = .

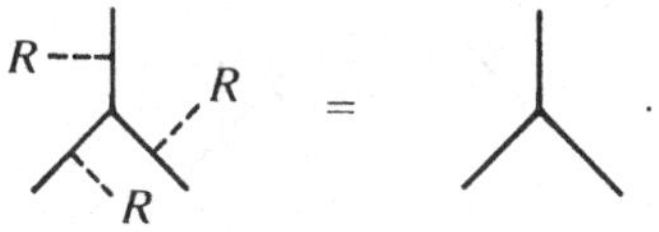

Comment

Note that for our definition of the 3jm (as opposed to the 2jm) symbol, the group operators are applied in a clockwise sense. In general, the *n*jm nodes are invariant, i.e. are unchanged by a joint rotation of all labels by any group operator in either a clockwise or anticlockwise sense. The sense (clockwise or anticlockwise) depends on the definition of each *n*jm symbol, and is reversed by taking complex conjugates. Such points will be spelled out in detail in §2.5.

Problem 2.4

Find the consequence of Clebsch–Gordan unitarity for the 3jm symbol.

Solution

J * = $\check{J}$

Σ_J $\hat{J}$ * = .

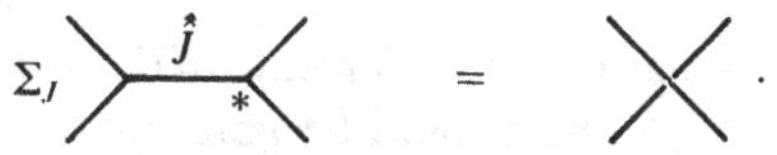

Comment

These might be termed *pseudo*unitarity conditions on account of the normalisation factors.

2.3 Invariant networks; Jucys–Levinson–Vanagas reduction theorems

The combination of *n*jm symbols in a *network* (by which we mean simply a linked set of nodes) will itself have a similar invariance property, provided the 3jm nodes are linked only to 2jm nodes and vice versa. For example:

Problem 2.5

Show that the network

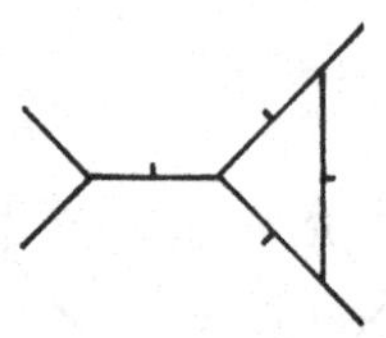

is an invariant network.

Solution

Use the 3jm invariance condition, applied clockwise (Problem 2.3) with a group element R, say, to each 3jm vertex. Then use the 2jm (anticlockwise) invariance condition with a group element R^{-1}. The internal group lines cancel ($RR^{-1}=E$).

In general, we may divide the component nodes into two classes: those with left invariance (such as a 3jm symbol, or a complex conjugated 2jm symbol) and those with right invariance (a complex conjugated 3jm, or a 2jm symbol). The network must be bipartite: left invariant nodes may link only with right invariant nodes, and vice versa, if it is to be invariant as a whole. This is guaranteed in a network of 3jm symbols if one 2jm symbol is placed on each internal line.

The *Great Orthogonality Theorem* (GOT for short) for SO(3) states that:

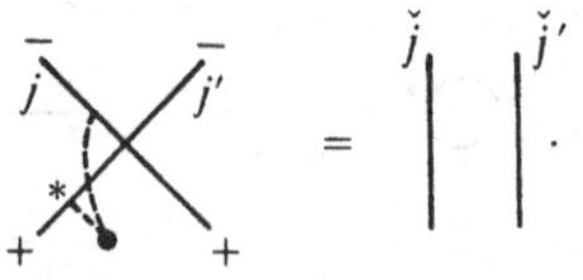

The dot denotes an average over the elements of the group. The parities are for future reference (§2.5), and should be ignored at a first reading.

There are various routes to the proof of this theorem. An extended exercise which is not spelled out in this book but which may interest the reader is the transliteration of the standard proofs of the Great Orthogonality Theorem in, say, Tinkham (1964) or in Leech and Newman (1969), into diagram form. We shall employ an alternative route.

Problem 2.6

Prove the *group summation relations*:

(*a*)  $j = \delta_{j0}$

(*b*)

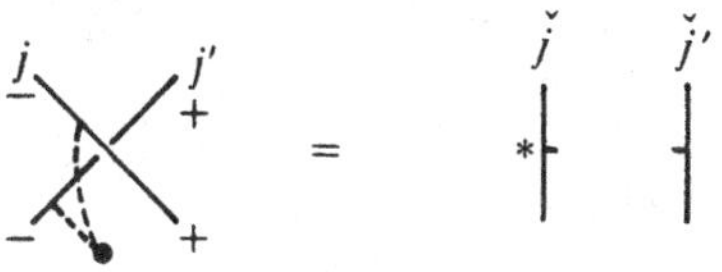

(*c*)

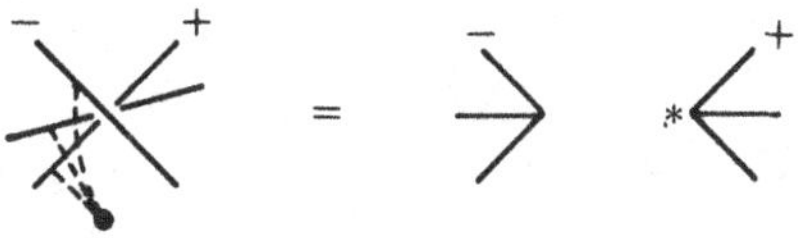

(*d*) $= \Sigma_J$

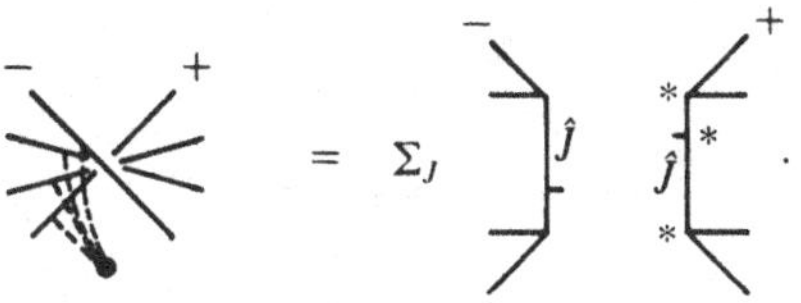

Solution

(a) We use a simple argument here; a more formal and accurate argument is given in the generalisation of §3.2 (see Problem 3.1). The rotational average of any expression which has no rotationally invariant part must vanish. Since this expression is otherwise arbitrary, the appropriate average of the rotation matrix element which performs the averaging must itself vanish.

(b) The Great Orthogonality Theorem may be proved as follows:

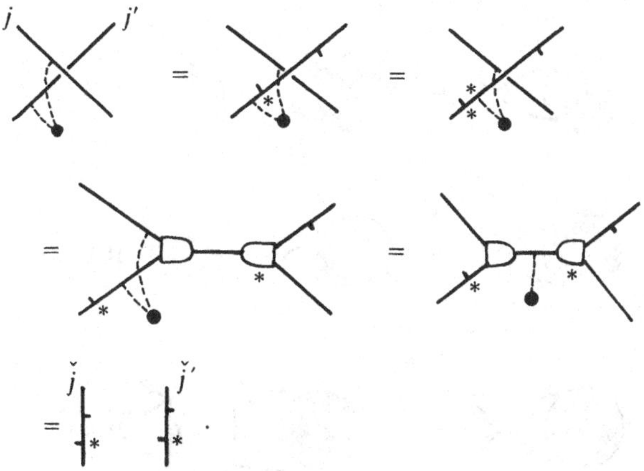

The result stated in the problem, in which one group operator is not complex conjugated and in which 2jm symbols appear on the right side, would follow similarly if we had omitted the stubs and asterisks in the first step. Alternatively, it may be derived as a corollary:

(c)

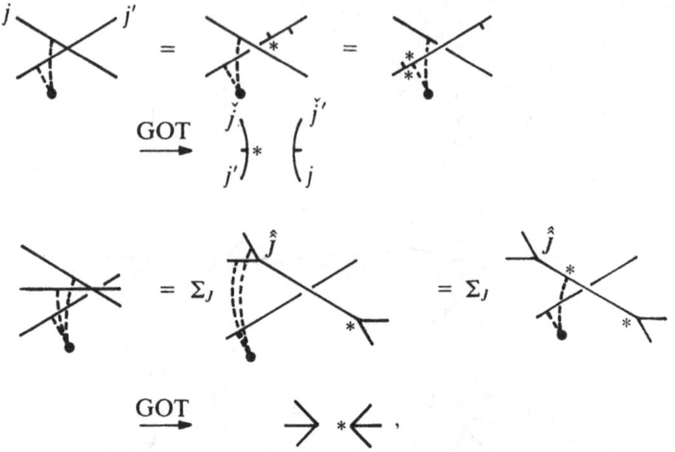

etc.

The last relationship is known as *Gaunt's theorem.*

If a network I_n with n external lines is invariant, the action of n group operators on these lines, when averaged over the group, leads to the basic results, called the *Jucys–Levinson–Vanagas theorems* or JLVn theorems:

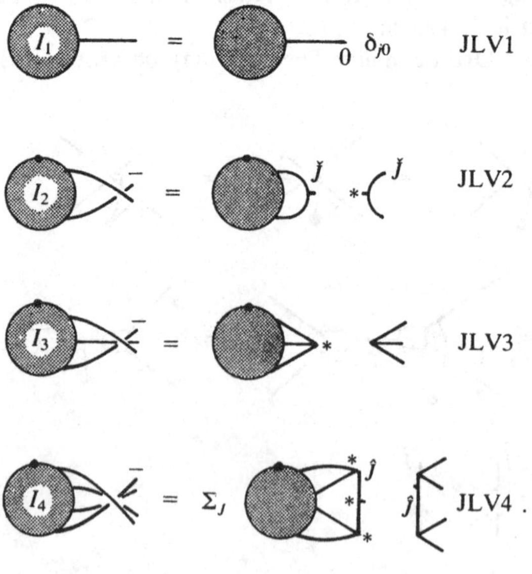

δ_{j0} JLV1

JLV2

JLV3

Σ_J JLV4 .

Problem 2.7
Prove these relations.

Solution
Either: If any diagram (D say) is invariant when both external lines are subjected to the action of any single group element g ($D \to O_g D = D$, say), it is equal to an average over the group of such transformed diagrams (so that $D = (1/|G|)\Sigma_g O_g D$). The part of the diagram in which the group average is found may be re-expressed using the appropriate group summation relation, to give the stated results.

Or: (for JLV2, JLV3, JLV4) we may use the unitarity condition for Clebsch–Gordan symbols to pinch two external lines together, and then commute the group operator nodes through a Clebsch–Gordan symbol to reduce their number, and then apply a lower-rank JLVn theorem. (Better still: do both for practice.)

Comment

(a) In broad terms, made precise in §2.5, *Schur's lemma* corresponds to the second, and the *Wigner–Eckart theorem* to the third, of these results.

(b) A first pedantic point. (Such points are not emphasised here, but will be considered fully in §2.5.) For definiteness, it is assumed in the above diagrams that the invariant node I_n is left invariant, i.e. invariant under clockwise application of the rotation matrix:

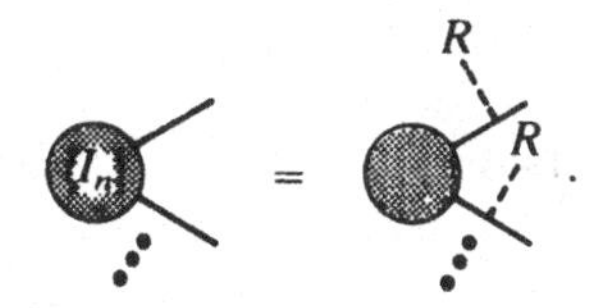

For right invariant nodes (invariant under anticlockwise application) one should ideally complex conjugate all 2jm and 3jm vertices on the right sides of the above results.

(c) A second pedantic point: of related vertices in JLV3, JLV4 (those with the same triad of j values), one is starred and one not. The numerical value is not affected by the choice of which node is asterisked. However, it is wise even now to form good housekeeping habits, and insert asterisks in such a way as to form a bipartite diagram, in which linked nodes always have opposite rotational invariance character (§2.5).

(d) These theorems are illustrated in §1.3 in vector analysis. In particular, we may make the equivalence:

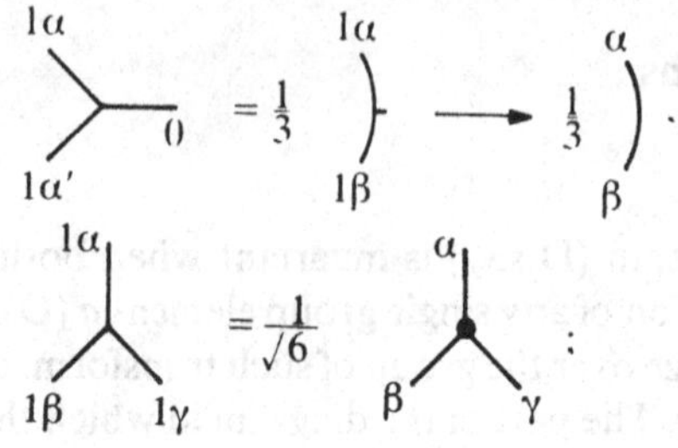

i.e.

$$\binom{1}{\alpha} = \delta_{\alpha\alpha'},\ \begin{pmatrix} 1 & 1 & 1 \\ \alpha & \beta & \gamma \end{pmatrix} = (1/\sqrt{6})\varepsilon_{\alpha\beta\gamma}.$$

This may be verified in detail when the 3jm symbols in the customary spherical (SO(3)→SO(2)) basis (Rotenberg *et al.* 1959) are transformed to a cartesian (SO(3)→D_2) basis. The necessary transformation matrix is the contragredient transformation of Fano and Racah (1959):

$$\begin{pmatrix} |x\rangle \\ |y\rangle \\ |z\rangle \end{pmatrix} = \begin{pmatrix} -1/\sqrt{2} & - & 1/\sqrt{2} \\ i/\sqrt{2} & - & i/\sqrt{2} \\ - & 1 & - \end{pmatrix} \begin{pmatrix} |1\rangle \\ |0\rangle \\ |-1\rangle \end{pmatrix}$$

and amounts to a different phase convention for spherical harmonics.

(e) At this stage we may use JLV2 to prove finally and generally that the matrix B of Problem 2.2 is indeed proportional (by a sign) to the unit matrix.

Problem 2.8

Show that each of the networks

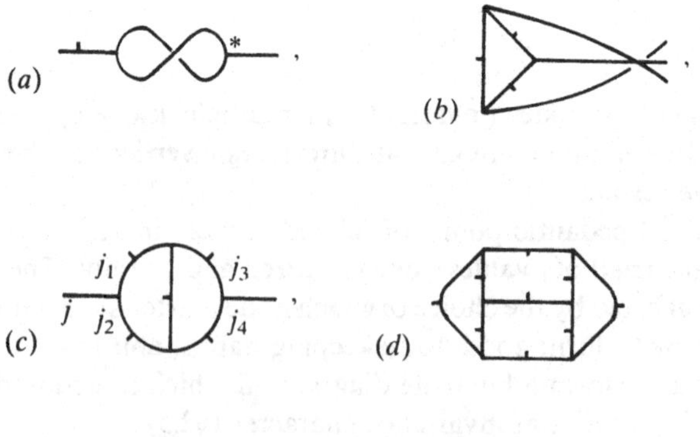

has an invariance property that permits the application of the Jucys–Levinson–Vanagas theorems. Hence simplify it.

Solution

Applicability of JLVn follows from the bipartite structure of each; explicit verification may be made by checking invariance or commutativity as in Problems 2.1, 2.2. The senses are: right invariant, (a); left invariant, (b)–(d). Hence:

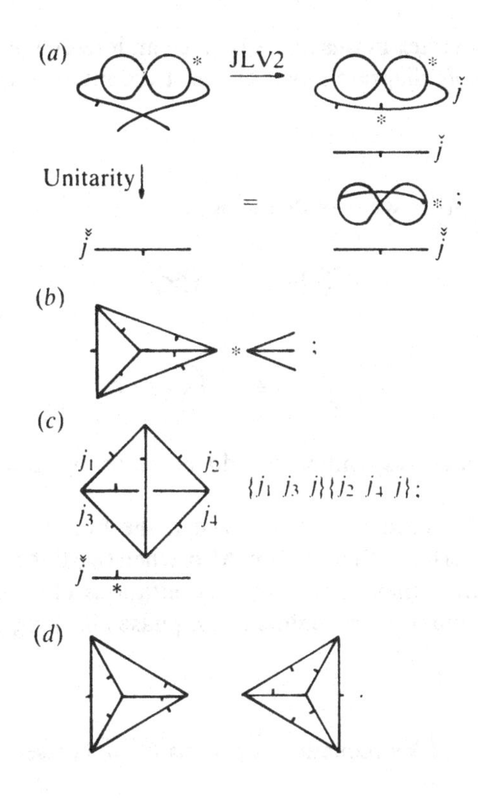

Comments

(a) The tetrahedral figures have a special importance (§2.4).

(b) JLVn is useful whenever a network consists of two subnetworks connected by n lines. These lines are essentially pinched together.

(c) The need to cross lines before pinching may be inconvenient. An adaptation to uncrossed lines follows shortly.

Problem 2.9

Show that the 3j phase can be represented by

$$\{Jjj'\} \leftrightarrow$$

Solution

The solution to Problem 2.8(a) is consistent with the pseudounitarity relationships for the 3jm symbol on the left side only if

$$\ast = 1.$$

Hence if one 3jm vertex in this figure is given an interchange permutation, the resulting *bubble* diagram must represent the interchange phase.

Comments

(a) There are other ways of depicting it:

$$\ast, \qquad \ast,$$

while

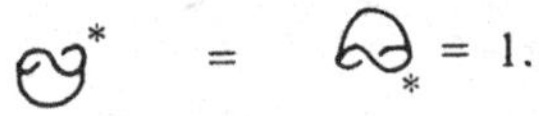

$$\ast = \ast = 1.$$

the above figures corresponding to odd permutations and even permutations, respectively.

(b) The j values need not be specified if the bubble is kept near an appropriate 3jm vertex in the diagram. Also, when two bubbles for the same triad come together, they burst (even permutations of a 3jm symbol are symmetry operations). These points make phase checking less laborious.

Problem 2.10

Find forms of the JLVn reduction theorems for uncrossed lines.

Solution

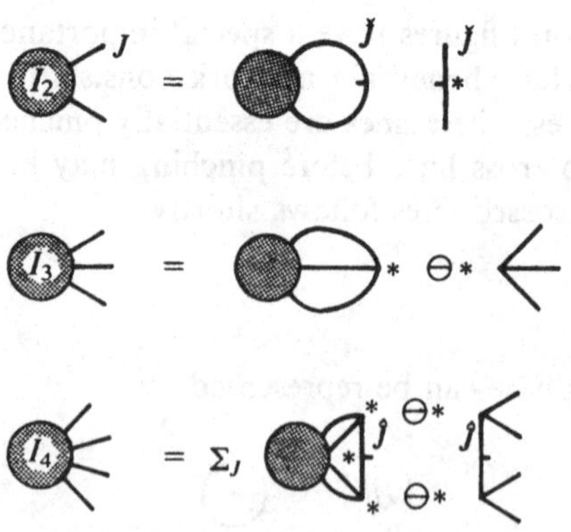

$$I_2 \; J = J \; J \ast$$

$$I_3 = \ast \; \ominus\ast$$

$$I_4 = \Sigma_J \; \ast \; J \; \ominus\ast \; \ast \; \ominus\ast \; J$$

Comments

(a) The necessary 3j phases may always be found after the main part of the calculation (in which lines are pinched ignoring phases), simply by comparing the ordering of the three *j* values at the two vertices formed in the pinching operation. If the ordering (on an anticlockwise convention) is the same, no 3j phase is needed. If not, a bubble is required.

(b) The disposition of asterisks in some results such as the Great Orthogonality Theorem or in Gaunt's theorem, for example, is arbitrary; the theorem is true whichever of two vertices is starred. When, however, these results are transferred to the present applications, it is a valuable habit (see §2.5) to add asterisks in just such a way as to preserve a bipartite structure, i.e. so that only terminals of opposite parity are connected. The asterisks in the above figures are chosen for the case when the subdiagrams I_n are left invariant networks; for right invariant networks, the complex conjugate of all the other nodes should be taken. For example, in the group summation relations of Problem 2.6, the leading terminals (in an anticlockwise sense) on a group operator node have positive parity, while the trailing terminals have negative parity.

(c) When writing out JLV*n* theorems, it is worth remembering that with *un*crossed lines the 2jm vertices (stubs) all face the *same* way, while with *crossed* lines the stubs face *opposite* ways. Which way any one stub faces is immaterial; reversing two corresponding stubs in the JLV2–JLV4 theorems introduces two 2j interchange phases which cancel.

2.4 *n*j symbols

The characteristically polyhedral figures encountered in pinching diagrams correspond to oft-recurring summations over 3jm symbols. (The summations are over all component labels, so that the results are basis-independent; hence the following nomenclature.) We define them as:

(a) The *6j symbol*

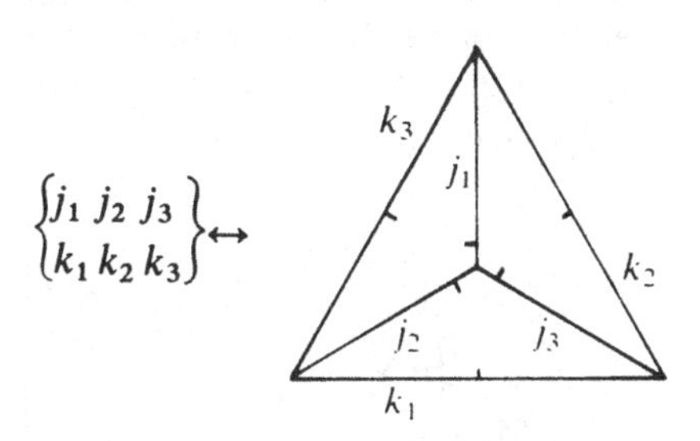

(b) *9j symbol*

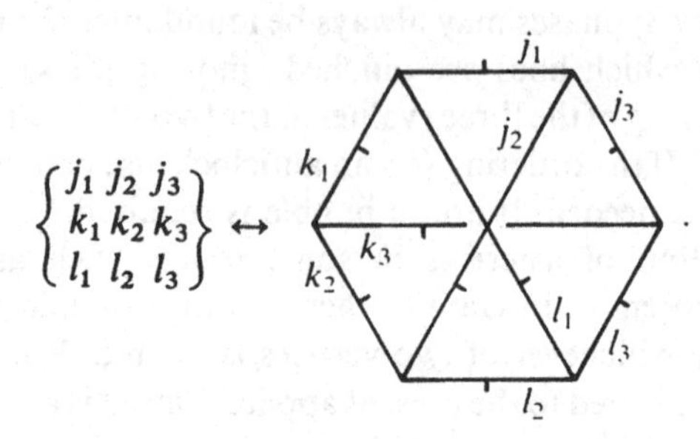

Note the anticlockwise ordering, and the opposition of the labels j_i, k_i.

One may similarly define *n*j symbols for higher *n* values. Their diagram forms are very helpful in displaying their symmetries (Jucys and Bandzaitas 1964, ElBaz and Castel 1972). All, including the 9j symbol, may be reduced to sums of products of 6j symbols using the JLV*n* theorems. We shall not discuss them further here.

We consider the *symmetries* of the 6j symbol.

First, it is unchanged under cyclic permutation of the columns (j_ik_i). This is obvious from the diagram which, on rotation by $\pm 120°$, suffers merely a label change.

Second, it is real. Use of the conjugation lemma and of stub unitarity of the 2jm symbol (stub) triggers the sequence of simplifications:

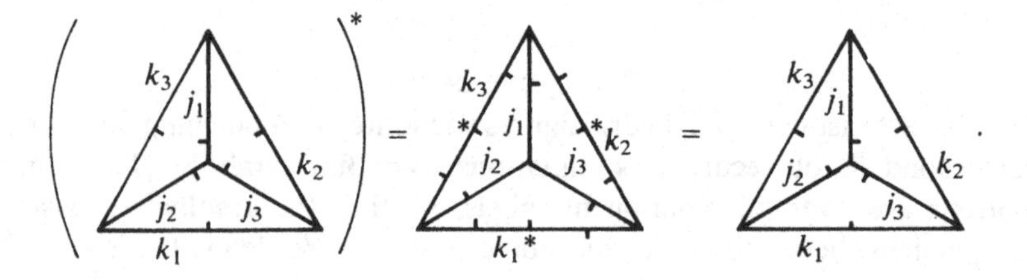

Third, one may interchange columns by exchanging two diagram vertices, so generating the product of four 3jm interchange phases, plus 2j phases which are needed to restore the orientations of the stubs. These phases conveniently cancel $(2(k_1+k_2+k_3)+j_1+j_2+j_3-(j_1+k_2+k_3)-(j_2+k_3+k_1)-(j_3+k_1+k_2)=0)$:

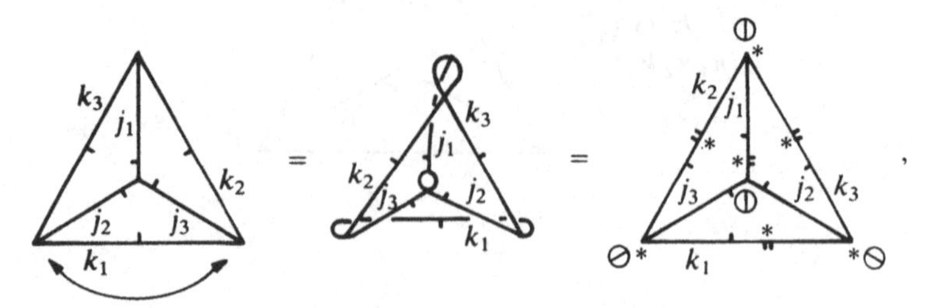

i.e.

$$\begin{Bmatrix} j_1 & j_2 & j_3 \\ k_1 & k_2 & k_3 \end{Bmatrix} = \begin{Bmatrix} j_1 & j_3 & j_2 \\ k_1 & k_3 & k_2 \end{Bmatrix}.$$

Fourth, we may 'pull the diagram inside out', transposing an exterior node to its interior. If we also swap two exterior notes as above, and use $\{j_1\}\{j_2\}\{j_3\} = 1$ etc. we have

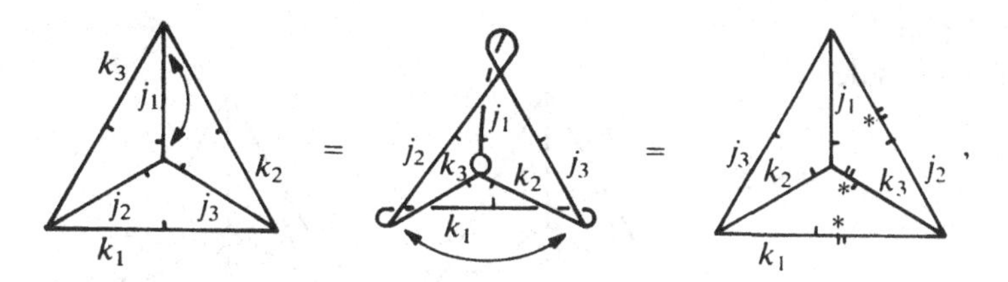

i.e.

$$\begin{Bmatrix} j_1 & j_2 & j_3 \\ k_1 & k_2 & k_3 \end{Bmatrix} = \begin{Bmatrix} j_1 & k_2 & k_3 \\ k_1 & j_2 & j_3 \end{Bmatrix}.$$

Problem 2.11

Prove the orthogonality relation for 6j symbols:

$$\Sigma_J \hat{J} \begin{Bmatrix} k_1 & j_2 & j_3 \\ J & j_1 & j_4 \end{Bmatrix} \begin{Bmatrix} k_2 & j_2 & j_3 \\ J & j_1 & j_4 \end{Bmatrix}^* = |k_1|^{-1} \delta_{k_1 k_2} \leftrightarrow$$

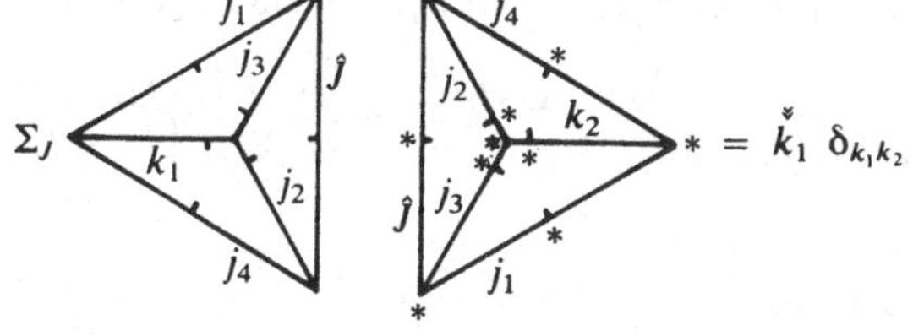

Solution

JLV4, followed by pseudounitarity, gives:

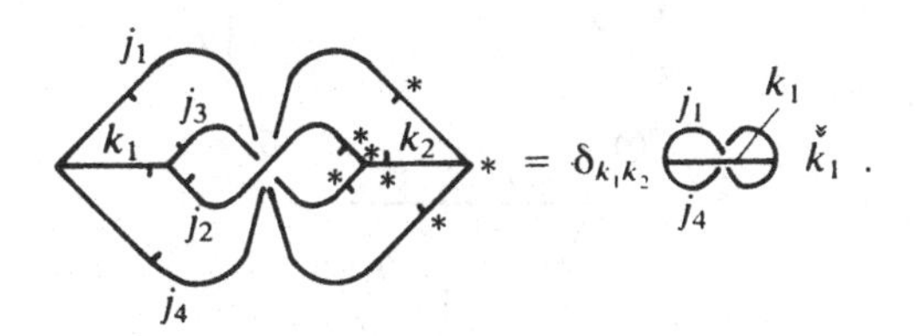

Problem 2.12

Prove the *Racah back-coupling* relation:

$$\Sigma_l \hat{\hat{l}}\{k_2\}\{k_1j_2k_3\}\{j_1j_2j_3\}\{j_1k_1l\}\begin{Bmatrix} j_2 & j_1 & j_3 \\ k_1 & k_2 & l \end{Bmatrix}\begin{Bmatrix} j_1 & k_1 & l \\ j_2 & k_2 & k_3 \end{Bmatrix} = \begin{Bmatrix} j_1 & j_2 & j_3 \\ k_1 & k_2 & k_3 \end{Bmatrix}.$$

Solution

The left side

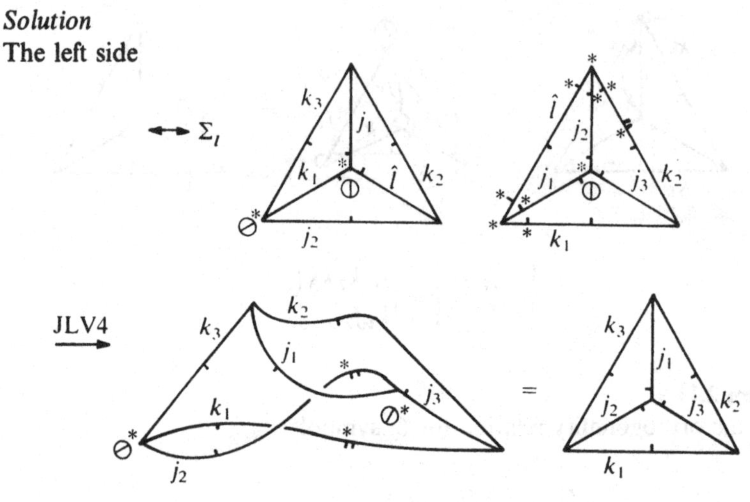

Problem 2.13

Prove the *Biedenharn–Elliott sum rule*:

$$\cdot\Sigma_J \hat{\hat{J}}\{j_1k_2k_3\}\{j_2k_3k_1\}\{j_3k_1k_2\}\{k_1l_1J\}\{k_2l_2J\}\{k_3l_3J\}\{l_2\}\{j_2\}$$
$$\times\begin{Bmatrix} k_2 & j_1 & k_3 \\ l_3 & J & l_2 \end{Bmatrix}\begin{Bmatrix} k_3 & j_2 & k_1 \\ l_1 & J & l_3 \end{Bmatrix}\begin{Bmatrix} k_1 & j_3 & k_2 \\ l_2 & J & l_1 \end{Bmatrix} = \begin{Bmatrix} j_1 & j_2 & j_3 \\ k_1 & k_2 & k_3 \end{Bmatrix}\begin{Bmatrix} j_1 & j_2 & j_3 \\ l_1 & l_2 & l_3 \end{Bmatrix}.$$

Solution

The left side becomes:

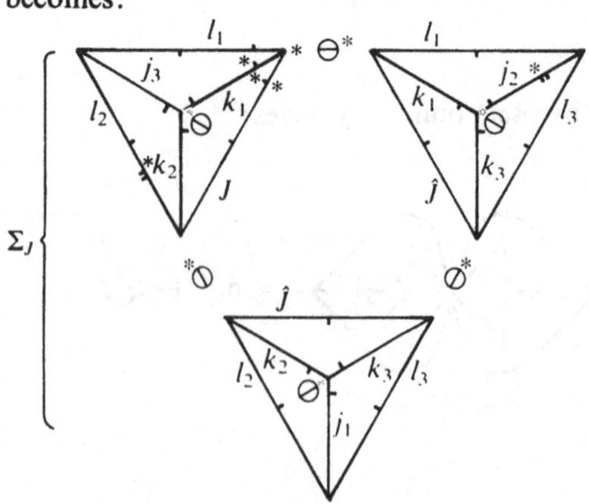

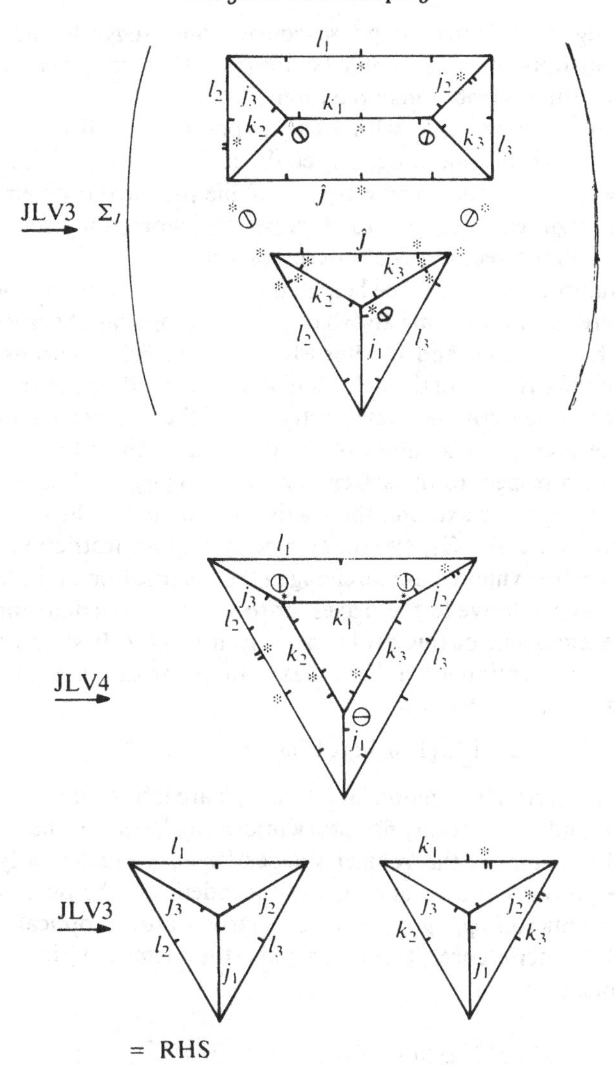

2.5 Diagram housekeeping: invariance, commutativity and parity rules

The reader interested in only the broad outline, and the reader who has a healthy contempt for keeping track of phases in a calculation, and especially of phase conventions, is wasting his or her time here. It is admittedly unpopular to raise such pedantic matters. Cvitanovic (1984) is able to claim that 'fixing a phase convention is a waste of time, as the phases cancel in summed over quantities'. Butler (1981), while giving explicit kets and spending much time on phase conventions, suggests that phase-dependent answers are rarely useful. Danos (1971) has similar comments in connection with 9j symbol manipulations.

I have enormous sympathy with such viewpoints. If one lived on a desert island, metaphorically speaking, one could achieve self-consistency fairly easily. However, it is often necessary in real life practical problems to get one's phases right when relating to other people's work, and necessarily so for phases with a direct physical interpretation.

A few examples of this may be helpful in making the point. Suppose that we are interested in knowing to what extent the optical absorption of a system in E1 coupling and for linearly polarised light depends on the direction of polarisation of the light. We would normally assign x, y, z axes to the system for describing its symmetry, and if the symmetry group were say C_{3v} interpret the generators of the group as a three-fold rotational symmetry with respect to the system z axis and as σ_y, a mirror reflection plane containing the z axis and the x axis. We might ask the question: to what extent would the C_{3v} system symmetry impose restrictions on the variety of spectra exhibited as one changed the polarisation of the light, e.g. from z to x to y, relative to these axes, or to some intermediate direction? The full development of this problem is given in §6.2. It shows that the intensity of absorption of a light beam of polarisation $\mathbf{e}$ involves a geometrical factor of the form

$$\Sigma_\mu \langle 1^- \rho | 1^- \varphi\lambda\mu \rangle \langle 1^- \varphi'\lambda\mu | 1^- \sigma \rangle e_\rho e_\sigma{}^*$$

where ρ, σ are cartesian component labels; λ, μ are labels for C_{3v} irrep and component; and φ, φ' are any necessary branching labels for the reduction $O(3) \downarrow C_{3v}$. In calculating this geometry dependence, one may clearly run up against the problem of determining any dependence of the phase of these matrix elements on φ, φ', μ etc. Similarly natural optical activity ($E1-M1/E2$ interference, §6.3) demands the determination of such combinations as

$$\langle 2^+ \rho\sigma | 2^+ \varphi\lambda\mu \rangle \equiv \Sigma_M \langle 2^+ M | 2^+ \varphi\lambda\mu \rangle \begin{pmatrix} 2 & 1 & 1 \\ M & \rho & \sigma \end{pmatrix}$$

in the calculation of which the relative phases of the 3jm symbol and the matrix element it multiplies must be consistent. In electric linear dichroism one requires the further combination

$$\Sigma_\mu \langle 2^+ \rho\sigma | 2^+ \varphi\lambda\mu \rangle \langle 1^\pm \varphi'\lambda\mu | 1^\pm \tau \rangle,$$

which obviously requires further consistency in phase choice.

It is essential that diagram methods have all the capabilities of the algebraic formalism. When Jucys *et al.* (1960) first published, their diagrams did not keep track of phases; it was not for nothing that their later work (Jucys and Bandzaitas 1964) remedied this.

All that is necessary to keep track of phases is a tidy, 'good housekeeping' approach. The ground work for this has already been laid, in Problem 2.7 and in a variety of Comments on earlier work. A few simple rules: make the diagram bipartite; insert a 3j phase bubble if triads do not match etc.; make it possible to get the correct phase in the answer even if the phase is ignored in the calculation! It is similar to the situation in relativity. While one may ignore the distinction between contravariant and covariant suffices, once the necessity of care is understood and accepted, the practical complications are minimal: simply sum each contravariant suffix against a covariant one. The connection with relativity in fact runs very closely. Much of the discussion will centre on judicious use of the 2jm symbol, which is the counterpart to the metric tensor of relativity, relating tensors that transform in complementary manner. Since good habits cannot be started too early, we spend this section in an explanation of these and of related points.

Many of the necessary points have already been made in our problems. The diagram technique of Jucys *et al.* depends for its utility on the property of invariance – or of commutativity – with respect to group operations, both for all constituent nodes, and for any network into which these nodes are combined.

Our first task is to emphasise the distinction between invariance and commutativity.

Invariance

A node is right invariant (or left invariant) if the action of any one group operator simultaneously on the external lines in an anticlockwise (or clockwise, respectively) sense leaves it unchanged:

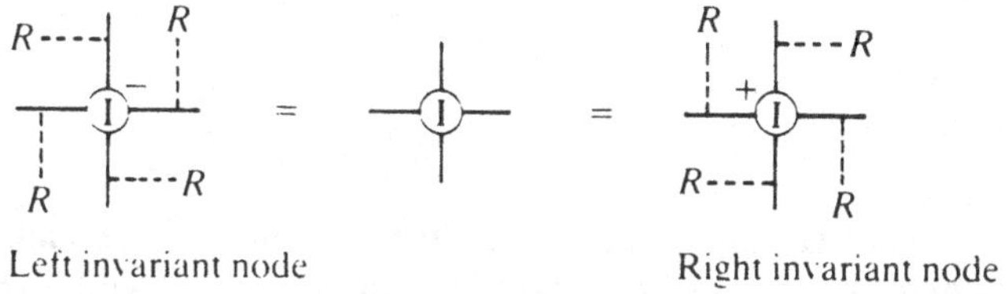

R R R R − I = I = R R + I R R

Left invariant node Right invariant node

Comments

(a) For invariant nodes, all lines are rotated similarly.

(b) An example of a left invariant node is a 3jm symbol, and an example of a right invariant node is the 2jm symbol (§2.2). The distinction is purely one of convention, since the taking of complex conjugates reverses the parity (see Comment (c)). If the *n*jm symbol is real, as in the Condon and

Shortley (1935) phase choice, it has both left and right invariance. However, we ignore this possible simplification in this section; it makes housekeeping easier to have a precise rule, and the transition to the general case is much less painful.

(c) From the unitarity of the rotation matrix, it follows that complex conjugation converts left invariance into right invariance:

(d) We assign a parity +1 to (all the lines of) a right invariant node, and −1 to a left invariant node.

Commutativity

A node is commutative if its terminals (or line–node junctions) may be divided into two classes, such that the action of any group operator simultaneously on all lines of the first class in an anticlockwise sense has the same effect as the action of the same group operator simultaneously on the lines of the second class in a clockwise sense:

Comments

(a) In the case of a commutative node, the two kinds of terminals, called above first and second class (those operated on the left and right sides of the above diagram equation), may be thought of as row and column labels, respectively, for a matrix which commutes with the matrix group operator:

$$D^{(j)}_{m'm}N_{mn}=N_{mn'}D^{(j)}_{nn'}.$$

(b) These terminals may be assigned opposite parities (+, −, respectively) accordingly. The invariant (or commutative) networks required (as in Problem 2.5) for the applicability of the JLVn theorems are then defined in full generality by the requirement of *bipartite construction*: all internal lines must connect terminals of opposite parity.

(c) An example of a commutative node is a Clebsch–Gordan coefficient (§2.2).

(d) If an invariant node is unchanged when group operators act simultaneously on all lines, a commutative node is unchanged when group operators act simultaneously on all lines, with the qualification that group operators acting on lines of one class are complex conjugated. This is readily proved using unitarity to take all group operators to one side of the equation. For example:

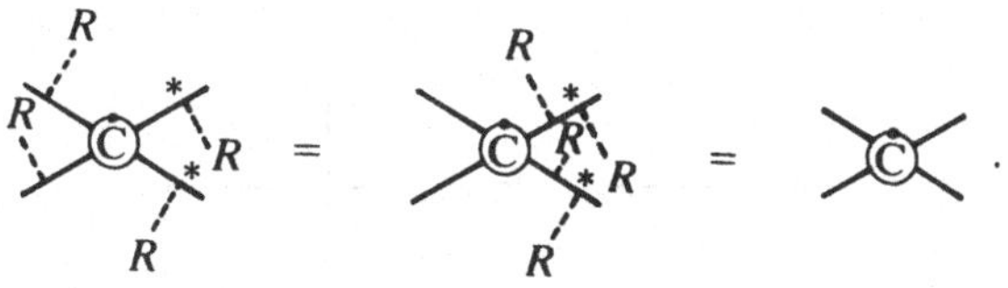

(e) An invariant node may be turned into a commutative node and vice versa, by adding stubs to one set of lines. For example, a Clebsch–Gordan coefficient and a 3jm symbol are so related (§2.2). Since the stub is itself an invariant node, if it is added to another invariant node of opposite parity the result will be a commutative node:

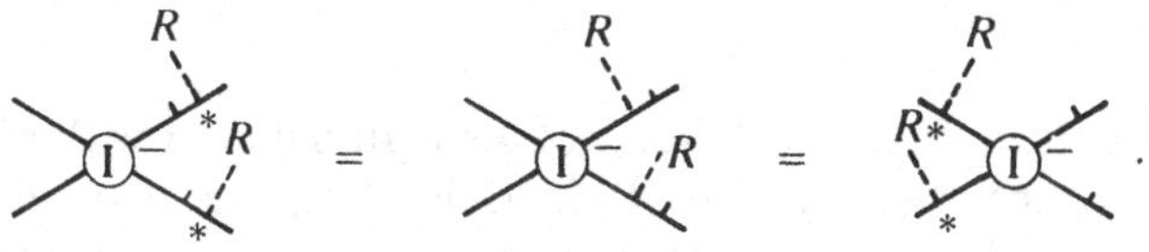

Examples of this include the commutative matrices A, B of Problem 2.2.

(f) The group summation relations and the JLVn theorems may be adapted once more and for commutative nodes using this property; see the following problems. This gives Schur's lemma directly.

(g) The matrix product of commutative nodes is itself commutative:

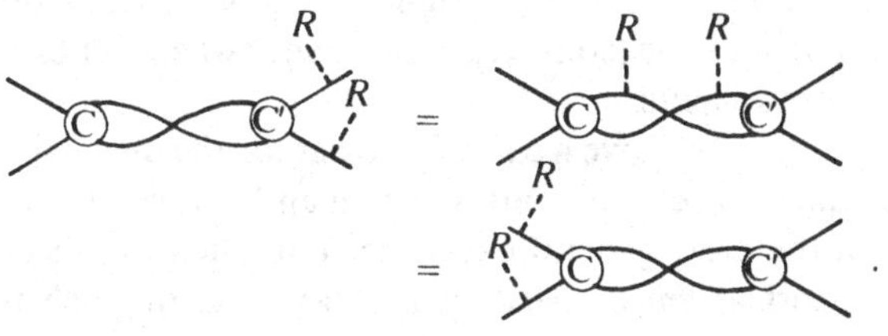

This is an example of the general construction of invariant networks (see Comment (b)).

(h) The group operator node (§2.2) is one node which is not itself either invariant or commutative in the irrep labels. We shall overcome this deficiency in §3.7.

(i) The problems already given, and those to follow, will aid in building up expertise in manipulating and combining the transformation characters of component nodes in a network, adding stubs judiciously to diagram equations where necessary, so as to construct invariant or commutative networks. For example, a group operator matrix element can be commuted through a stub at the price of a complex conjugation:

Throughout this work, we have placed asterisks and stubs etc. on all results in such a manner as to ensure left or right invariance of the networks obtained on combining them. As mentioned in Comment (b) of Problem 2.10, the application of the group summation relations gives a searching test of one's abilities in handling parities consistently, since in the right side of such a summation relation, either of two *njm* symbols may apparently be complex conjugated; however, only one of those choices preserves the canonical structure.

Let us as an illustration rewrite some of the material in §1 with more care. As noted in the discussion of Rule 6 in §1.2, we may distinguish contravariant and covariant vector components by adding a sign (parity) in the spirit of the present section. For example

$$a^i \leftrightarrow \textcircled{a}\overset{+}{\underset{i}{—}}, \quad a_i \leftrightarrow \textcircled{a}\overset{-}{\underset{i}{—}}.$$

One may standardise the nodal sign choice on, say, the covariant node (this conforms with the negative parity assigned to kets, cf. §2.7), omitting the minus sign, and include the (contravariant) metric whenever discussing the contravariant components.

When, for example, we have a cartesian basis for the discussion of SO(3) vectors, the distinction between contravariant and covariant components is unnecessary. In the more general basis choice implicit in this chapter, for example the spherical tensor basis, it is vital to distinguish these components, and the raising/lowering metric components may be identified with the 2jm symbol and its complex conjugate, respectively:

$$g^{ij} \leftrightarrow \quad , \qquad g_{ij} \leftrightarrow \quad .$$

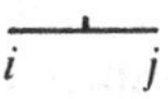

The metric character of the 2jm symbol arises since (for any basis) within SO(3) g_{ij} is an invariant tensor and so is proportional to the 2jm matrix through JLV2. Hence, as illustrated above for a vector, a contravariant component of a tensor is obtained by adding suitable 2jm transformations to the covariant component. (A fuller analysis is given in §§3.6, 5.3.)

The necessity of introducing a metric in turn demands further precision in defining combinations of vectors. The scalar product should be written

$$\mathbf{a}\cdot\mathbf{b} = a^i b_i = a_i g^{ij} b_j \leftrightarrow \quad = \quad ,$$

while the vector product is written

$$(\mathbf{a} \times \mathbf{b})_k \leftrightarrow \quad = \quad .$$

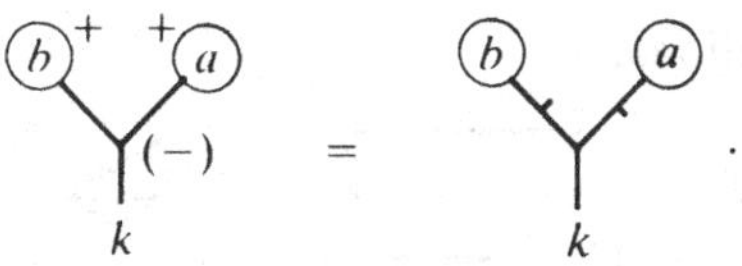

These diagrams now satisfy the requirement of bipartite structure.

The *conjugate node*, denoted by adding a tilde, is defined by transforming by both complex conjugation and the addition of 2jm symbols on each external line:

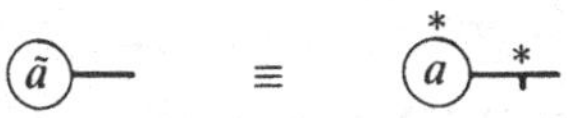

$$\tilde{a} \equiv a^*$$

The 2jm symbols have the effect of restoring the original parity, and in the spirit of this subsection afford a clearer comparison with the original node. The conjugate node is not necessarily equal to the original node. This illustrates the fundamental distinction mentioned at the end of §1.2, and treated more fully in §5.3. In general, the node conjugate to that representing the covariant components of a tensor is the complex conjugate of the node representing the contravariant components: $\tilde{a}_i = (a^i)^*$.

If the conjugate node is equal to the original node, or to its negative, as for the *n*jm symbol of §2.2, we call the node *self-conjugate* or *-anticonjugate*, respectively. When this equivalence is restricted to one sense only of the application of the stubs – anticlockwise or clockwise but not both – we term the node *right-* or *left-conjugate* accordingly.

A network may possess a conjugation property unshared by the component nodes; for example, whatever the conjugation character of a vector **a**, the coupled tensorial product $[\mathbf{a}\tilde{\mathbf{a}}]^J$ of itself and its conjugate will have the conjugacy parity $(-1)^J$ (i.e. the coupled node will be self-conjugate or -anticonjugate as J is even or odd, respectively):

$$([\mathbf{a}\tilde{\mathbf{a}}]^J)^{\sim} \leftrightarrow$$

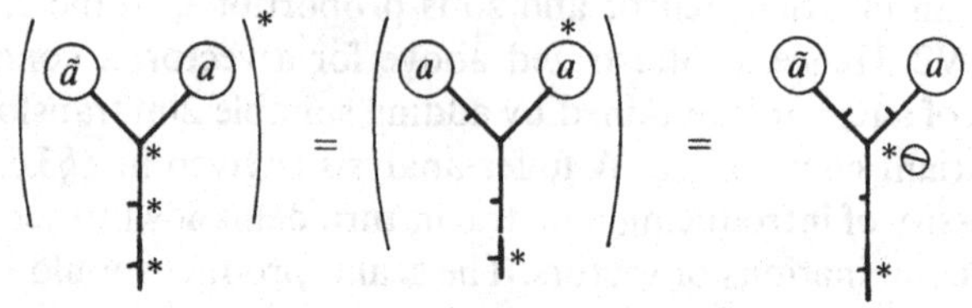

$$\leftrightarrow [\mathbf{a}\tilde{\mathbf{a}}]^J\{1\}\{1\,1J\}.$$

2.6 Schur's lemma, Wigner–Eckart theorem, and matrix elements of irreducible tensor operators

Problem 2.14
Prove the relations:

(a)

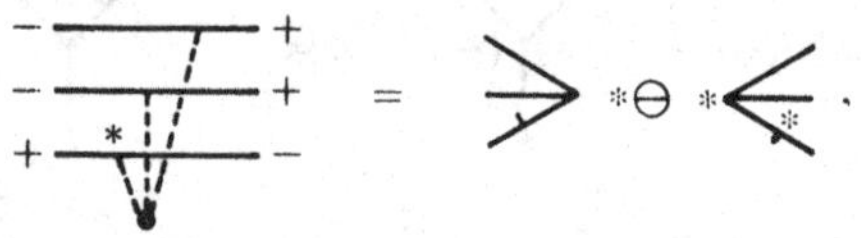

(b) $= \Sigma_j$

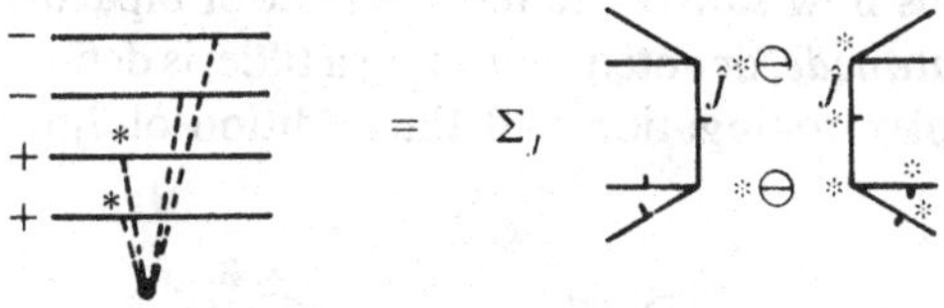

and deduce the (adapted) JLV2–4 theorems in the form:

(a′)

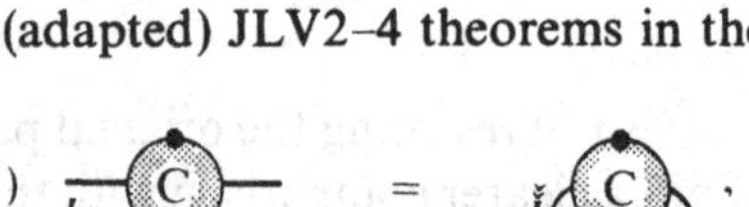

(b′)

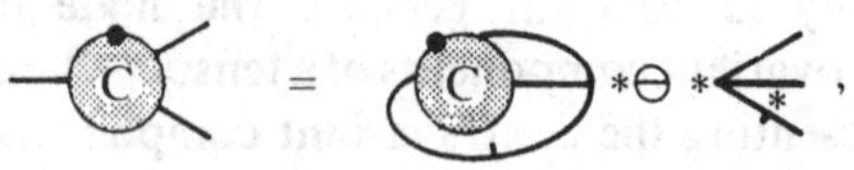

(c′) $= \Sigma_j$

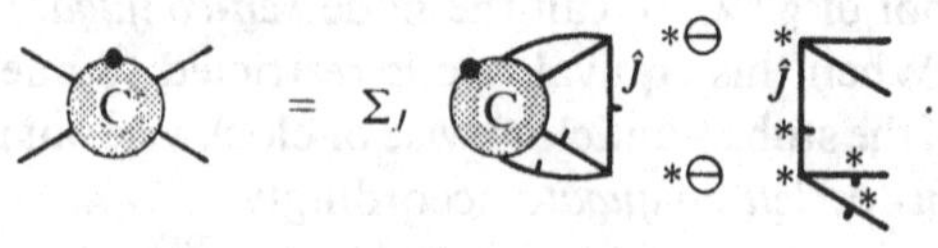

Solution

For example: (a)

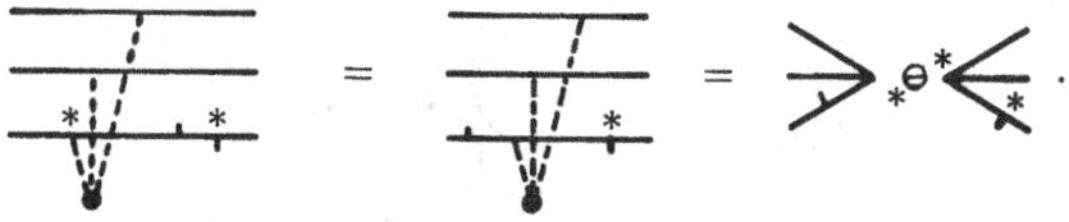

For JLVn, proceed as for Problem 2.7. We have added parities to help the reader of §2.5.

Comments

(a) We have now constructed a result, figure (a′), which corresponds to *Schur's lemma II*: a matrix acting within an irrep which commutes with a group operator must be a multiple (itself determined from the trace) of the unit matrix.

(b) The second figure (b′) leads to the *Wigner–Eckart theorem*, which states that the matrix elements of an irreducible operator in an irreducible basis are proportional to the appropriate Clebsch–Gordan coefficient. (Apart from a trivial normalisation, the Clebsch–Gordan coefficient may be recognised in the open diagram on the right side of (b′), as in §2.2.) In this subsection we use our present notation to work entirely in terms of matrix elements; in the next, we give an economical approach to working directly with operators and state vectors.

An *irreducible tensor operator* (ITO) $\mathbf{X}^j$ may be defined as one whose matrix elements correspond to commutative nodes:

$$\langle JM|X^j_m|j'm'\rangle \leftrightarrow$$

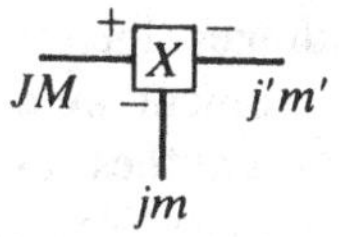

From JLV3, this matrix element is proportional to a Clebsch–Gordan coefficient. This proves the Wigner–Eckart theorem for SO(3):

$$\langle JM|X^j_m|j'm'\rangle = \langle J||X^j||j'\rangle\langle JM|jmj'm'\rangle \leftrightarrow$$

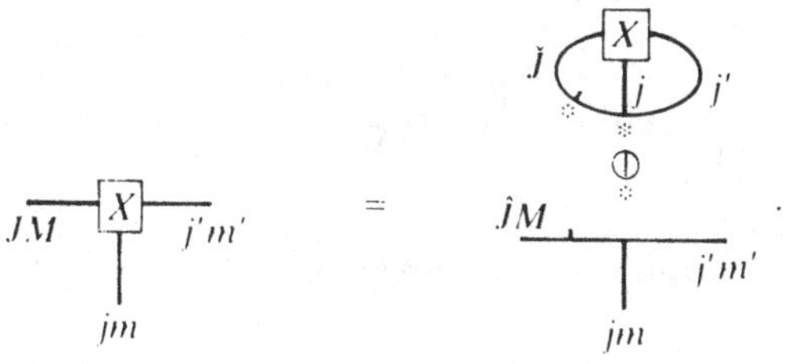

(c) The *reduced matrix element*, i.e. the ratio of a matrix element of an ITO to the corresponding Clebsch–Gordan coefficient, is in diagram form the closed diagram generated by JLV3:

$$\langle J \| X^j \| J' \rangle \leftrightarrow$$

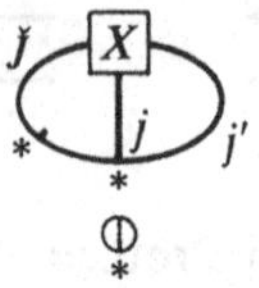

(d) Adding a stub to the node for an irreducible tensor operator matrix element gives a left-invariant node, distinguished from its commutative counterpart by a circular symbol in place of the square:

Details of this will be given in the next section.

2.7 Basis functions and operators

2.7.1 Representations of kets and operators

We demonstrate in this subsection that our fundamental notational technique, based as it is on matrix elements, is adequate for denoting Hilbert space kets in a particular 'representation' (in the sense discussed by Butler 1981). In the next subsection we introduce a more abstract notation which is compatible with the earlier notation.

Consider the Schrödinger eigenfunctions $\{\psi_i(\mathbf{r}) | i=1, 2, \ldots\}$ for a Hamiltonian $H(\mathbf{r})$; for the moment we consider a spatial dependence only. In Dirac notation we represent these as the inner product of the Hilbert space ket or state vector $|i\rangle$ with the eigenfunction $|\mathbf{r}\rangle$ of the position operator $\mathbf{r}$:

$$\psi_i(\mathbf{r}) = \langle \mathbf{r} | i \rangle.$$

In general we may find a representation $\langle e | i \rangle$ of a ket $|i\rangle$ using a complete set $\{|e\rangle\}$ of eigenfunctions of a suitable operator on the space.

The natural diagram representation for a matrix element forces a notation like

$$\langle e | i \rangle \leftrightarrow$$

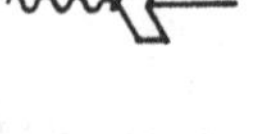

$$\langle i | e \rangle = \langle e | i \rangle^* \leftrightarrow$$

for a ket representation and its conjugate. (In earlier papers, such as Stedman 1976a, we used a circle for the node; however, it is advisable to have a logo which is not reflection symmetric so as to distinguish ket from bra labels, and this particular shape is chosen with an eye to the extended notation of §2.7.2.) For example, ElBaz and coworkers write a spherical harmonic as

$$Y^l_m(\theta,\phi)\leftrightarrow \quad \underset{lm}{\longrightarrow\!\!\!\dashv}\!\!\text{----}_{\theta\phi} \; .$$

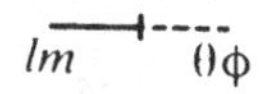

Orthonormality and completeness give:

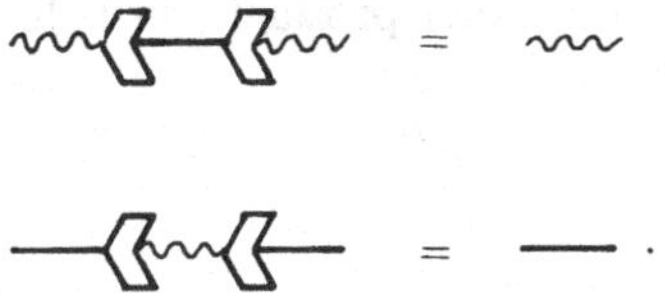

The wavy line is the unit operator on the Hilbert space:

$$\Sigma_e|e\rangle\langle e|=1\leftrightarrow \;\sim\!\!\sim\!\!\sim \; .$$

while a projection operator onto a subspace is defined as usual by limiting the summation range of the state labels:

$$\mathrm{P}^{\mathscr{S}}\leftrightarrow\Sigma_{i\in\mathscr{S}} \; .$$

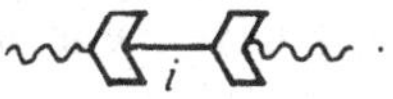

One advantage of this notation is that we can define the action of a group element in Hilbert space as a line on a wavy line, oriented in such a way as to make a ket left invariant. Indeed, all the basic operations may be handled in this format, and time reversal leads to a result analogous to the conjugation lemma (Stedman 1976a). We defer a discussion of this till after the development of the new notation which follows.

2.7.2 Abstract ket notation

It is valuable to have a basis-independent depiction of a state vector. We suggest the definitions:

$$|\alpha jm\rangle\leftrightarrow \; jm \; .$$

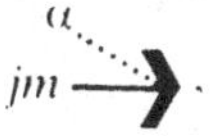

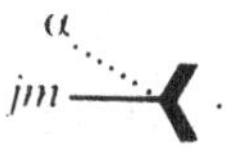

$$\langle\alpha jm|\leftrightarrow \; jm \; .$$

Comments

(a) The chevrons, or angular brackets, serve to distinguish bra from ket, and are therefore preferable to a square end, say.

(b) The angles are chosen in this particular way to make a mnemonic connection with the angular brackets in the usual algebraic notation.

(c) The angular notation is compatible with the notation of §2.7.1 in the sense that:

$\langle e|i\rangle \leftrightarrow$ 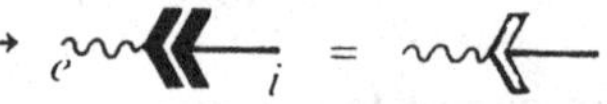

(d) Orthonormality under inner products on Hilbert space give

(e) Completeness also gives the forms

 $\leftrightarrow \mathbb{1}$

 $\leftrightarrow \mathrm{P}^{\mathscr{S}}$

for the unit operator and for a projection operator onto a subspace $\mathscr{S}$.

(f) In general, operators have the form

$\mathrm{O} = \Sigma_{ab}|a\rangle\langle a|\mathrm{O}|b\rangle\langle b| \leftrightarrow$ 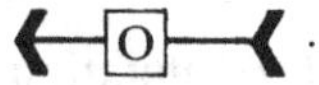

(g) Hermitian conjugation would complex conjugate all nodes and reverse the sense of the chevrons:

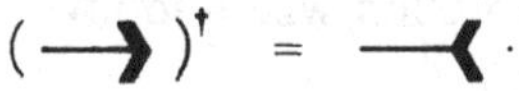

(h) Equation (2.1) has the diagram form

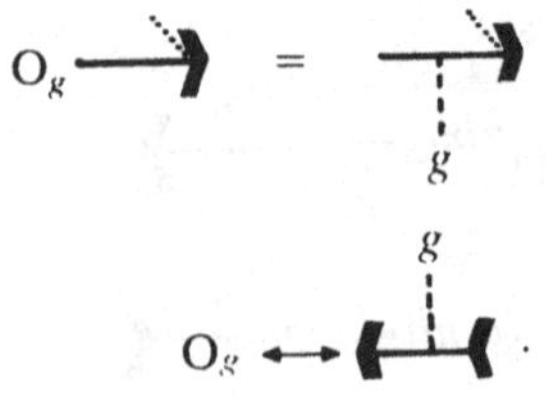

2.7.3 Irreducible tensor operators, the Wigner–Eckart theorem, and hermitian conjugate operators

The Wigner–Eckart theorem, as proved at the end of §2.6, relates to matrix elements of irreducible tensor operators (ITOs for short). These operators correspond to commutative nodes in which first class terminals (of positive parity) correspond to bra labels, and second class terminals (negative parity) to the operator and ket labels. If a stub is added explicitly to the bra label line, the residual node is of negative parity, and so left (rather than right) invariant.

The hermitian conjugate of an irreducible tensor operator is not itself in the same standard irreducible form. The action of hermitian conjugation interchanges ket and bra character, and we restore the original pattern only if we also introduce a stub on the operator line. We may see this problem and resolve it by the methods of either §2.6 or §2.7.

Problem 2.15

Show that the hermitian conjugate $(Q^j_m)^\dagger$ of an ITO Q^j_m is transformed into an ITO by the addition of a 2jm symbol:

$$\tilde{Q}^j_m \equiv \begin{pmatrix} j \\ -m \end{pmatrix}^* (Q^j_{-m})^\dagger.$$

Solution

We give two solutions. The first works in terms of matrix elements and uses an invariant form of the node corresponding to the tensor. The second works in terms of operators, and uses a commutative form of the tensor node.

(*a*)

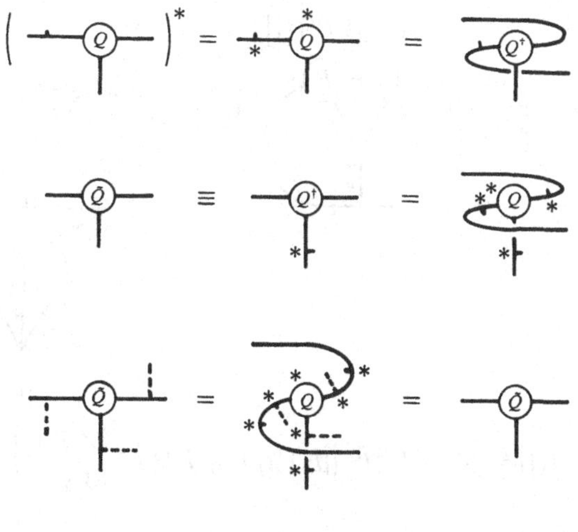

(b)

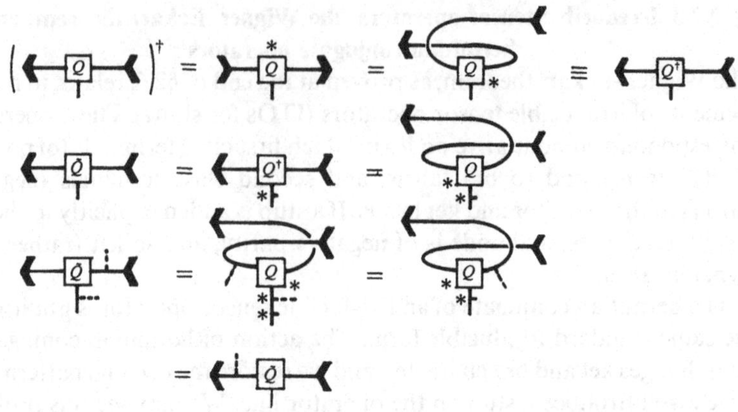

Comments

(a) A creation operator, creating a state $|jm\rangle$ with irreducible transformation properties from the vacuum, is *ipso facto* an ITO with these same irreducible labels. Hence an appropriately rephased annihilation operator will also be an ITO. We shall use this in §2.8.

(b) The reduced matrix elements of a creation/annihilation ITO are related to the *fractional parentage coefficients* of Racah algebra (Judd 1967), which more generally couple the product of a one-particle and an n-particle state to a symmetrised $(n+1)$-particle state.

Problem 2.16

Find the relationship between the reduced matrix element of a coupled tensor operator and those of its component ITOs.

Solution

$$\langle j_1||[A^j B^{j'}]^J||j_2\rangle \leftrightarrow$$

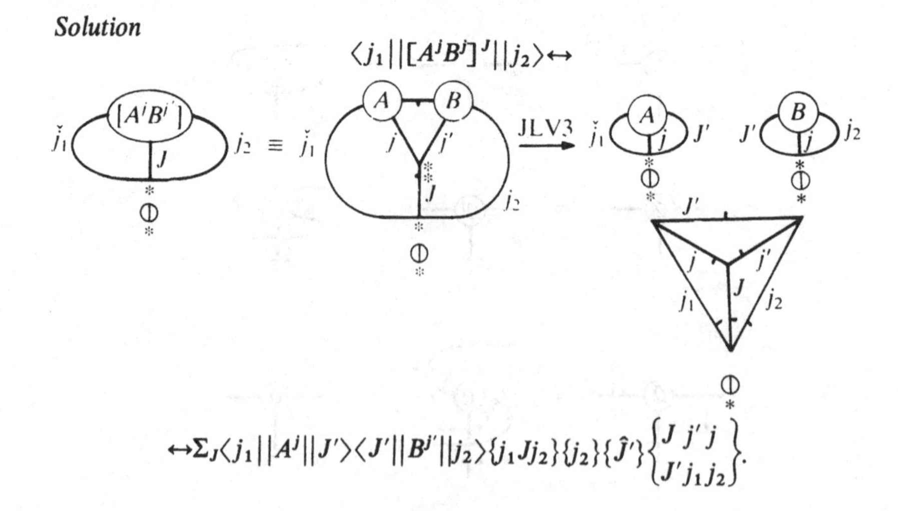

$$\leftrightarrow \Sigma_J \langle j_1||A^j||J'\rangle\langle J'||B^{j'}||j_2\rangle\{j_1 J j_2\}\{j_2\}\{\hat{J}'\}\begin{Bmatrix} J & j' & j \\ J' & j_1 & j_2 \end{Bmatrix}.$$

2.7.4 Time-reversed kets and operators

Time reversal commutes with spatial operators: $\overline{R}=R$ (we use an overbar for time reversal throughout this book). However, it is antilinear, and under time-reversal equation (2.1) (including a parentage label) reads:

$$R|\overline{\alpha jm}\rangle=[D^{j}_{m'm}(R)]^{*}|\overline{\alpha jm'}\rangle.$$

Hence, the time-reversed ket transforms by the complex conjugate representation matrices. A 2jm transformation will turn these complex conjugated matrices into the standard matrices for the complex conjugate irrep (see Comment (c) on the definition of the 2jm symbol in §2.2), which in the case of SO(3) is the same irrep (j is 'real'). Hence

$$|\overline{\alpha jm}\rangle=\varepsilon\begin{pmatrix} j \\ & m \end{pmatrix}|\alpha^{*}j-m\rangle,$$

where ε is a conventional phase inserted for historical reasons (reflecting an inconsistency between the phase choice of Condon and Shortley (1935) for spherical harmonics and the Wigner choice of phase for 3jm symbols of SO(3)):

Comments

(a) Obviously time reversal reverses the angular momentum projection!

(b) We have added a stub to the parentage line (a matrix performing the conjugation $\alpha\rightarrow\alpha^{*}$) to the diagram, in case it should be necessary to distinguish the set of state vectors from the time-reversed set. Such transformations will be even more necessary in §3.

(c) If we choose to regard time reversal (of kets) as similar to complex conjugation (of numbers) in the definitions of §2.5, and are careful to include '2jm-conjugation' of parentage labels as in Comment (b), we may interpret this result as the statement that, for a basis choice in which $\varepsilon=1$, *kets* (and bras) *are* (say) *right-conjugate.*

(d) For half-integral angular momentum, the 2j phase is not unity, and we cannot also have left-conjugacy at the same time. Double time reversal gives for the time-reversal signature of the state vector the product of 2j phases on the group and on the parentage lines:

If this ket describes a multielectronic state of real angular momentum j, since odd-electron states, with half-integral angular momentum, are of spinor type (transform with negative eigenvalue under T^2), and since even-electron states with integral angular momentum correspond to true irreps of SO(3), the phase factor under double time reversal is $(-1)^{2j}$, and is equal to that from the group label 2j phase by itself. This indicates that in this case the conjugation matrix $\alpha \rightarrow \alpha^*$ does not contain any essential phase information.

(e) Notice that on each side of the equation defining the time-reversal conjugate diagram, external lines still have the same labels, in conformity with the rules of §1.

(f) However, the node terminals to which these external lines are attached now have opposite parity to those in the original diagram. To this extent, the application of time reversal has changed the character of the external labels. The necessity of a parity change is best seen by noting that the time-reversed ket must transform by the complex conjugated group matrices. The 2jm symbol nicely averts the threat of an inconsistency between this essential parity change and our previous rule that a ket node has negative parity in all circumstances.

For physical operators O, time reversal induces (a) the possibility of an overall sign τ_o for time-odd operators (momentum, angular momentum, spin, magnetic field and velocity- or magnetism-related vectors, etc.); (b) a departure from the standard ITO phasing as for any antilinear operation on an ITO. This may be remedied by either a 2jm transformation, which was used above to restore the equivalent problem posed by hermitian conjugation, or by hermitian conjugation itself.

To see this, we take the fundamental relation (Messiah 1962)

$$\langle a|\mathrm{O}|b\rangle^* = \langle \bar{a}|\bar{\mathrm{O}}|\bar{b}\rangle \leftrightarrow$$

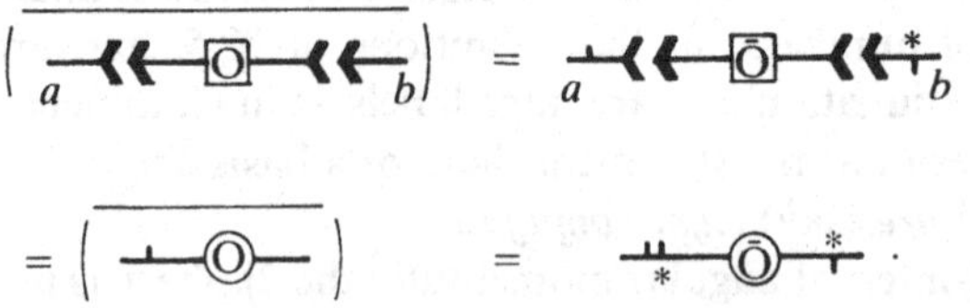

Hence

In short, the *conjugate node* (as defined in §2.5) describes matrix elements of the *time-reversed* operator.

We reserve full details of the interconnections between an irreducible tensor operator, its time reverse and its hermitian conjugate for the more general discussion in §3.6.4.

2.8 Quasispin – an example of second quantised spherical tensors

The second quantisation analysis of §1.6 will be extended to discuss in diagram form the elegant concept of quasispin. This concept is used in nuclear and atomic physics (Bell 1959; Lawson and Macfarlane 1965; Judd 1967, 1985, 1987b; Parikh 1978; Cheng-tian Feng and Judd 1982; Arima 1983; Rudzikas and Kaniauskas 1984; Judd, Lister and Suskin 1986), in ligand field theory (Watanabe 1966, Wybourne 1973), in particle–hole conjugation theory in organic systems (Koutecky, Paldus and Čížek 1985) and in thermopower theory (Stedman 1987, Stedman and Kaiser 1987) for example.

There are many other ways of coupling fermion and/or boson operators to give operators with the commutation relations of spin operators; we mention isospin (Problem 1.18) and drones (e.g. McKenzie and Stedman 1976). Quasispin is particularly powerful where the time-reversed states of the fermion have a useful symmetry, as in an atomic system where they correspond to opposite projections m, $-m$ of the angular momentum j in a shell.

We *define* the set of operators

$$A_m \equiv \left(f_m^\dagger, \begin{pmatrix} j \\ m \end{pmatrix} f_{-m}\right)$$

as a *quasispin tensor* of rank $\frac{1}{2}$, for reasons which will emerge. The 2jm phase is included so that both operators have the same tensorial character under ordinary angular momentum (Problem 2.15 verifies this), and thus quasispin and real spin decouple. The two operators are so defined to have opposite quasispin projection. In diagrams:

$f_m^\dagger \leftrightarrow$ 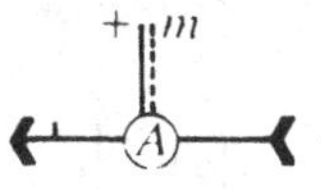

$\begin{pmatrix} j \\ m \end{pmatrix} f_{-m} \leftrightarrow$ 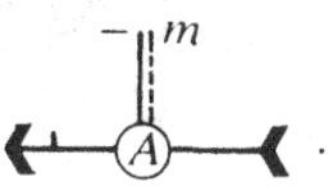

This has been written out for the operators. For a change, we write the next proof in terms of matrix elements. The matrix elements of the anticommutation relations take the form:

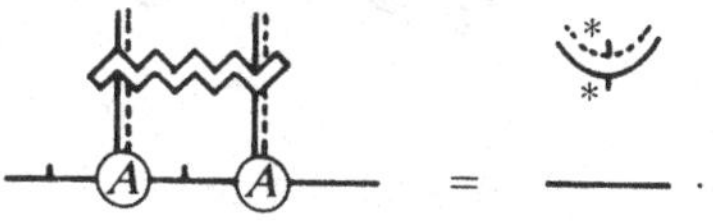

Comments

(a) The dotted line also stands for angular momentum (and projection) labels, and that the corresponding angular momentum j will be half integral, as befits a fermion state. Hence its 2j or interchange phase $\{j\} = -1$.

(b) The antisymmetric symbol $\varepsilon_{\lambda\mu}$ of §1.6 is now to be recognised as the 2jm symbol for (quasi)spin $\frac{1}{2}$.

(c) The above-mentioned interchange symmetry is needed to ensure the validity of this expression for all projection choices.

We define the *generators of quasispin* by coupling these quasispin-$\frac{1}{2}$ operators to rank 1. Their matrix elements have the form:

$Q_\alpha \leftrightarrow -\frac{1}{2}\sqrt{(\frac{3}{2})}\mathrm{i}$ 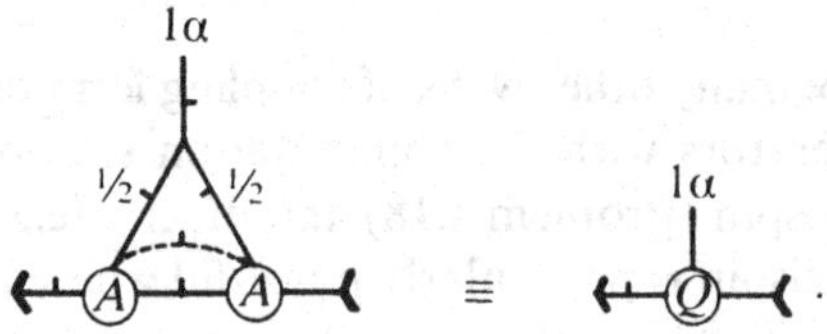

Problem 2.17

Show that these operators have the commutation relations of angular momentum.

Solution

$$\langle [Q_\alpha, Q_\beta] \rangle \leftrightarrow$$

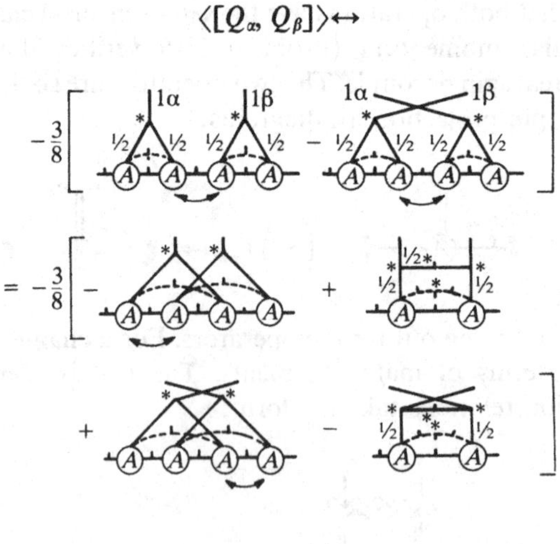

$= -\frac{3}{8}\Big[$ ... $-$ ... $+$ ...

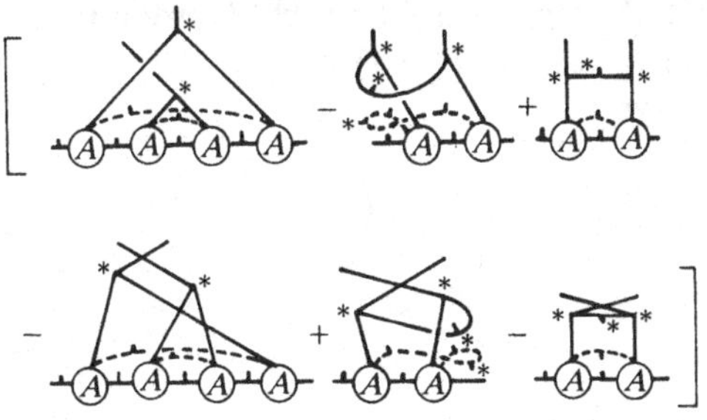

$-$ ... $+$ ... $-$ ... $\Big]$

$=\frac{3}{4}$ 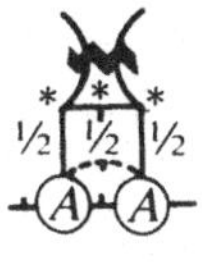 ... $=\frac{3}{4}$ ...

$=\frac{9}{2}$ 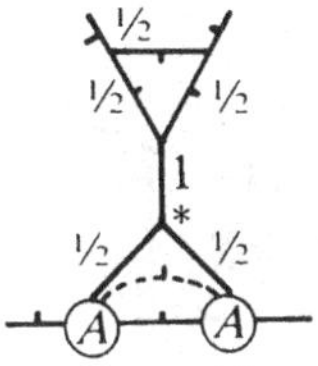 ... $= 3\sqrt{6}\mathrm{i}$ ...

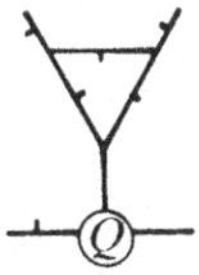

$=3\sqrt{6}\mathrm{i}$ ...

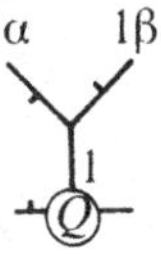

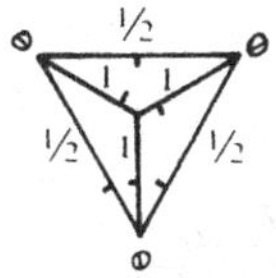

$\leftrightarrow \mathrm{i}\varepsilon_{\alpha\beta\gamma}\langle Q_\gamma\rangle. \longleftrightarrow -3\mathrm{i}\varepsilon_{\alpha\beta\gamma}\ Q_\gamma \begin{Bmatrix} 1 & 1 & 1 \\ ½ & ½ & ½ \end{Bmatrix}$

Comments

(a) Steps 1–3 follow from the anticommutation relations.

(b) In the fourth line, passing the symmetriser through two 3jm vertices has the effect of turning it into the negative of an antisymmetriser, since the uncrossed case has the same sign, and the crossed case involves the reversal of a stub on an internal line with $j=\frac{1}{2}$.

(c) In the fifth step, we use Clebsch–Gordan unitarity, with the result that only $j=1$ appears in $[\frac{1}{2}\times\frac{1}{2}]_+$.

(d) The 6j symbol

$$\begin{Bmatrix} 1 & 1 & 1 \\ \frac{1}{2} & \frac{1}{2} & \frac{1}{2} \end{Bmatrix} = -\tfrac{1}{3}.$$

Problem 2.18

Show that $\{Q_\alpha\}$ and A_m have the commutation relations of a spin-1 and a spin-$\frac{1}{2}$ tensor.

Solution

For example:

$$\langle[Q_\alpha, A_{mq}]\rangle \leftrightarrow -\tfrac{1}{2}\sqrt{\left(\tfrac{3}{2}\right)}\,\mathrm{i}\left[\ \cdots\ -\ \cdots\ \right]$$

$$= -\tfrac{1}{2}\sqrt{\left(\tfrac{3}{2}\right)}\,\mathrm{i}\left[-\ \cdots\ +\ \cdots\ +\ \cdots\ -\ \cdots\ \right]$$

$$= -\sqrt{\left(\tfrac{3}{2}\right)}\,\mathrm{i}\ \cdots\ \leftrightarrow -\sqrt{\left(\tfrac{3}{2}\right)}\,\mathrm{i}\begin{pmatrix} 1 & \frac{1}{2} & \frac{1}{2} \\ \alpha & q' & -q \end{pmatrix}^* \begin{pmatrix} \frac{1}{2} \\ -q \end{pmatrix}^* A_{mq'},$$

which in the case $\alpha=0$ (z) is proportional to $q\langle A_{mq}\rangle$ in the appropriate manner.

Problem 2.19

Show that quasispin and angular momentum commute.

Solution
First we note that the angular momentum operator may be depicted, up to a constant, by

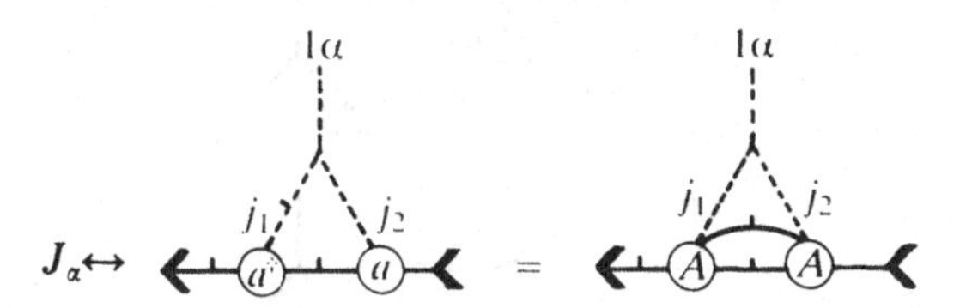

Hence,

$\langle [Q_\alpha, J_\beta] \rangle \leftrightarrow$ α β A A A A − A A A A

≡ [A A − A A

− A A + A A]

= 0 .

2.9 Summary

Rotation matrix:

$D^j_{mm'}(R) \leftrightarrow$ R, jm, jm' ; =

Clebsch–Gordan coefficient:

$\langle JM | jmj'm' \rangle \leftrightarrow$ JM, $j'm'$, jm = $\hat{J}M$, $j'm'$, jm

=

= .

3jm symbol:

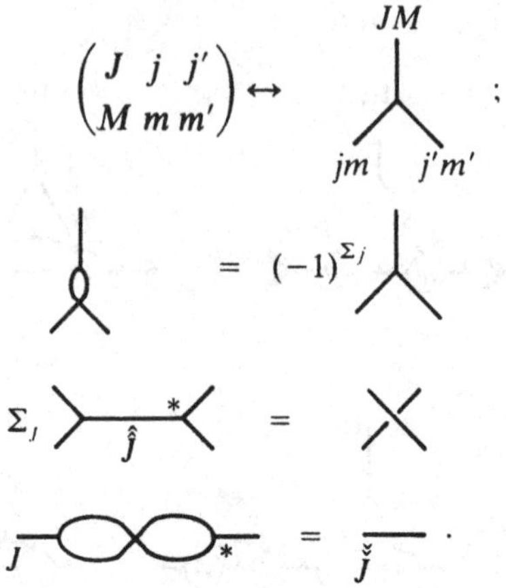

$$\begin{pmatrix} J & j & j' \\ M & m & m' \end{pmatrix} \leftrightarrow$$

JM

jm j'm'

;

$$= (-1)^{\Sigma j}$$

$$\Sigma_j \quad \hat{\hat{j}} \quad * \quad =$$

$$J \quad * \quad = \quad \check{\check{J}} \quad .$$

2jm symbol:

$$\begin{pmatrix} j \\ m \end{pmatrix} \leftrightarrow$$

jm = $\hat{j}m$ * 00 ;

$* = \quad , \qquad j \; * = (-1)^{2j}$,

$= *$.

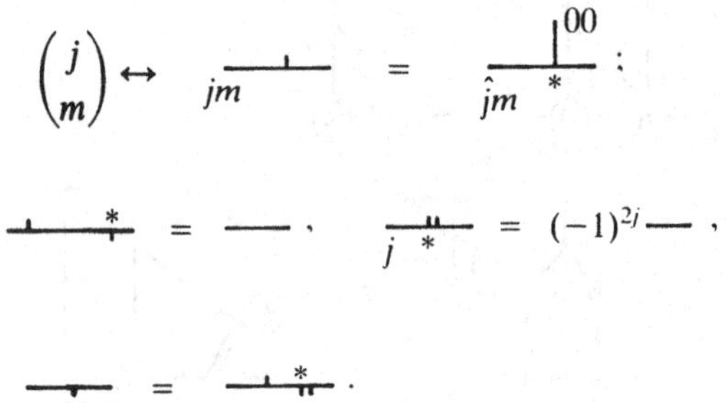

Complex conjugation:

$* = \quad = \quad .$

Invariance and commutativity:

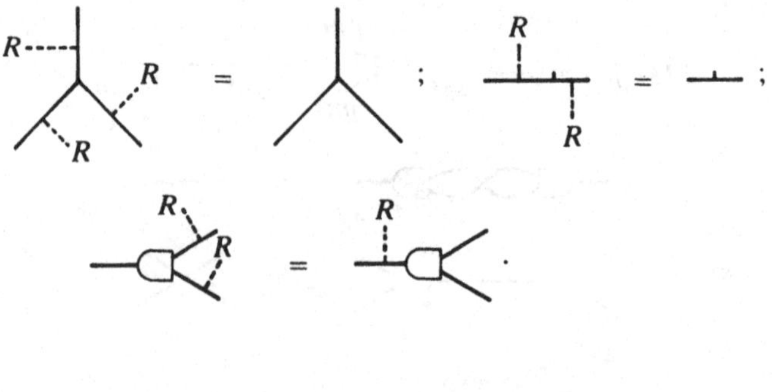

R R R = ; R R = ;

R R = R .

Great Orthogonality Theorem and Gaunt's theorem:

J * = $\check{j}$; = * .

JLVn:

I J $= \delta_{J0}$ I 0

I J = I $\check{j}$ $\check{j}$ *

I = I *

I = Σ_J I * $\check{J}$ * * $\check{J}$.

Wave functions and operators:

$|\alpha jm\rangle \leftrightarrow$ α jm , $\langle \alpha jm| \leftrightarrow$ α jm

$\mathrm{O} \leftrightarrow$ O .

Time-reversed state:

$\overline{|\alpha jm\rangle} = \varepsilon$ α * jm * .

3

Extension to compact simple phase groups

What is not, perhaps, adequately appreciated, is the extent to which we actually depend on visual images rather than algebraic manipulation in thinking about physics. A deplorable trend in applied mathematics – perhaps an excess of zeal by the followers of Descartes in revolt against the geometric proofs of Newton's *Principia* – was the deliberate banishment of diagrams from the exposition of physical theory. The literature of physics is now recovering from this impoverishment of the imagination, but there are still inhibitions, in the formal primary literature, against the pictorial representation of physical ideas.

Ziman (1985)

3.1 Introduction to basic representation theory

We review standard representation theory for finite groups, or more generally compact groups (those with a bounded parameter space; these have finite-dimensional unitary irreducible representations (irreps)). Even in joint reviews (e.g. Killingbeck 1970) this has traditionally been separated from diagram techniques. Our goal is to introduce a diagrammatic form (Agrawala and Belinfante 1968, Lulek 1975; Stedman 1975, 1976a; Kibler and ElBaz 1979; Nencka-Ficek and Lulek 1982) for all standard results. Since this is not a text on group theory *per se*, we refer the reader to Butler (1981), also such books as Wigner (1959), Heine (1960), Hamermesh (1962), Tinkham (1964), Leech and Newman (1969), Lax (1974), Elliott and Dawber (1979) and Cornwell (1984) for fuller accounts. We work in terms of subgroups G of O(3) (the group of rotations and reflections in 3-space) i.e. the point groups, in view of their importance in condensed matter theory. Various accounts are available of the Racah algebra and the associated 3jm symbols of the point groups (Griffith 1962, Koster *et al.* 1963, Damhus, Harnung and Schaffer 1984). We take the work of Butler (1981), with a slight modification discussed by Reid and Butler (1982), as definitive. We also use the notation of Butler (1981) with minor variations as described below.

The main alterations to the material of §2 are: the need to include a repetition or multiplicity index for repeated irreps in a Kronecker product; the inequivalence of complex conjugate irreps; and perhaps the need to include a duality transformation for the repetition label in analogy to the 2jm symbol for the group label. The 'good housekeeping' habits of §2.5 have been introduced, in anticipation of such complications, to remove much of the sting of these alterations.

We label the *irreps* of the group G by Greek letters λ, μ, ν, . . . For dimensional factors we use carets: $|\lambda|^{\frac{1}{2}}=\hat{\lambda}$, $|\lambda|^{-\frac{1}{2}}=\check{\lambda}$, $|\lambda|=\hat{\hat{\lambda}}$, $|\lambda|^{-1}=\check{\check{\lambda}}$. The *identity* irrep will be designated 0. The *components* of the irrep λ will be labelled by the corresponding Latin letter l (and l', etc.). *Basis functions* are then represented by $|\alpha\lambda l\rangle$. α is any extra parentage label needed to specify the ket uniquely; in a group–subgroup scheme, this will include the irrep labelling of the state in higher groups. α will be suppressed or expanded (to include, say, j^{π} when denoting spin j and parity π) as desirable without warning throughout this chapter. We use the notation $\{\lambda^{l}{}_{l'}(g)\}$ for the matrices *representing* an irrep λ; the suffix disposition will be explained with the fuller definition of equation (3.1). The complex conjugate matrix elements obviously represent another irrep, called the *complex conjugate irrep* λ^*, which is called *real* or *complex* as it is isomorphic, or not isomorphic, respectively, to λ. In either case, a unitary transformation T (§3.3) links the two irrep representation matrices:

$$\mathbf{T}\boldsymbol{\lambda}(g)^*\mathbf{T}^{\dagger}=\boldsymbol{\lambda}^*(g).$$

3.2 Representation matrices and basic theorems

Solid lines will carry group (irrep and component) labels. The group operators O_g, for all g in G, act on these basis functions to give linear combinations, the coefficients $\lambda(g)^{l'}{}_{l}$ being elements of the *matrix representation* of this irrep:

$$O_g|\alpha\lambda l\rangle=\Sigma_{l'}|\alpha\lambda l'\rangle\lambda(g)^{l'}{}_{l}. \tag{3.1}$$

We have superscripted the leading label, which stands for the row label of the corresponding matrix, in anticipation of a fuller analysis (§§3.7, 5.3) which highlights the conjugate transformation characters of these two labels. The diagram form of the representation matrix can be made into a commutative, rather than an invariant, node. Equation (3.1) suffices to indicate that l has the (covariant) character of a ket label, but that the sum over a covariant (ket) label l' *against* the row label produces an invariant. We can expand on this notation (e.g. Butler 1975); the complex conjugate matrix element may be denoted by reversing the raising and lowering, so that the first suffix is now lowered:

$$\lambda_{l}{}^{l'}(g)\equiv[\lambda^{l}{}_{l'}(g)]^*.$$

This preserves the association of subscripts and covariant or ket transformation character.

Following §2.2, the matrix elements are represented in diagram form by

$$\lambda^{l}{}_{l'}(g) \leftrightarrow \quad \text{[diagram: horizontal line labelled } \lambda l \text{ and } l'\text{, dotted vertical line up to } g]$$

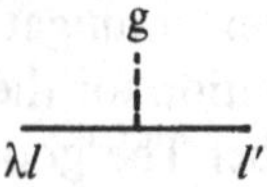

and the inverse operation is, by unitarity,

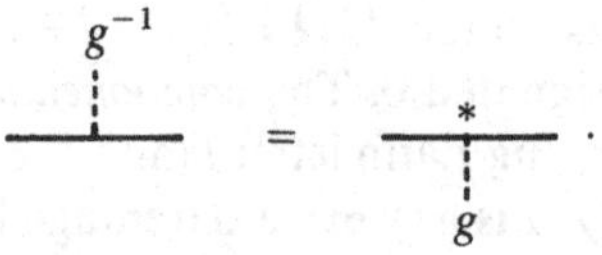

$$\text{[diagram: line with dotted vertical to } g^{-1}\text{ above]} \quad = \quad \text{[diagram: line marked } * \text{ with dotted vertical to } g \text{ below]} \; .$$

The trace of the representation matrix is basis-independent and will be called the *character* of the irrep. It is represented diagrammatically by

$$\chi^{\lambda}(g) \leftrightarrow \quad \text{[diagram: circle labelled } \lambda \text{ with dotted line up to } g] \; .$$

The *group average* ($\Sigma_g O_g/|G|$ for a finite group, and for a continuous but compact group the analogue (a normalised integral over the group element parameter space)) will be denoted by a heavy *dot*.

Problem 3.1
Show that the group average of a representation matrix element $\lambda(g)$ vanishes unless λ is the identity irrep 0, when the average is unity:

$$\text{[diagram: line labelled } \lambda \text{ with dotted vertical ending in heavy dot]} \quad = \quad \delta_{\lambda 0} \; .$$

Solution
When the group operator O_g acts on the state $|\alpha\lambda l\rangle$, and the result is averaged over the group element label g, we obtain the ket $|A_l\rangle = |G|^{-2}\Sigma_g O_g|\alpha\lambda l\rangle$. Consider the action of another group operator O_h on this ket. By the definition of group multiplication, we obtain O_{hg} inside the summation, which by the group rearrangement theorem may be replaced within the sum by O_g. Hence $|A_l\rangle$ is invariant under a group operation. However, the ket $|A_l\rangle$ forms part of a basis set for the irrep λ, since it is formed from one such ket by the action of the group itself. By the definition of an irrep, $|A_l\rangle$ must be inextricably mixed with other states in

this basis set via the action of the group operators. Hence either $\lambda=0$ or $|A_l\rangle$ vanishes for all choices of l. If $\lambda=0$, the group average normalises to unity. Otherwise $\langle\alpha\lambda l'|A_l\rangle$ vanishes for all l, l'.

Comment

(a) This amounts to a special case of Schur's lemma I.

(b) Many standard texts imply that this result is a consequence of the great orthogonality theorem. However, we prefer to take it as a starting point from which to deduce the orthogonality theorems, Schur's lemmas, etc. The corresponding diagram theorems form a natural sequence.

(c) This leads almost trivially to a *theorem* (*JLV1*):

Any invariant network with just one external group line, labelled with the irrep-component labels λ, l, vanishes unless $\lambda=0$.

$$(I)\!\!-\!\!_{\lambda l} \;=\; \delta_{\lambda 0}\, (I)\!\!-\!\!_{0}\,.$$

The proof follows since, if it is invariant under any group operator, it is invariant under an average of group operators.

3.3 Coupling of representations

3.3.1 Clebsch–Gordan coefficients

The set (for all l, m) of product kets of the irrep basis kets $|\lambda l\rangle\mu m\rangle$ (we suppress parentage labels for simplicity) spans the basis of a new representation. The corresponding representation matrices may be reduced to the standard form of the irreps of G. The elements $\langle\Lambda L|\lambda l, \mu m\rangle$ of the matrix that perform this block diagonalisation correspond to the expansion coefficients of the product state in irreducible basis functions, and are generalised Clebsch–Gordan coefficients.

For most continuous groups and the point groups K, O and T, a given irrep will appear more than once in the decomposition of the Kronecker product $\lambda\times\mu$; the various occurrences will be given an (arbitrary) *repetition label* r $(=0, 1, \ldots)$. This label was unnecessary in SO(3), where all Kronecker products contain any j at most once – SO(3) is simply reducible.

A natural and convenient extension of our earlier notation is to add a fourth line at a Clebsch–Gordan or 3jm node to carry the repetition label. The summations over repetition labels essential in the generalised Wigner–Eckart theorem etc. are conveniently transliterated into diagrams: corresponding lines are joined. When cyclic symmetry does not hold, the different character of this line from the three group theoretical lines (those carrying irrep and component labels) can be used, as we show below, to fulfil the function of a conventional point on the node in determining the canonical order of indices.

We have then the following definitions, as natural generalisations of those in §2 for SO(3):

A *Clebsch–Gordan coefficient* will be represented by

$$\langle r\Lambda L|\lambda l\mu m\rangle \leftrightarrow$$

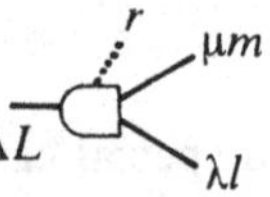

Comments

(a) The Clebsch–Gordan coefficient is unitary:

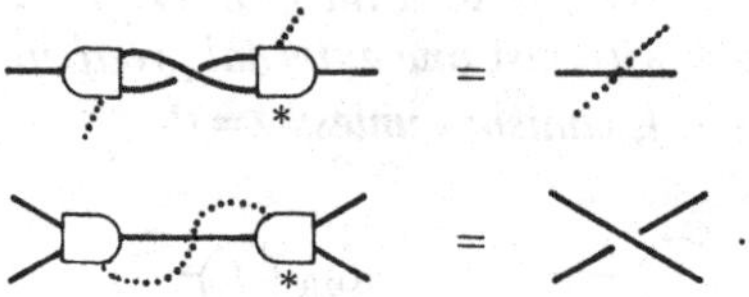

(b) The Clebsch–Gordan coefficient is a commutative node:

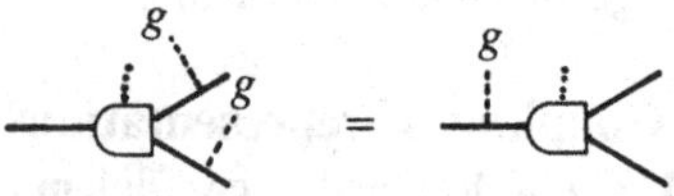

3.3.2 *nj*m symbols and *nj* phases

A *2jm symbol* will be defined as:

$$\begin{pmatrix}\lambda\\ l\end{pmatrix} = \hat{\lambda}\langle\lambda l\lambda^* l'|00\rangle \leftrightarrow$$

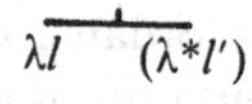

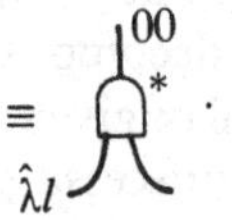

Comments

(a) The two irrep labels are necessarily complex conjugate. From character theory (Problem 3.2; see §3.4), $\lambda\times\mu$ contains the identity once at most (hence a repetition label is unnecessary) and then only if $\lambda=\mu^*$.

(b) The use of a stub rather than an arrow to represent the 2jm symbol is particularly useful in this section, where λ and λ^* are not equivalent; the vertex appearance distinguishes the labels on each side more clearly. In addition the connection with the 3jm symbol is implied in this depiction of the symbol. As always, leading labels are related to the orientation of the stub in an anticlockwise convention.

(c) The 2jm symbol is right invariant:

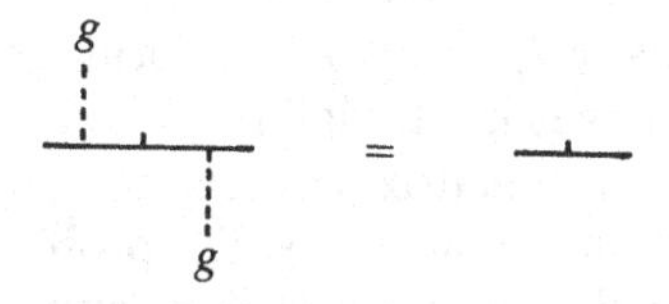

since the Clebsch–Gordan node is commutative and since a rotation has no effect on scalar labels.

(d) The 2jm symbol is a unitary matrix in the *bra* labels l, l':

Because of the invariance and normalisation of the 2jm symbol (cf. §2.2), application of Schur's lemma shows that it can be identified with the matrix $\mathbf{T}^{\dagger}$ of §3.1:

g g
(λ) (λ*)

(To prove this, we may as in Problem 2.2 prove a restricted form of Schur's lemma II first.) In general, the 2jm symbol need not be real. Butler (1981) shows that it can be chosen to be real. We prefer in any case to allow for the 2jm symbol being complex, in order to implement the 'good housekeeping' rules of §2.5, and insist on bipartite network construction.

The *2j phase* $\{\lambda\}$ is defined as formerly:

$$\{\lambda\} \leftrightarrow \quad \lambda$$

Comments

(a) Again as before, the invariance of the 2jm symbol shows that this is a commutative matrix. Hence from Schur's lemma the 2j phase is just a multiple of the unit matrix. From the unitarity of the 2jm symbol, this factor has modulus unity.

(b) It amounts to a permutation matrix for the 2jm symbol:

(c) However, the 2j phase is not obviously basis-independent, and in fact so only for real irreps. Even real irreps may be subdivided into two types, named here (following Butler 1981) *orthogonal* and *symplectic*, respectively, for which the 2j phase is, respectively, +1 and −1.

(d) This indicates that a 2jm matrix transformation **T** between complex conjugate representations is nontrivial in general; even in SO(3) for real representations a rotation matrix and its complex conjugate do not transform an irrep basis in the same way. The problem is not eradicable by changing basis from say the Condon and Shortley basis to the Fano–Racah basis.

A *3jm symbol* is defined, again in parallel with §2, by

$$\begin{pmatrix}\lambda & \mu & \nu\\ l & m & n\end{pmatrix}^r = \check{\lambda}\begin{pmatrix}\lambda^*\\ l'\end{pmatrix}^* \langle r\lambda^* l' | \mu m \nu n\rangle \leftrightarrow$$

and conversely

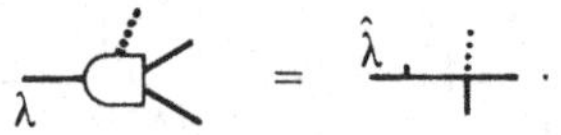

Comments

(a) The major change is to introduce a fourth line carrying the repetition labels r distinguishing the repetitions of 0 in $\lambda\times\mu\times\nu$. This amounts also to the distinguishing of the repetitions of λ^* in the Kronecker product $\mu\times\nu$, or equivalently the repetitions of μ^* in $\lambda\times\nu$ etc. The line wearing repetition labels is useful in other ways, since there is no guarantee in general that this symbol has any interchange or even cyclic symmetry; hence it gives a reference point for defining an initial label set, with the others determined by anticlockwise ordering (§1.2).

(b) Clebsch–Gordan and 2jm symbol unitarity give the pseudounitarity constraints:

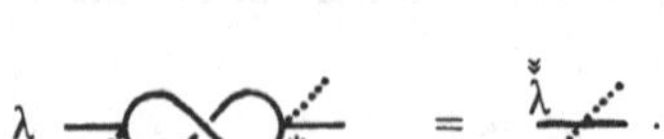

(c) Hence

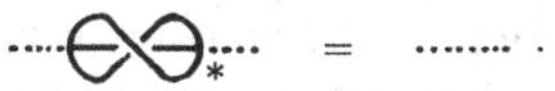

(d) In general, we have the *conjugation lemma*:

The complex conjugate of a 3jm symbol is related to the original symbol by a 2jm transformation on each line:

$$\left[\begin{pmatrix}\lambda & \mu & \nu \\ l & m & n\end{pmatrix}^{r}\right]^{*} = \begin{pmatrix}\lambda^* & \mu^* & \nu^* \\ l^* & m^* & n^*\end{pmatrix}^{r}\begin{pmatrix}\lambda^* \\ l^*\end{pmatrix}\begin{pmatrix}\mu^* \\ m^*\end{pmatrix}\begin{pmatrix}\nu^* \\ n^*\end{pmatrix}$$

Comments

(a) We shall not give a proof; see Butler (1975, 1981) for its background, under the name of the Derome–Sharp lemma (Derome and Sharp 1965). For this to hold for all groups, we should include a matrix acting on the repetition labels. This matrix may be thought of as the analogue of a 2jm symbol for the repetition label. However, we have for simplicity omitted this matrix in the above statement of the lemma. This means that we do not distinguish r and r^*, nor introduce a conjugation phase for the label r. In making this simplification, we are assuming the property of *quasiambivalence* (Butler 1981), itself satisfied by the continuous and point groups of our applications, though not by certain pathological groups. Where quasiambivalence holds, the product of 2j phases of the three irreps is unity:

$$(\{\lambda\}\{\mu\}\{\nu\} = 1) \leftrightarrow$$

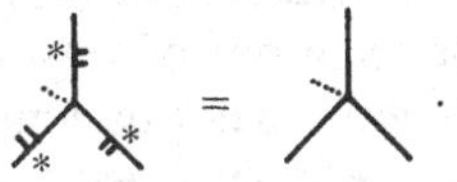

(This relation may be seen to follow from the above simplified form of the theorem, simply by using the theorem twice.)

(b) It follows that, as shown in the diagram, the 2jm transformations may be applied in either rotational sense.

(c) From one viewpoint, however, it is unhelpful to introduce quasiambivalence: it breaks the symmetry of our development, because it makes the resulting diagrams no longer bipartite. For this reason, we shall in our diagrams leave the complex conjugation star on vertices of even parity, rather than replacing the star by stubbing all lines.

(d) We note, however, that in fact all 2jm symbols in all point groups may be chosen to be real. The same does not hold for 3jm symbols in all point groups. The freedom in the choice of phase for some primaeval kets may certainly be exploited to make as many 3jm symbols as possible to be real. One may, in many groups such as most point groups, choose phases so that all 3jm symbols are real (Bickerstaff 1984, 1985, Bickerstaff and Damhus 1985).

Permutation matrices need to be defined, in general terms, for the 3jm symbol. A trivial manipulation gives the corresponding diagram:

$$\begin{pmatrix} \lambda & \mu & \nu \\ l & m & n \end{pmatrix}^r = m_\pi(\lambda\mu\nu) \begin{pmatrix} \pi(\lambda) & \pi(\mu) & \pi(\nu) \\ \pi(l) & \pi(m) & \pi(n) \end{pmatrix}^r$$

$m_\pi(\lambda\mu\nu) \leftrightarrow$ [diagram]

Comments

(a) For SO(3), the cyclic permutations all are trivial, and the interchange permutations are all the same, giving merely a phase factor. This generalises to all simple phase groups – namely, those satisfying the condition $\Sigma_g[\chi^\lambda(g)]^3 = \Sigma_g \chi^\lambda(g^3)$ (Butler 1981). All point groups are simple phase (Stedman 1975). As before, we call the interchange phase factor the *3j phase*:

$\{\lambda\mu\nu r\} \leftrightarrow$ [diagram]

(b) These permutation matrices may be included in a calculation where necessary as an afterthought, simply by the expedient of checking the compatibility of the ordering of vertices in conjugate pairs (see §2.5). This helps to take the sting out of network manipulations in which lines are crossed under diagram deformations.

(c) The 3j phase is determined from character theory when two irreps coincide (see Problem 3.3).

3.3.3 nj symbols

A *6j symbol* is defined by

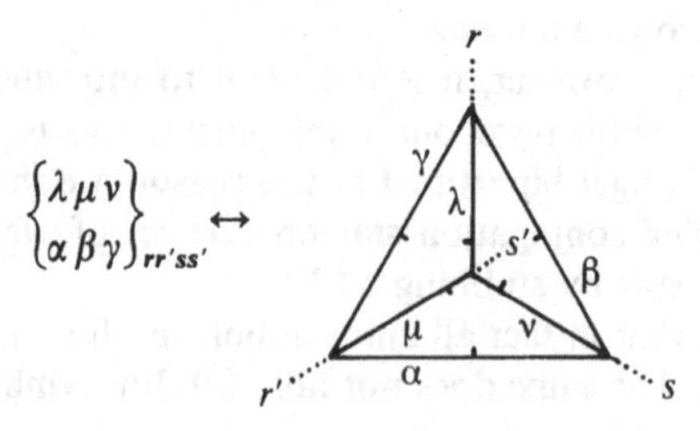

and a *9j symbol* similarly:

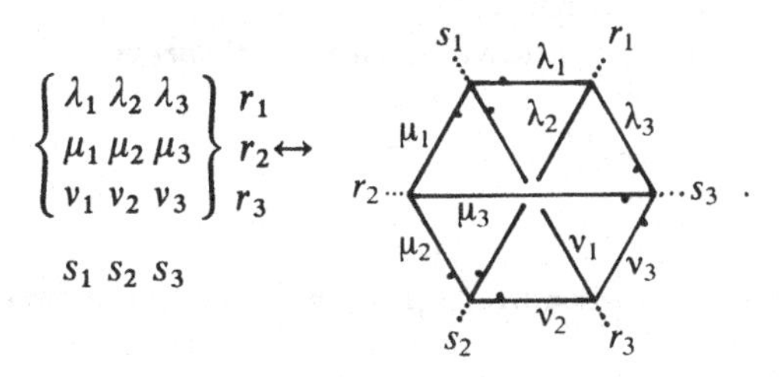

$$\left\{\begin{matrix} \lambda_1 & \lambda_2 & \lambda_3 \\ \mu_1 & \mu_2 & \mu_3 \\ \nu_1 & \nu_2 & \nu_3 \end{matrix}\right\} \begin{matrix} r_1 \\ r_2 \\ r_3 \end{matrix} \leftrightarrow$$
$$s_1 \; s_2 \; s_3$$

The symmetries of these symbols, suitably generalised, together with examples of their origin in diagram manipulation, may be treated in exact analogy to the case of SO(3) in §2. Detailed formulae are given in Butler (1981).

3.4 Network reduction theorems

With the orthogonality relations for 3jm symbols, we may build on Problem 3.1 to obtain a hierarchy of results for group averages of n matrix elements, in either of two forms (one of which mixes complex conjugated matrix elements):

λ $= \delta_{\lambda 0}$

λ μ $=$ $\check{\lambda}$ $\check{\mu}$

$=$ $*$

$= \Sigma_\lambda$ $\hat{\lambda}$ $*$ $*$ $\hat{\lambda}$ $*$.

Comments

(a) The second form is the *Great Orthogonality Theorem*, which leads (by looping diagonally opposite ends) to the *Character Orthogonality Theorem*:

λ $*$ μ $= \delta_{\lambda\mu}$.

This result in turn permits us to verify some useful and familiar results from standard representation theory:

(b) The third relation is known as *Gaunt's theorem.*

Problem 3.2

Show that 0 occurs just once in $\lambda \times \mu$ if $\mu = \lambda^*$, and zero times otherwise.

Solution

$$n_0 = (1/|G|)\Sigma_g \chi^{\lambda \times \mu}(g)\chi^0(g)^* \leftrightarrow$$

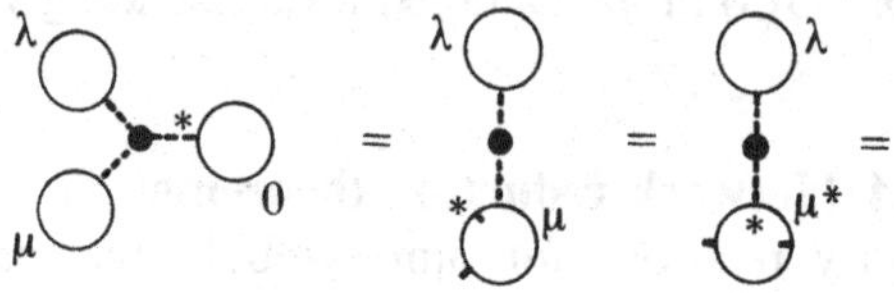

$$= \delta_{\lambda\mu^*}.$$

Comment

We have used the standard character theoretic formula (Tinkham 1964), derived from the character orthogonality theorem, for the number of occurrences of an irrep in any representation. We then use the results that: the trace of a direct product of matrices is the product of the individual traces; $\chi^{(0)}(g) = 1$; the unitarity of the stub, which permits us to complex conjugate one character with the associated irrep label.

Problem 3.3

Show that the characters of the symmetric (antisymmetric) part of the Kronecker square of an irrep are given by

$$\chi^{[\lambda \times \lambda]\pm}(g) = \tfrac{1}{2}[\{\chi^\lambda(g)\}^2 \pm \chi^\lambda(g^2)],$$

respectively.

Solution

The basis functions representing the symmetrisation or antisymmetrisation, respectively, may be taken as

$$|ll'\eta\rangle \equiv (\psi_l^\lambda \phi_{l'}^\lambda + \eta \psi_{l'}^\lambda \phi_l^\lambda)/N_{ll'\eta} \leftrightarrow \frac{1}{N_{ll'\eta}} \qquad (l \geq l')$$

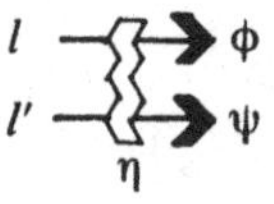

$N_{ll'\eta}$ is determined by normalisation:

$$\langle ll'\eta | mm'\zeta \rangle = 1 \leftrightarrow$$

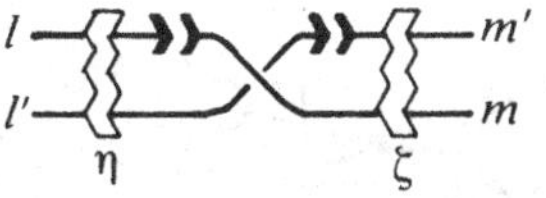

l l' η ζ m' m $/(N_{ll'\eta}\, N_{mm'\zeta})$

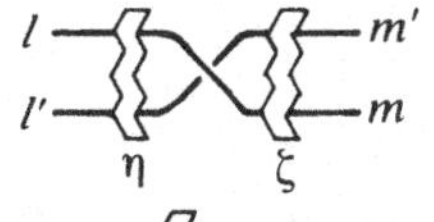

$=$ l l' η ζ m' m $/(N_{ll'\eta}\, N_{mm'\zeta})$

$= 2\delta_{\eta\zeta}$ $(+)$ $/(N_{ll'\eta}\, N_{mm'\eta})$.

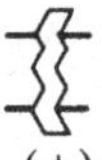

The factor $\delta_{\eta\zeta}$ ensures orthogonality of the different symmetry types. For the case of unequal labels: $l > l'$, the uncrossed term in the expansion of the last symmetriser necessarily vanishes, and we obtain the orthogonality conditions $\delta_{lm}\delta_{l'm'}$ in the answer together with the normalisation condition $2/[N_{ll'\eta}]^2 = 1$, so that $N_{ll'\eta} = \sqrt{2}$ for $l \neq l'$. When $l = l'$, we obtain the factors $\delta_{\eta+}$, and with an extra factor of two from the generalised Kronecker delta obtain $N_{ll+} = 2$.

The group matrix element is diagonal in the symmetry labels η, ζ for a similar reason. The diagonal matrix element is:

$$\langle ll'\eta | \mathrm{O}_g | mm'\eta \rangle \leftrightarrow 2$$ g l g l' η m' m $/(N_{ll'\eta}\, N_{mm'\eta})$.

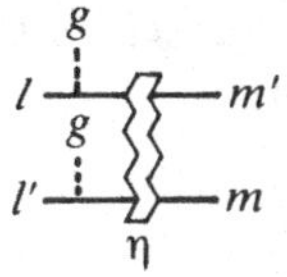

Finding the character of the representation spanned by these symmetrised/antisymmetrised kets amounts to taking the trace of this matrix, i.e. choosing $l = m$, $l' = m'$, and summing over l, l' with $l \geq l'$. We can remove the last restriction by including a factor $\frac{1}{2}$ in the sum for the case $l \neq l'$; this compensates the difference in the normalisation factors for the two cases where l is, respectively, equal and unequal to l'. We have:

$$\chi^{[\lambda\times\lambda]\pm}(g) \leftrightarrow \tfrac{1}{2}$$ g η g $= \tfrac{1}{2}\Big[$ g g $+ \eta$ g g $\Big]$

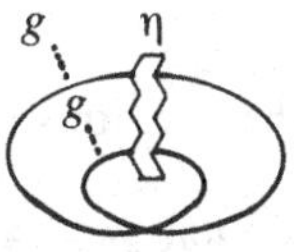

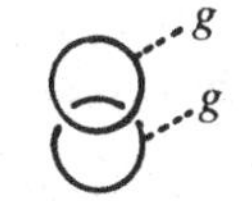

$$\leftrightarrow \tfrac{1}{2}[(\chi^{\lambda}(g))^2 \pm \chi^{\lambda}(g^2)].$$

Comment

This has the consequence that as μ_r (the rth occurrence of μ) in $\lambda \times \lambda$ is in $[\lambda \times \lambda]_{\pm}$, the interchange 3j phase $\{\lambda\lambda\mu^{*}r\}$ is ± 1, respectively:

$n(\mu \in [\lambda \times \lambda]_{\pm}) \leftrightarrow$.

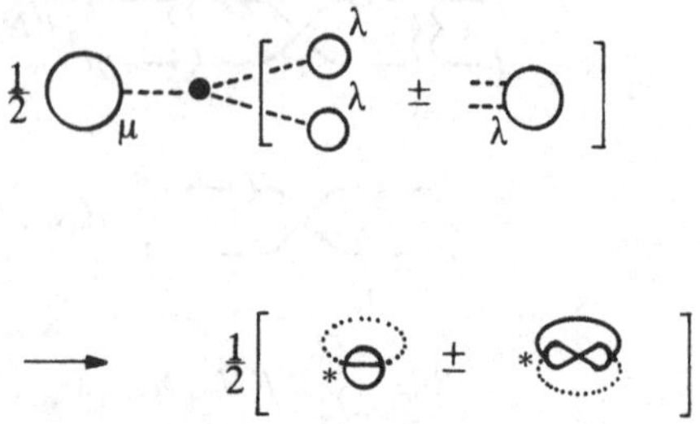

(the last step follows from Gaunt's theorem) which is 1 or 0 accordingly.

This may be summarised by writing

$$\mu_r \in [\lambda \times \lambda]_{\{\lambda\lambda\mu^{*}r\}},$$

a result to be used in Problem 3.12.

We now apply the theorems for group averages given at the commencement of this subsection to determine the group average of an invariant diagram or (using the alternative form of the above results) of a commutative diagram with say n external lines. We obtain the *generalisations* (to a general compact group) *of the JLVn theorems*, which may be written in '*crossed*' or '*uncrossed*' form:

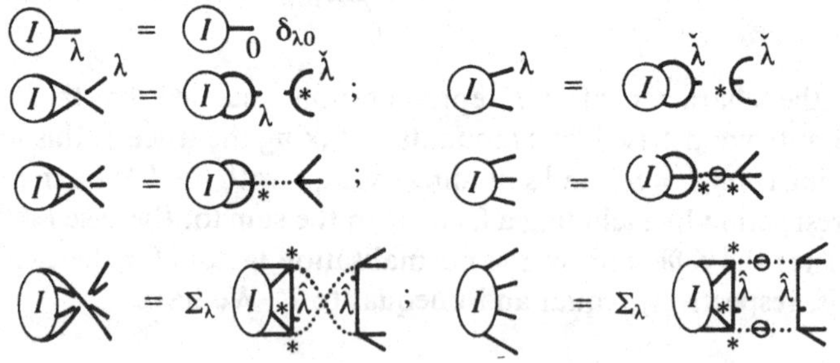

Comments

(a) The proofs of these results parallel the previous analysis for SO(3) (§2.4–§2.6).

(b) We shall assume that by now the reader has gained facility in picking out the most useful of the above alternative forms for any purpose, and shall not distinguish the alternative forms of JLVn in the following discussion.

(c) The general forms of Schur's lemma II and of the Wigner–Eckart theorem are implied by JLV2, JLV3, respectively, (in the form for commuting nodes, to be precise). We may use this in turn to prove a number of results assumed earlier. Details are deferred to the problems.

Problem 3.4
Show that the Frobenius–Schur invariant

$$c_\lambda = \Sigma_g \chi^\lambda(g^2)$$

is equal to the 2j phase for a real irrep, and is zero for a complex irrep.

Solution

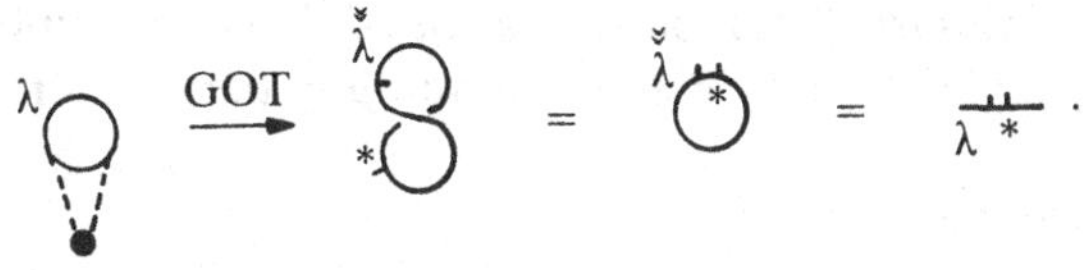

Comment
Since, from the Character Orthogonality Theorem, the identity occurs once only in λ^2 if λ is real, and not at all otherwise, this occurrence of the identity irrep 0 must be in either the symmetric or antisymmetric part of the Kronecker square. If we add to this the results of Problems 3.1–3.2, we may conclude that the Frobenius–Schur invariant is $+1$, -1 as $0 \in [\lambda \times \lambda]_\pm$, respectively; and again that the Frobenius–Schur invariant vanishes for a complex irrep. Hence the 2j phase is $+1$ for an orthogonal irrep and -1 for a symplectic irrep. This is an alternative way of distinguishing orthogonal and symplectic irreps.

3.5 Group–subgroup reduction

Take the functions $\{\phi^\lambda{}_l | l = 1,2, \ldots |\lambda|\}$ to span the irrep λ of a group G; the corresponding representation matrices are $\{\lambda^{l'}{}_l(g) | g \in \mathrm{G}\}$. If H is a subgroup of G, the $\{\phi^\lambda{}_l\}$ span a representation of H. In general this is reducible. Suppose that λ reduces under group–subgroup reduction to several irreps μ of H, perhaps repeated, and with their repetitions distinguished by a label $r = 1,2, \ldots, r_\mu$:

$$\lambda(\mathrm{G}) \downarrow \Sigma_\mu r_\mu \mu(\mathrm{H}).$$

(In this subsection only, λ denotes an irrep of a group G, and μ an irrep of a subgroup H.) The subset $\{\lambda^{l'}{}_l(h) | h \in \mathrm{H}\}$ of matrices belonging to operations of H can then be rendered block diagonal, so as to give r_μ blocks of $\mu^{m'}{}_m(h)$, by a suitable unitary transformation (B) whose elements we call $B^{\lambda l}{}_{r\mu m}$. We depict these by a suitable node:

$$B^{\lambda l}{}_{r\mu m} \leftrightarrow$$

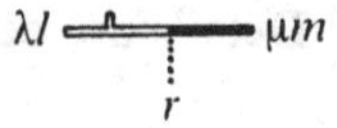

λl μm r

corresponding to the diagonalisation transformation in the form $(B^\dagger)(\lambda)(B) = (\mu)$, i.e.

$$B_{\lambda l}{}^{r\mu m}\lambda^l{}_{l'}(h)B^{\lambda l'}{}_{r'\mu' m'} = \delta_{\mu\mu'}\delta_{rr'}\mu_{mm'}(h) \leftrightarrow$$

h h

Note that an open bar carries irrep–component labels of G, while the solid bar as before carries irrep–component labels of H. The multiplicity label r distinguishing copies of μ(H) in λ(G) leads to μ in the standard anticlockwise sense.

Unitarity requires:

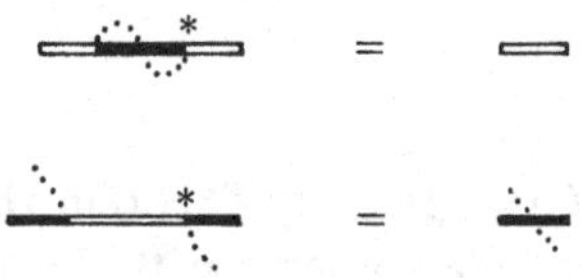

Problem 3.5
Show that the node as defined above is left invariant.

Solution
Since the matrix transformation performs the block diagonalisation for all h, yet does not depend on h, it is a commutative matrix. The stub added to the definition transforms parities in the requisite manner to convert a commutative to a left invariant node.

More directly: combining unitarity with the block diagonalisation condition gives

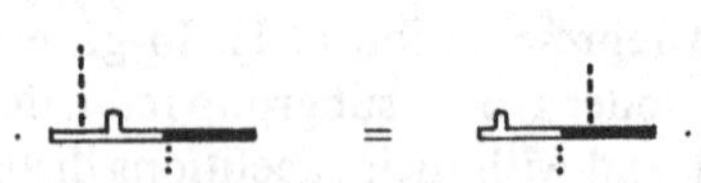

We note for use below that the matrix (B) amounts to a transformation of basis in which the subgroup irrep and component labels μ and m provide at least a partial (if r is trivial, a total) and purely group theoretic specification of the component label l of the irrep λ of G. In effect, the transformation by B of the rotation matrix expresses it in a subgroup basis.

Problem 3.6

Show that the 2jm symbols in group and subgroup are related by the *2jm factor*, which is defined as a stub or parity transformation on the multiplicity label by

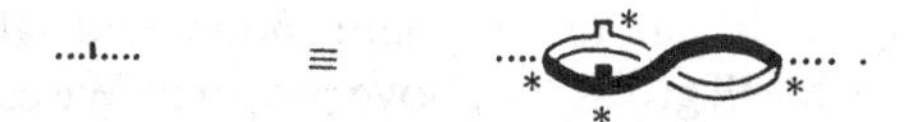

Solution 1

This definition, coupled with the unitarity of the basis transformation, shows that the basis transformation node defined above satisfies an analogue of the conjugation lemma of §3.3.2, viz. complex conjugation is equivalent to adding a stub to each line in either sense:

This enables us to write a relation between the 2jm symbols:

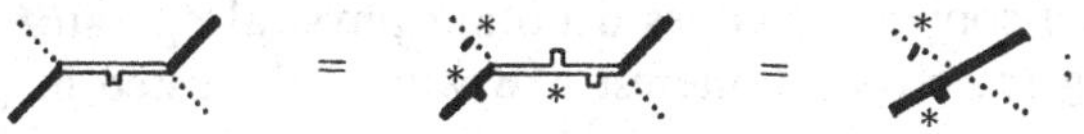

i.e. the 2jm symbol of the group is a product of the 2jm symbol of the subgroup and the 2jm factor.

Solution 2

If the 2jm symbol of G is expressed in the subgroup basis afforded by H via the transformation (B) as a label transformation, the resulting diagram is right invariant, and JLV2 may be applied to create the necessary factorisation:

Comments

(a) The 2jm symbols of any group irrep may thus be written as products of 2jm factors in a subgroup chain (Butler 1981).

(b) The 2jm factor is unitary:

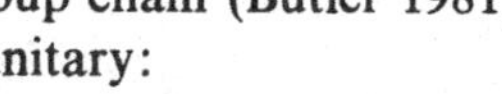

This links the above two solution methods.

(c) This is a general method of construction (whether we call the results stubs, arrows, parity transformations, or conjugacy transformations, or metric tensors, or 2jm symbols) which is not limited to group–subgroup bases; any representation of a group may be decomposed into its irreducible parts by a commutative unitary transformation, and so permits the definition not only of an invariant node, but also of a parity transformation which will give it the above property. We call such relations the *generalised conjugation lemma*:

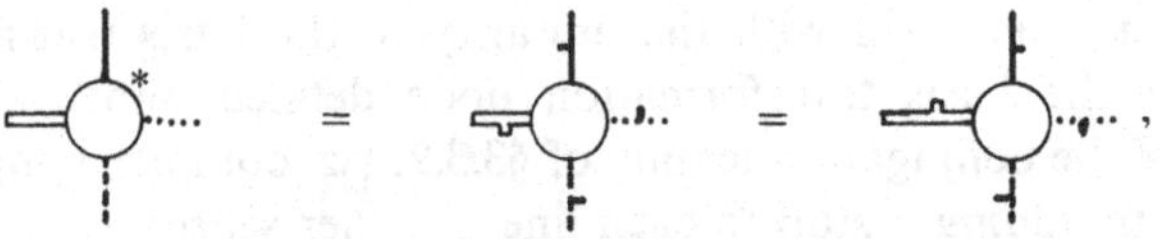

and refer to this property of any node as *self-conjugacy*. Its particular manifestation for the reducible representation constructed as a Kronecker product of irreps was the subject of §3.3; its application to the representation formed from a higher group G containing the group H under consideration is the topic of this section. Its analogues for the (not necessarily *self*-conjugate) nodes denoting physical operators or expressions are of great physical interest and will be discussed in §3.6. Nodes which on complex conjugation are equivalent to the original node with stubs on all lines *times a minus sign*, will be called *anticonjugate*.

(d) One consequence of self-conjugacy is that the double application of complex conjugation must be an invariance operation, and hence the product of '2j phases' on each line at a vertex must be unity:

$$\{j\}\{a\}\{\lambda\}=1\leftrightarrow$$

where $\{a\}$ is the (counterpart to a) 2j phase for the branching multiplicity label. (The irrep $\lambda\equiv\frac{3}{2}$ of the group C_3 is an example of a case where $\{a\}=-1$, since while from Problem 3.4 $\{\frac{3}{2}\}$ is necessarily 1 for a real and one-dimensional irrep, $j=\frac{3}{2}$ is a spin irrep. This has some interesting if pathological consequences; see Comment (f) of Problem 3.12, and §8.4.5.) Again, this product phase relation is of general application wherever any representation or invariance space of a group is irreducibly decomposed.

Problem 3.7

Derive the *Racah factorisation lemma* and hence give a diagram form for the isoscalar factor, or 3jm factor.

Solution

We use the second method of solution from the previous problem, but apply it to an invariant node of degree 3 rather than 2. If the 3jm symbol of G is expressed in the basis choice afforded by H, the resulting diagram is invariant with respect to H, and JLV3 affords a factorisation:

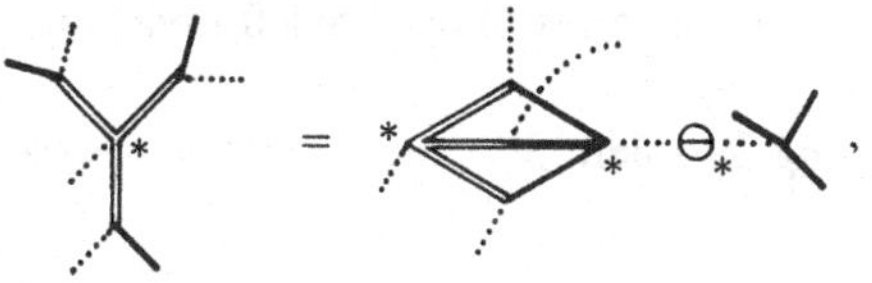

in which the 3jm symbol of G is seen to be a linear combination of the 3jm symbols of H. The coefficients in the linear combination are *3jm factors.*

Comment

The isoscalar factor differs from the 3jm factor by relating the Clebsch–Gordan symbol (rather than the 3jm symbol) in group and subgroup. By now, we assume that the reader is familiar with the necessary adjustments of definition, i.e. the judicious addition of stubs to convert invariant to commutative nodes (§2.5).

3.6 Wavefunctions and operators

3.6.1 Basic definitions

A ket symmetrised with respect to the group G will be written as $|\alpha j^\pi a\lambda l\rangle$ where the various labels represent parentage (α, j, a), irrep (λ) and component (l) labels. It may be defined from the O(3) ket $|\alpha j^\pi m\rangle$ (§2.7) via the basis transformation of Problem 3.4, and may be depicted by:

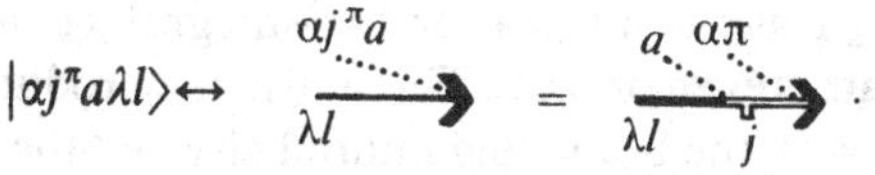

$|\alpha j^\pi a\lambda l\rangle \leftrightarrow$

Comments

(a) Given some sufficiently detailed group chain of the form $A \supset O(3) \supset \hat{G} \supset \ldots G \supset H \ldots$, the parentage label α corresponds to (an) irrep label(s) of some higher group(s) A. The spin–parity labels j^π are those of irreps of the full orthogonal group O(3) corresponding to rotations/reflections in 3-space. The labels a may possibly be irrep labels of one or more groups $\hat{G}$ intermediate between O(3) and the group G under consideration; failing that possibility, they might be defined as eigenvalues of some suitable labelling operator. The component labels l are similarly irrep labels of lower groups H. The full parentage labels will be included only in expressions where such precision and clarity is helpful.

(b) It is purely a matter of taste and convenience as to how far one includes these labels on the node or on the lines, or suppresses them entirely. For simplicity we shall usually write only a (which implies a particular choice of j, λ) for all parentage labels.

(c) The stub in the definition is to restore left invariance character to the irrep labels.

(d) It follows, using Problem 3.5, that the time-reversed states have the form:

$$\overline{|\alpha j a \lambda l\rangle} = \varepsilon \begin{pmatrix} j \\ a \\ \lambda \end{pmatrix} \begin{pmatrix} \lambda \\ l \end{pmatrix} |\alpha j a^* \lambda^* l^*\rangle; \leftrightarrow$$

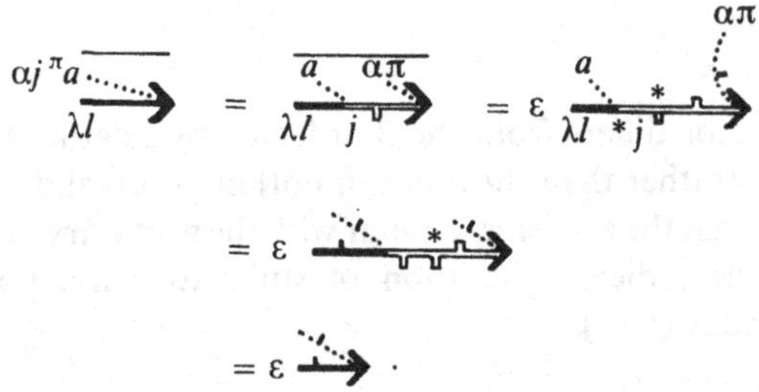

$= \varepsilon$ · $= \varepsilon$ · $= \varepsilon$.

The time-reversed state involves 2jm-symbol conjugation of irrep and of parentage labels, since it is essentially determined by the 2jm symbol of O(3) rather than of G. ε is a conventional phase, as before. Similarly, we may regard this relation as the statement that (for a suitable choice of ε) the node is right-conjugate.

Under double time reversal (when the conventional phase ε cancels) the state returns to itself modulo a phase τ_j equal to the O(3) 2j phase $\{j\}$ (which is $(-1)^{2j}$ and is ± 1 as j is integral or half-integral, i.e. as the number of electrons in the state is even or odd). This is also the product $\{a\}\{\lambda\}$ of the 2j phases associated with the branching multiplicity and the subgroup labels.

(e) The usual orthonormality and completeness relations take the form

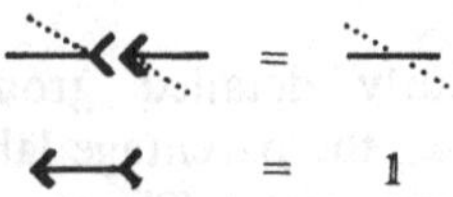

=

= 1

as in §2.7.2.

(f) Similarly operators may be written

$O(a,a') \leftrightarrow$

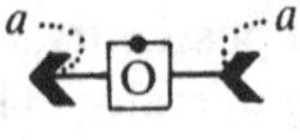

the labels a,a' summarising parentage information in bra and ket action, including the G-irrep labels.

(g) Hermitian conjugation reverses the sense of the chevron (bra↔ket) and as an antilinear operator complex conjugates all nodes. For example,

$$\mathrm{O}(a,a')^{\dagger} \leftrightarrow \quad \xrightarrow{R_{2\pi}}$$

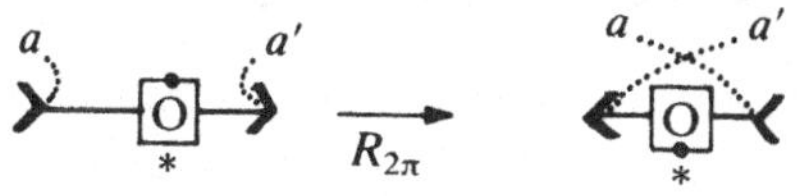

$$\leftrightarrow \mathrm{O}^{\dagger}(a',a). \; =$$

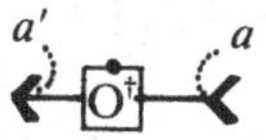

(h) The standard relation for the effect of a group operator on an irreducible basis member (Butler 1981)

$$\mathrm{O}_g|\alpha a\lambda l\rangle = \Sigma_{l'}\lambda^{l'}{}_{l}(g)\,|\alpha a\lambda l'\rangle \leftrightarrow$$

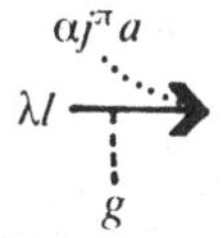

This gives the representation

$$\mathrm{O}_g \leftrightarrow \quad , \quad \mathrm{O}_g^{\dagger} \leftrightarrow$$

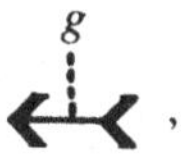

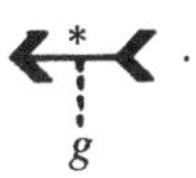

For bra vectors:

$$\langle \alpha a\lambda l|\mathrm{O}_g \leftrightarrow \quad = \quad .$$

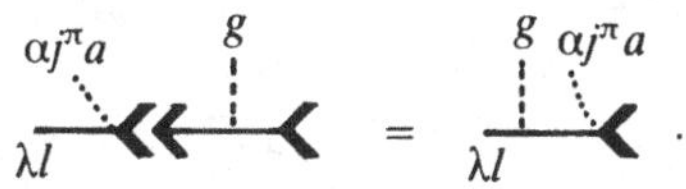

(i) This notation is offered here as representing an improvement on that of Stedman (1976a). The more abstract version (compare §§2.7.1, 2.7.2) and the use of a chevron has a greater mnemonic value and elegance than our old notation.

Problem 3.8

Verify that the operator

$$\mathrm{P}^{\lambda} \equiv (|\lambda|/|\mathrm{G}|)\Sigma_g \chi^{\lambda}(g)^{*}\mathrm{O}_g$$

projects onto the irreducible subspace associated with λ.

Solution

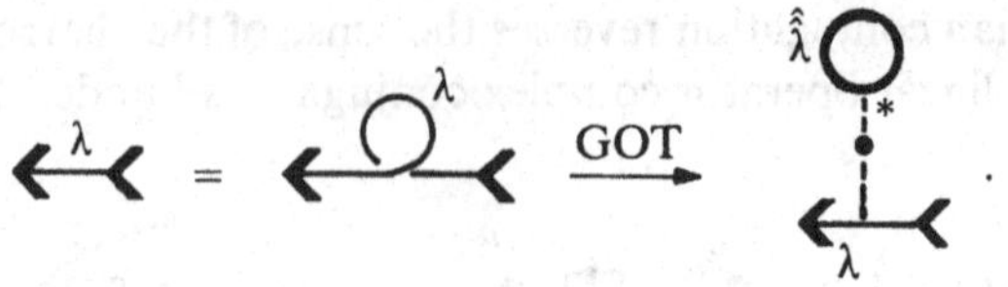

λ λ GOT $\hat{\hat{\lambda}}$ * λ

Comment

A fuller analysis is given in §3.8.

3.6.2 Irreducible tensor operators and the Wigner–Eckart theorem

An *irreducible tensor operator* is defined to have the property

$$\mathrm{O}_g T^{\mu}_{m} \mathrm{O}_{g^{-1}} = \Sigma_{m'} T^{\mu}_{m'} \mu_{m'm}(g).$$

In diagrams, this means that the corresponding node is then commutative (or, if modified by a stub, invariant):

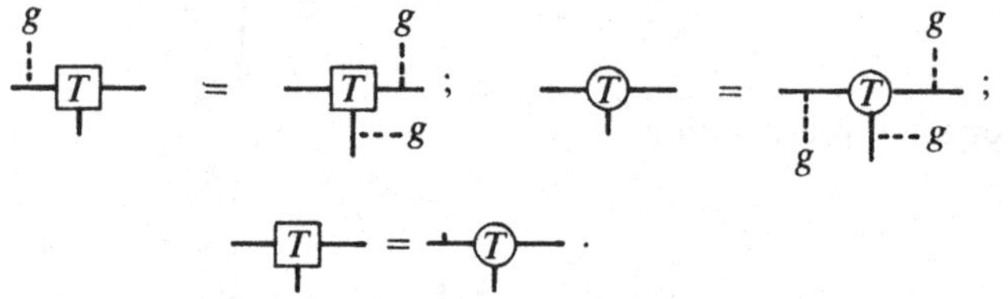

g T = T g ; g; T = T g ; g g

T = T .

Problem 3.9

Verify the above claim and derive the Wigner–Eckart theorem in the form:

$$\langle a\lambda l | T^{\mu}_{m} | a'\lambda' l' \rangle = \Sigma_r \check{\lambda} \langle a\lambda \| T^{\mu r} \| a'\lambda' \rangle \langle r\lambda l | \mu m \lambda' l' \rangle \leftrightarrow$$

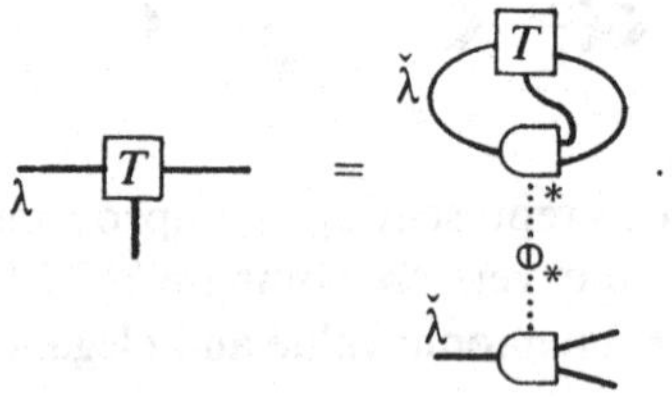

λ T = $\check{\lambda}$ T * Φ * $\check{\lambda}$.

Solution

$\mathrm{O}_g T^{\mu}_{m} \mathrm{O}_{g^{-1}} \leftrightarrow$ g T g^{-1} μm = gg^{-1} T g μm

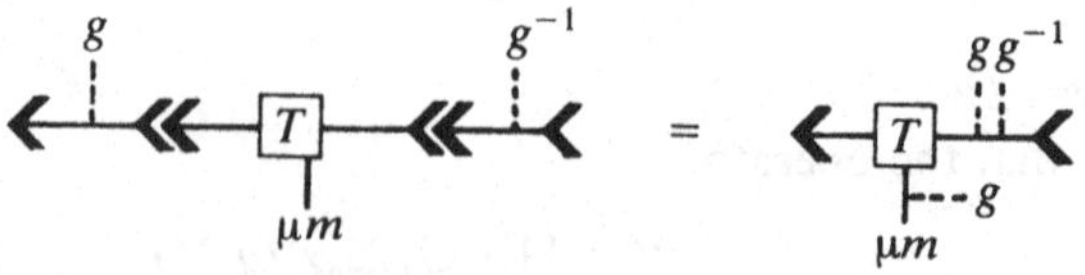

$$\leftrightarrow \Sigma_{m'} T^{\mu}_{m'} \mu_{mm'}(g).$$

Hence JLV3 gives the Wigner–Eckart result in the above form or equivalently as:

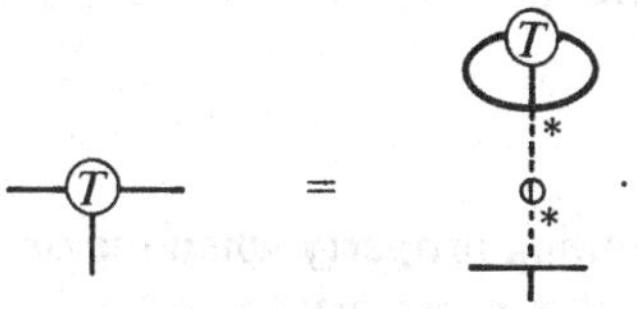

Comments

(a) Note the summation over repetition labels in the theorem, and the natural diagram form of this, and hence of the reduced matrix element:

$$\langle a\lambda \| T^{\mu r} \| a'\lambda' \rangle \leftrightarrow$$

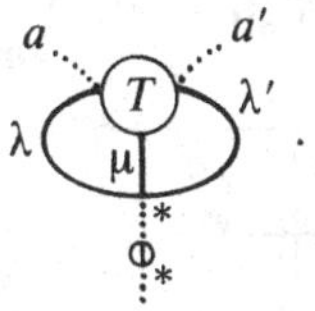

(b) We use a circle for invariant, and a box for commutative, nodes, as in Comment (e) of Problem 2.14.

3.6.3 Unit tensor operators; time reversal and hermitian conjugate tensor operators

The *unit tensor operators* are defined by

$$U^{\mu}_{m}(a,a',r) = \Sigma_{ll'} |a\lambda l\rangle\langle a'\lambda' l'| \begin{pmatrix} \lambda \\ l \end{pmatrix} \begin{pmatrix} \lambda^* & \mu & \lambda' \\ l^* & m & l' \end{pmatrix}^r \leftrightarrow$$

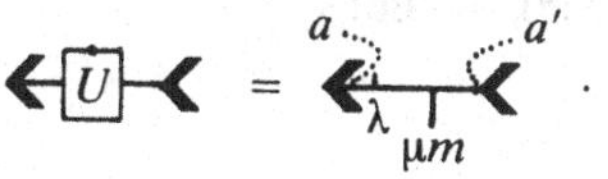

These are irreducible tensor operators with unit reduced matrix elements, since their structure is precisely the kernel left when applying the Wigner–Eckart theorem. In detail, its reduced matrix element is

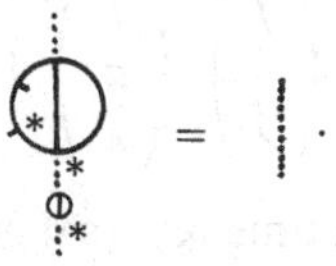

Problem 3.10

Convert the hermitian and time-reversal conjugates $(U^{\mu}_{m})^{\dagger}$, $\overline{U^{\mu}_{m}}$ of the unit tensor operator into irreducible tensor operators, and find their reduced matrix elements.

Solution

To obtain the corresponding property when the operator is conjugated by an antilinear operator, it is necessary to add a 2jm symbol. We define

$$V^{\mu}_{m}(a,a',r) \equiv \begin{pmatrix} \mu \\ m \end{pmatrix}^{*} [U^{\mu^*}_{m^*}(a',a,r)]^{\dagger} \leftrightarrow$$

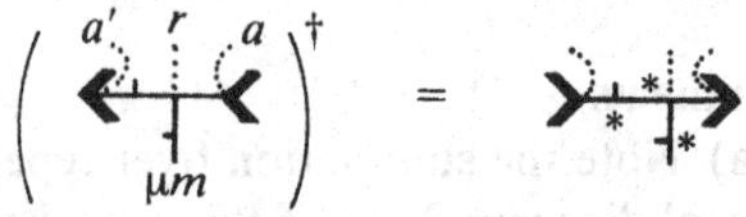

$$W^{\mu}_{m}(a,a',r) \equiv \begin{pmatrix} \mu^* \\ m^* \end{pmatrix}^{*} \overline{[U^{\mu^*}_{m^*}(a,a',r)]} \leftrightarrow$$

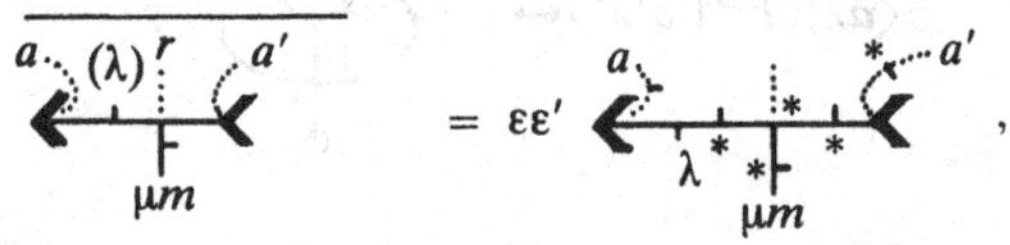

where the bar denotes a time reversal of the bras and kets. Then one may check similarly that these are irreducible tensor operators according to the definition. Alternatively, we can relate these directly to the unit tensor operators:

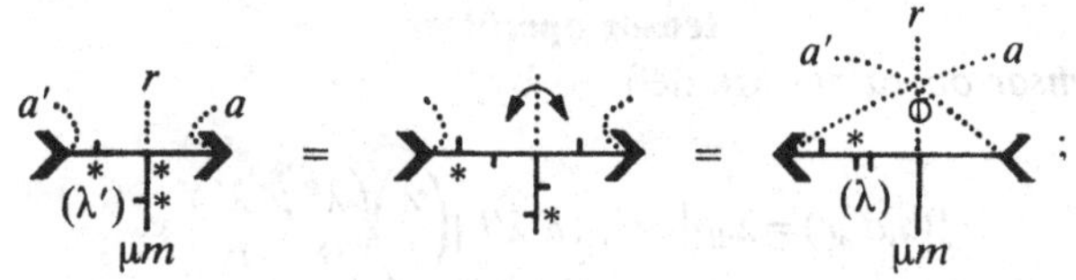

i.e. $V^{\mu}_{m}(a,a',r) = \{\lambda\}\{\lambda^*\mu\lambda' r\}\, U^{\mu}_{m}(a,a',r)$, while:

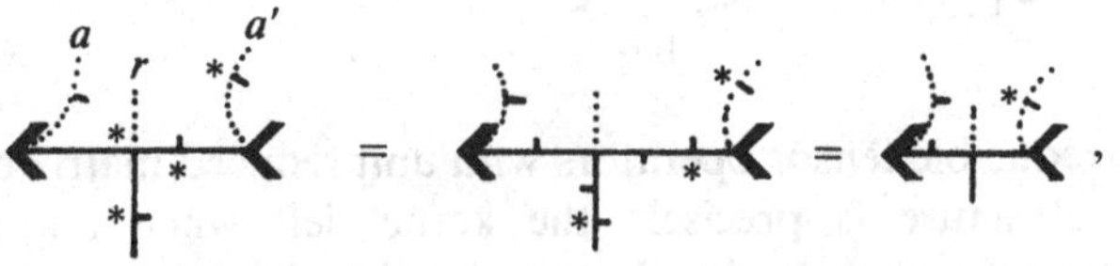

so that

$$W^{\mu}_{m}(a,a',r) = \varepsilon\varepsilon' \begin{pmatrix} j \\ a \\ \lambda \end{pmatrix} \begin{pmatrix} j' \\ a' \\ \lambda' \end{pmatrix}^{*} U^{\mu}_{m}(a^*,a'^*,r),$$

where the phases ε, ε' arise from time reversal of the corresponding ket and bra. Apart from these phases, this relation amounts to the statement that the unit tensor nodes are (say) right conjugate.

Comments

(a) We pause at this (somewhat arbitrary) point to reiterate and illustrate the advice of §2.5. It is good practice in 'good housekeeping' to see in diagram terms the reasons for the particular arrangement of suffices in each of the above expressions. One fundamental rule is that all external labels must match on each side, thanks to the presence of 2jm transformations where necessary. Every reader will have his or her own way of checking this, and the following discussion is just one of the possible logical routes. Take, for example, the definition of V^{μ}_{m}. The reversal of the order of the parentage labels on the right side of that definition reflects the effect of hermitian conjugation in interchanging bra and ket. If (by convention) this switch is performed in the definition, the order is restored in the final expression. This interchange has the further effect of complex conjugating the irrep labels near the invariant node (by which we mean the 3jm node in the above case of unit tensors, or, in the case of a more general operator, the circular node as opposed to square node). This is because when constructing an irreducible tensor operator from an invariant node a stub is needed on just the bra labels. In turn, this means that to preserve the triad structure $\{\lambda^*\mu\lambda' r\}$ the operator irrep label μ must be conjugated as well, so that the invariant node in V^{μ}_{m} corresponds to the fully conjugated triad $\{\lambda'^*\mu^*\lambda r\}$. This is one way of seeing why, in order that the label μ be attached to an external line on each side of the diagram equation defining V^{μ}_{m}, it is necessary to introduce a 2jm symbol on the labels μm. Fortunately, this also has the effect of achieving a rotational invariance appropriate for an irreducible tensor operator. All these felicities are consequences of the antilinear effect of hermitian conjugation of the 3jm vertex – namely complex conjugation – plus the conjugation lemma of §3.3.2, itself a further generalisation of the conjugation lemma of §2.5, which replaces the conjugated 3jm vertex by the unconjugated vertex (as required, if we are to have a left invariant kernel for an irreducible tensor operator) and a stub on each irrep line. One of the latter cancels (apart from a 2j phase, discussed below) the stub which was previously present on the bra label and which is no longer necessary, since it now corresponds to a ket label; the other two stubs have the necessary effect, as described above, on the new bra and the operator labels. The orientation of the stub on the operator labels μm is arbitrary, and may be adjusted by introducing another 2j phase; if two 2j phases were to be generated in this way, since the three 2j phases at any vertex cancel, we could replace the two 2j phases $\{\lambda\}\{\mu\}$ by the third, $\{\lambda'\}$.

(b) Similarly, for the time-reversed operator, the conjugation lemmas perform the necessary conjugations of vertices and labels. The reason for conjugating the group labels in the 2jm symbol when defining the conjugate operator in the time-reversed case is for convenience rather than necessity. First, it eliminates some phases in the $W^{\mu}_{m} \leftrightarrow U^{\mu}_{m}$ relation. Second, it ensures that under sequential hermitian conjugation and time reversal these 2jm symbol transformations will cancel. If

$$X^{\mu}_{m}(a,a',r) \equiv \overline{[U^{\mu}_{m}(a',a,r)]^{\dagger}},$$

this is an irreducible tensor operator standing in relation to V^{μ}_{m} as W^{μ}_{m} does with respect to U^{μ}_{m}, and also standing in relation to W^{μ}_{m} as V^{μ}_{m} does with respect to U^{μ}_{m}. Hence,

$$X^{\mu}_{m}(a,a',r) = \varepsilon\varepsilon' \begin{pmatrix} j \\ a \\ \lambda \end{pmatrix} \begin{pmatrix} j' \\ a' \\ \lambda' \end{pmatrix}^{*} \{\lambda\}\{\lambda^{*}\mu\lambda r\} U^{\mu}_{m}(a^{*},a'^{*},r).$$

3.6.4 Tensorial expansion of a physical operator; hermitian and time-reversal conjugation

Suppose that an operator V of physical interest is expanded in terms of a standard set of irreducible tensor operators O^{μ}_{m} (for example, these may be such well-known operators as position, momentum, etc., or the operator equivalents of angular momentum). As a physical operator, V is hermitian ($V = V^{\dagger}$) and of some definite time-reversal signature ($\overline{V} = \tau_{V} V$, where $\tau_{V} = \pm 1$ as V is time-even or time-odd; e.g. as V represents an electronic operator associated with electric or magnetic effects, respectively).

We ask now: what conditions are imposed on the standard operators' reduced matrix elements, and on the coefficients in the expansion, by these hermiticity and time-reversal properties?

As the first part of the answer, the standard operators within states of given symmetry and parentage are of arbitrary normalisation; hence the expansion coefficients c^{μ}_{m} in the fundamental expansion

$$V = \Sigma_{aa'\mu m r} c^{\mu}_{m}(a,a',r) O^{\mu}_{m}(a,a',r) \leftrightarrow$$

are independent in the labels μ, m, r; similarly for the associated reduced matrix elements $\langle a\lambda l \| O^{\mu r} \| a'\lambda' l' \rangle$.

By the Wigner–Eckart theorem, a standard tensor operator is closely related to the unit tensor operator:

$$O^{\mu}_{m}(a,a',r) = \langle a\lambda \| O^{\mu r} \| a'\lambda' \rangle U^{\mu}_{m}(a,a',r)$$

so that, if $u^{\mu}_{m}(a,a',r)$ are the coefficients in the expansion of V in unit tensor operators, then

$$c^{\mu}_{m}(a,a',r) = [\langle a\lambda \| O^{\mu r} \| a'\lambda' \rangle]^{-1} u^{\mu}_{m}(a,a',r).$$

Now we define the operators H^{μ}_{m}, T^{μ}_{m} and P^{μ}_{m} associated with the hermitian, time-reversal and double conjugates of O^{μ}_{m} in the same way that the tensor operators V^{μ}_{m}, W^{μ}_{m} and X^{μ}_{m} are related to U^{μ}_{m} in §3.6.3:

$$H^{\mu}_{m}(a,a',r)=\begin{pmatrix}\mu\\ m\end{pmatrix}^{*}[O^{\mu*}_{m*}(a',a,r)]^{\dagger}$$

$$T^{\mu}_{m}(a,a',r)=\begin{pmatrix}\mu^{*}\\ m^{*}\end{pmatrix}^{*}\overline{[O^{\mu*}_{m*}(a,a',r)]}$$

$$P^{\mu}_{m}(a,a',r)=\overline{[O^{\mu}_{m}(a',a,r)]^{\dagger}}.$$

Because of the relationship between the unit tensor operator and the standard tensor operator,

$$\langle a\lambda||H^{\mu r}||a'\lambda'\rangle=\{\lambda\}\{\lambda^{*}\mu\lambda' r\}\langle a'\lambda'||O^{\mu r}||a\lambda\rangle^{*}$$

$$\langle a\lambda||T^{\mu r}||a'\lambda'\rangle=\varepsilon\varepsilon'\begin{pmatrix}j\\ a\\ \lambda\end{pmatrix}\begin{pmatrix}j'\\ a'\\ \lambda'\end{pmatrix}^{*}\langle a^{*}\lambda^{*}||O^{\mu r}||a'^{*}\lambda'^{*}\rangle^{*}$$

$$\langle a\lambda||P^{\mu r}||a'\lambda'\rangle=\varepsilon\varepsilon'\begin{pmatrix}j\\ a\\ \lambda\end{pmatrix}\begin{pmatrix}j'\\ a'\\ \lambda'\end{pmatrix}^{*}\{\lambda\}\{\lambda^{*}\mu\lambda' r\}\langle a'^{*}\lambda'^{*}||O^{\mu r}||a^{*}\lambda^{*}\rangle.$$

Problem 3.11

Verify these relations by diagram methods.

Solution

For example, from the definition,

$$\langle a\lambda l|H^{\mu}_{m}(a,a',r)|a'\lambda' l'\rangle=\begin{pmatrix}\mu\\ m\end{pmatrix}^{*}\langle a'\lambda' l'|O^{\mu*}_{m*}(a',a,r)|a\lambda l\rangle^{*}\leftrightarrow$$

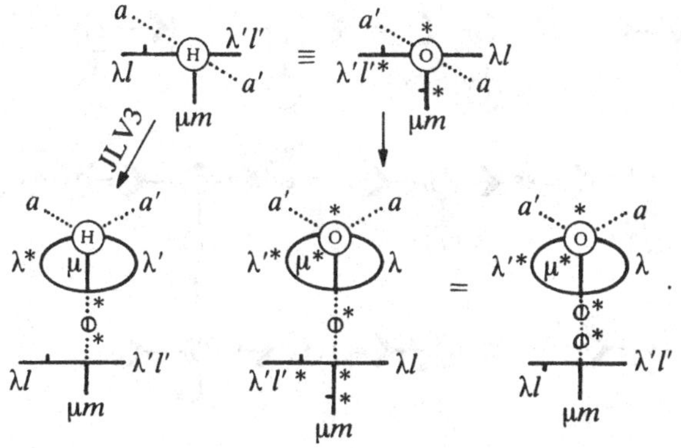

Problem 3.12

Show that, corresponding to these relations, one may write the physical interaction V in terms of the irreducible tensor operators H, T or P with coefficients h^{μ}_{m}, t^{μ}_{m} and p^{μ}_{m} respectively, which are then related by

$$h^{\mu}_{m}(a,a',r) = \begin{pmatrix}\mu\\ m\end{pmatrix}[c^{\mu^*}_{m^*}(a',a,r)]^* = \{\lambda\}\{\lambda^*\mu\lambda'r\}c^{\mu}_{m}(a,a',r)$$

$$t^{\mu}_{m}(a,a',r) = \tau_{V}\begin{pmatrix}\mu^*\\ m^*\end{pmatrix}[c^{\mu^*}_{m^*}(a,a',r]^* = \varepsilon\varepsilon'\begin{pmatrix}j\\ a\\ \lambda\end{pmatrix}\begin{pmatrix}j'\\ a'\\ \lambda'\end{pmatrix}^* c^{\mu}_{m}(a^*,a'^*,r)$$

$$p^{\mu}_{m}(a,a',r) = \tau_{V}c^{\mu}_{m}(a',a,r) = \varepsilon\varepsilon'\{\lambda\}\{\lambda^*\mu\lambda'r\}\begin{pmatrix}j\\ a\\ \lambda\end{pmatrix}\begin{pmatrix}j'\\ a'\\ \lambda'\end{pmatrix}^* c^{\mu}_{m}(a'^*,a^*,r).$$

Solution

These follow directly from the definitions. In diagrams (and omitting parentage lines for simplicity):

V = H (h) ≡ O * (h) = O * * (h)

= O (h) ↔ O (c)

= V * = O (c) * = O (c) * ;

V = O (c) = T (t) ≡ O (t)

= $\varepsilon\varepsilon'$ O * (t) = $\varepsilon\varepsilon'$ O * (t) = $\varepsilon\varepsilon'$ O (t) = τ_V V

= $\tau_V\varepsilon\varepsilon'$ V * = $\tau_V\varepsilon\varepsilon'$ O (c) * = $\tau_V\varepsilon\varepsilon'$ O (c) * .

Comments

(a) Arrows denote the comparisons that give the first two above results.

(b) The last relation has an important physical application when $a=a'^*$, i.e. when we are looking at matrix elements between manifolds which contain time-reversal conjugate partners (these can be the same manifold, with $a=a'=a'^*$ etc.). For the coefficients c^{μ}_{m} not to vanish,

$$\tau_{V}=\{\lambda\}\{\lambda^*\mu\lambda'r\}\begin{pmatrix} j \\ a \\ \lambda \end{pmatrix}\begin{pmatrix} j^* \\ a^* \\ \lambda^* \end{pmatrix}^* .$$

The last two 2jm factors give the 2j phase $\{a\}$ which with $\{\lambda\}$ gives $\{j\}=\tau_j$, according to Comment (d) on the definition in §3.6.1 (p. 86). Hence the interchange symmetries of the relevant 3jm symbols involved in the coupling are restricted by the time-reversal character of the operator and of the quantum states:

$$\tau_V\tau_j=\{\lambda^*\mu\lambda'r\}.$$

Hence from the Comment on Problem 3.3 (p. 82), the operator irrep labels are restricted to a certain interchange symmetry, itself dependent on the time-reversal character of both state and operator:

$$\mu^*_r\in[\lambda\times\lambda]_{\tau_V\tau_j}.$$

Applications of this in the theory of Raman scattering and of the Jahn–Teller effect will be given in §§6.5.2, 7.3.2, 8.4.1. For a more physical proof see Abragam and Bleaney (1970) or Stedman and Butler (1980).

(c) The first two conditions give restrictions on the real or imaginary character of the reduced matrix elements which will be both stringent and important in many physical situations (Stedman and Butler 1980, Stedman 1985).

(d) As illustrated in §2.7.4 for SO(3) operators, the effect of the extra phase τ_V from the physical operator transformation under time reversal is to replace self-conjugacy by anticonjugacy (as defined in §2.5) and vice versa, when the operator is time-odd.

In the present case we may see this by noting the appearance of τ_V in the relation between t^{μ}_{m} and c^{μ}_{m}, combining this with the results of Problem 3.11. This further generalisation of the conjugation lemma (§§2.5, 3.3.2, 3.5) is in strict accordance with, and directly derivable from, the fundamental relation $\langle a|V|b\rangle^*=\langle\bar{a}|\bar{V}|\bar{b}\rangle=\tau_V\langle\bar{a}|V|\bar{b}\rangle$.

(e) Under double time reversal, V as well as the complex coefficients in the operator expansions are unchanged. Hence the various conjugation phases appearing in the joint time-reversal operations must cancel. The 2jm factors in the SO(3)⊃G branching which appear in the relation between W^{μ}_{m} and U^{μ}_{m}, and hence in that between T^{μ}_{m} and O^{μ}_{m}, combine to give 2j phases of the final form $\{a\}\{a'\}\{\lambda\}\{\lambda'\}$, equal to $\{j\}\{j'\}$ or $\tau_j\tau_{j'}$. Hence, in general, *the time-reversal phases* τ_j *of bra and ket must be identical.* This may be termed a *superselection rule*: bosonic states are coupled only to bosonic states, and fermionic to fermionic.

(f) This does not mean that the 2j phases of the corresponding G-irreps are related, since the 2j phases $\{a\}$ in the branching labels may differ (see Comment (d) on Problem 3.6, also §8.4.5). However, it does mean that *the operator irrep* μ *corresponds to a true, rather than a spin, irrep of* O(3) (we write $\tau_\mu = +1$ rather than -1). If we enlarge our spatial operations to include 2π rotations $R_{2\pi}$ and distinguish them from identity operations so as to cover projective (spin) irreps of O(3), the connection $\tau_j \tau_{j'} \tau_\mu = 1$ (and the consequence $\tau_\mu = 1$) amounts to the invariance of the 3jm symbol under the double application of $R_{2\pi}$.

(g) Group 2jm symbols $\binom{\mu}{m}$ etc. cancel between the respective transformations of coefficient and operator (see the 2jm factors in the above equation for t^μ_m and that for T^μ_m).

3.7 Algebra of group operators

Some years ago, there was a resurgence of interest in level splitting algebras and similar tools for the bypassing or substitution of the traditional representation-theoretic methods of group theory in physical applications. Some of the relevant papers are referenced in Black and Stedman (1982), which also develops a diagram form of the dual analysis of group algebra, complementary to that discussed so far in this chapter. We review this dual extension here.

As defined in §2.1, the group operator matrix element in an irreducible representation space

$$\lambda^l{}_{l'}(g) \leftrightarrow$$

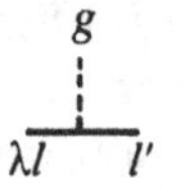

g
λl l'

does not correspond to an invariant or commutative node. We note immediately that since the parities of bra and ket labels are opposite, the hope might be for a commutative (rather than invariant) node. We therefore define it to be one by the expedient of taking the third line of the diagram more seriously and allowing for a new form of group element action, i.e. the action of a group element on itself:

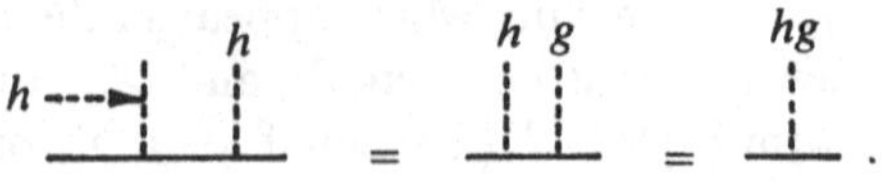

h --→ h = h g = hg .

From the group multiplication relation, this obviously requires the definition

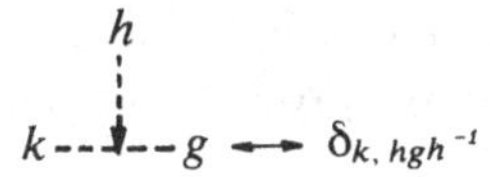

$$k \text{ - - }\overset{h}{\downarrow}\text{ - - } g \longleftrightarrow \delta_{k,\, hgh^{-1}}$$

for *the action of group element on group element*, i.e. the operation of conjugacy of g with respect to h.

Problem 3.13

Show that this new node is itself both left and right invariant.

Solution

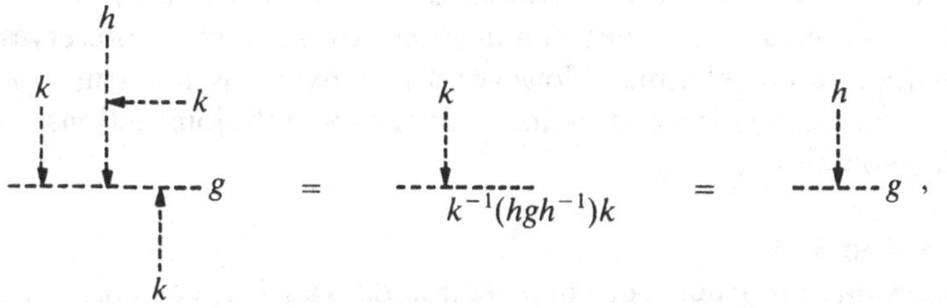

k, h, k, g, k $=$ k, $k^{-1}(hgh^{-1})k$ $=$ h, g ,

since, for example, $(k^{-1}hk)(k^{-1}gk)(k^{-1}h^{-1}k)=k^{-1}(hgh^{-1})k$.

Comments

(a) This is initial evidence that a new and elegant extension of all earlier work is possible.

(b) This in turn reflects the value of class sum theory in many physical contexts (see, for example, Lax 1974).

(c) We note here that this action of one group element on another keeps the element in the same class, just as the action of a group element on an irrep space is diagonal in irrep label, changing only the component; in this sense a group (class-) element line has very similar properties to an irrep component line, and is a further indicator of an extensive duality.

(d) The group element node has mixed parity: the leading terminal and trailing terminal have positive and negative parity, respectively. Hence the housekeeping of §2.5 has to be carried through terminal by terminal: the diagrams containing the group element node are not bipartite. Bipartite construction could be avoided by defining a new, left invariant, node as the group element with a (complex conjugated) stub on the leading terminal and re-expressing all following results in terms of this node. However, this takes us away from the group element *per se*, and we have not felt the change is worthwhile at this stage of the discussion.

A simpler node corresponding to group-algebraic coupling of three group elements is enshrined in the basic group multiplication relation (independent of basis). It is sensible to make this a symmetric definition dual to the 3jm symbol by requiring the product of the three group elements to be the group identity element:

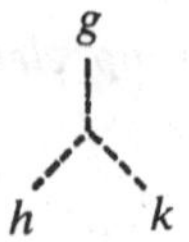

$$\leftrightarrow \delta_{ghk,E}.$$

This preserves cyclic symmetry in accordance with the rules of §1.2. It corresponds to the 3-K symbol of van Zanten and de Vries (1973). We shall call it the 3Kk vertex, to emphasise its dependence not only on the class K, but also the element k within the class (as for the 3jm nomenclature). Again, it is an invariant node; conjugation within a class preserves group multiplication relations. However, the converse is not true: the group element conjugation node is not invariant under the joint actions of the 3Kk node on all lines.

Problem 3.14

Show that the group element inverse node is analogous to the 2jm symbol, and check its invariance property.

Solution

Define

$$\delta_{h,k^{-1}} \leftrightarrow \quad \equiv$$

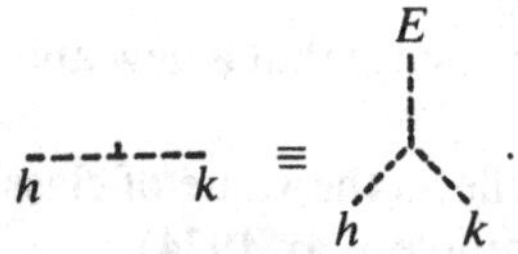

Clearly it is unitary. Under group conjugation

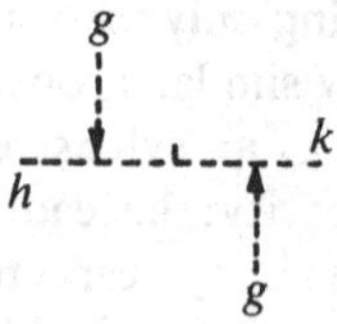

$$\leftrightarrow \delta_{g^{-1}hg,(g^{-1}kg)^{-1}} = \delta_{h,k^{-1}}.$$

Comments

(a) A logical choice for the dual of the 2jm symbol is the conjugation node that would make the group operator node self-conjugate, in the sense that:

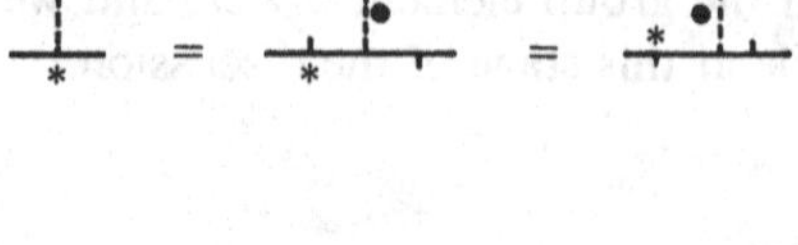

(we have to complex conjugate one 2jm symbol in each of the last two diagrams since the group operator is a mixed node, in which the irrep–component terminals have opposite parity). However, the right invariance of the stub, i.e. the 2jm symbol on the irrep–component line, immediately shows that this definition is trivially the Kronecker delta symbol.

(b) It is a pleasant if circuitous exercise to confirm this by using the definition of Comment (c) of Problem 3.6 for the dual 2jm symbol, to prove the unitarity etc:

$$\Sigma_\lambda \quad \hat{\lambda}/|G|$$

via the Great Orthogonality and Character Orthogonality theorems.

(c) Another possible choice is the node

$$\leftrightarrow \delta_{C_g, C_h}.$$

equating the classes, but not necessarily the elements.

A third node may be defined which relates three elements of the group, and which does so very symmetrically. We simply require that all group elements at the node be identical:

$$\leftrightarrow \delta_{gh}\delta_{hk}.$$

This is of use in some earlier work as well as the present. For example, the black dot of §2.3 is a weighted sum over this node.

As a generic symbol, we use

for any of these third-degree invariant nodes.

We may now form combinations of these nodes with the ones defined in earlier chapters to obtain invariant nodes with an arbitrary number of irrep–component or class–element lines. For example, if we join our old 3jm symbol and the group element (adapted with a stub to make it invariant), we obtain a duality transformation matrix between class–element and irrep–component labels, parametrised by the irrep and multiplicity associated with the Kronecker product in the 3jm symbol:

$$\chi^{g}_{\lambda l}(\mu r) \leftrightarrow$$

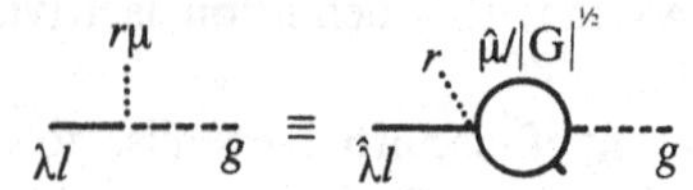

Note that we are using all the housekeeping rules of §2.5 to maintain compatibility of parities etc., joining only lines of opposite parity, i.e. left invariant only to right invariant node–line junctions. (We have also added carets indicating dimensional factors to simplify later results.) One may check for example that this mode is left invariant:

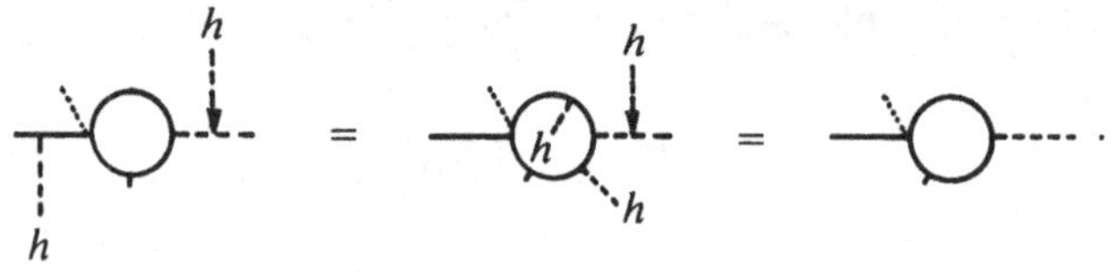

Similarly we can obtain other constructs with more of either type of line simply by adding 3jm or $3Kk$ vertices, for example.

Apparently (but only apparently) there is little relation between invariant nodes of the same external structure. Such a vertex as the above could have been formed, for example, from a network with one irrep–component and one class–element line by forming any of the diagrams

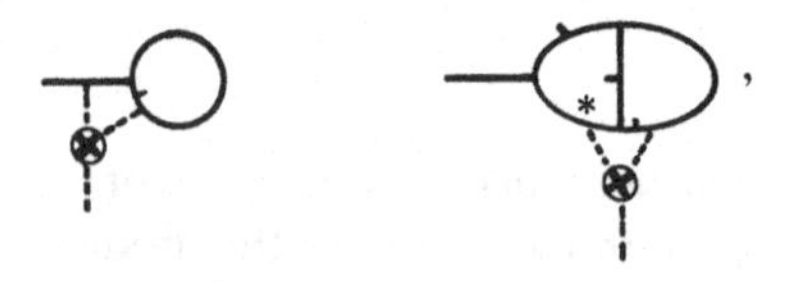

each with any of the previous definitions of the third-degree group element node, where relevant; just as one can manufacture an arbitrary number of networks with a given number of external lines using irrep–component lines alone.

The important result, whose proof will be indicated below, is that all such nodes, with the same degrees in irrep–component and in class–element lines, may be related by extensions of the JLV network theorems to certain standard nodes – in this case the duality transformation node – provided only they are invariant (or commutative; we shall not labour the distinction, but use invariant nodes for explicitness).

This arises since, in finite groups at least, the Great Orthogonality Theorem:

$$= |G|$$

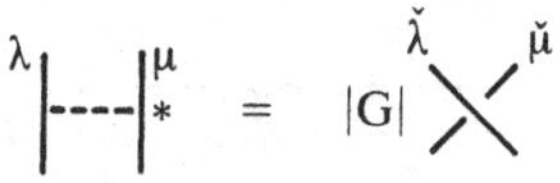

central to our development of the JLVn theorems, has a dual:

$$= |G| \text{------} .$$

Problem 3.15
Prove this result.

Solution
The matrix $U^{\lambda l l'}{}_g \equiv \lambda^l{}_{l'}(g)[|\lambda|/|G|]^{\frac{1}{2}}$ is a square matrix, thanks to the theorem that the irrep λ appears in the regular representation $|\lambda|$ times. The Great Orthogonality Theorem shows that it is also a unitary matrix ($UU^\dagger = 1$):

$$U^{\lambda l l'}{}_g \leftrightarrow \quad ;$$

$$= \quad .$$

Hence, unitarity ($U^\dagger U = 1$) gives

$$= \quad ,$$

and the stated result is its transliteration.

Comments
(a) Similarly the Character Orthogonality Theorem

$$= |G|\, \delta_{\lambda\mu}$$

can be interpreted as a unitarity condition ($UU^\dagger = 1$) on a square matrix $U^\lambda{}_C$ where C is a class label:

$$U^\lambda{}_C \equiv \chi^\lambda(C)[|C|/|G|]^{\frac{1}{2}} \leftrightarrow$$

$$= \quad .$$

$U^{\lambda}{}_{C}$ is also a square matrix since the number of irreps and classes in a finite group is the same. Hence unitarity gives also the dual theorem ($U^{\dagger}U = 1$)

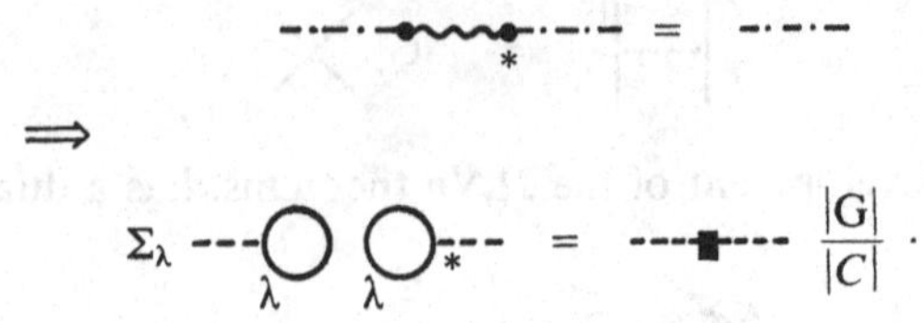

$\Rightarrow$

$$\Sigma_{\lambda} \quad = \quad \frac{|G|}{|C|} \, .$$

(b) We may check the result by noting that the following diagram may be reduced by either the original or dual form of the theorem:

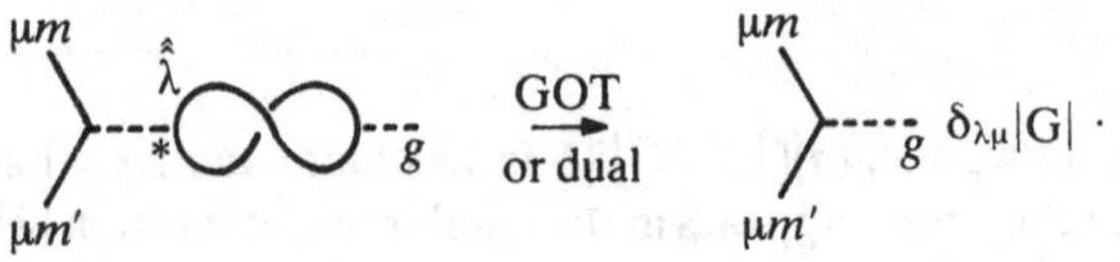

$$\xrightarrow[\text{or dual}]{\text{GOT}} \quad \delta_{\lambda\mu}|G| \, .$$

(c) Hence the duality transformation introduced before has a unitarity property:

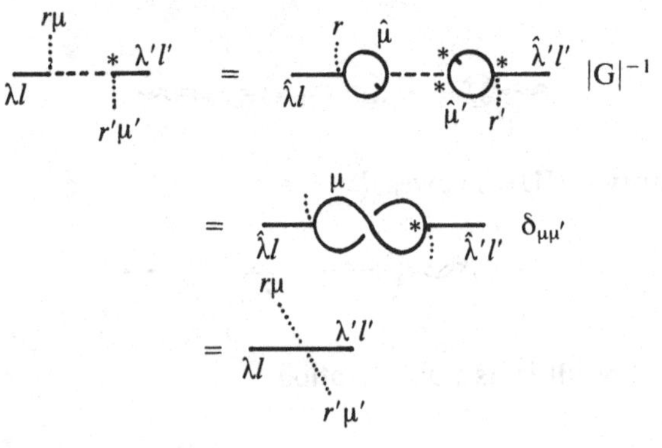

$$= \quad |G|^{-1}$$

$$= \quad \delta_{\mu\mu'}$$

$$=$$

and vice versa:

$$= \quad .$$

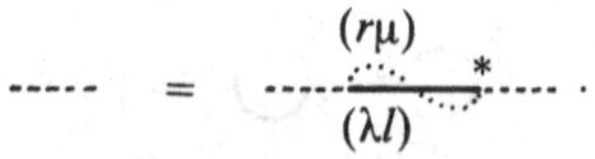

We can apply this unitarity condition for any set of class–element labels on any external lines of an invariant diagram, then apply the standard JLV*n* theorem to the irrep–component lines of the diagram to which all the external lines are now connected in tree fashion. For example, while it is possible to find invariant nodes which are alternatives to the duality transformation in that they have a different structure, but the same set of external lines, they may each be written as a linear combination of the duality transformation nodes:

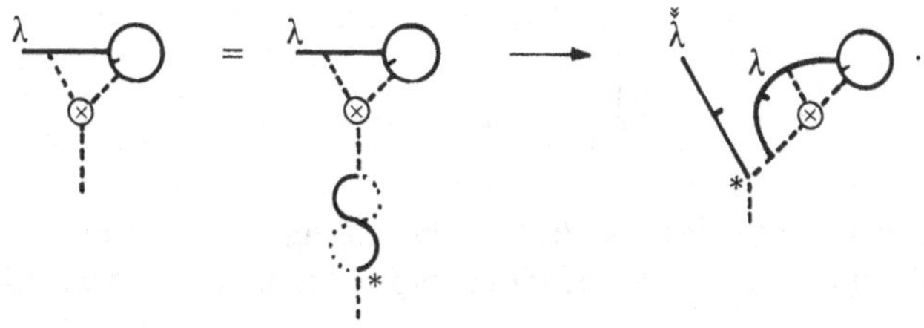

Therefore, just as an invariant node of third degree in irrep–component legs can be written as a sum over 3jm symbols with suitable coefficients, which we might call 'reduced matrix elements' in analogy with the Wigner–Eckart theorem, or JLV3, any invariant node – say $N(m,n)$ – with an arbitrary number of external lines – say m class–element lines and n irrep–component lines – can be similarly decomposed, by the theorem $T(m,n)$ say, in terms of some standard vertices, $M(m,n)$ say. One tabulation (using left invariant nodes for definiteness) is given below:

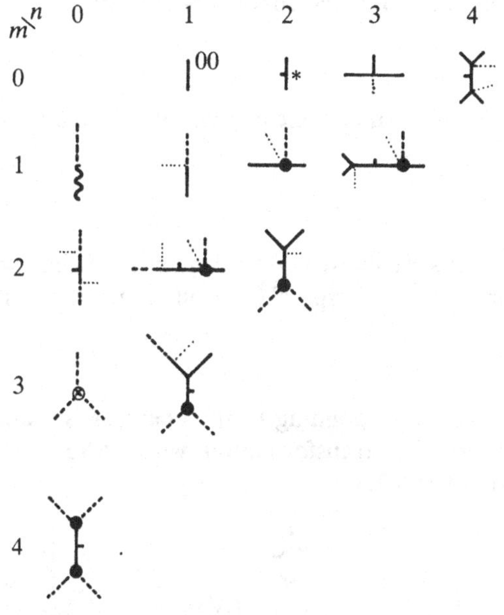

For elegance in this we have introduced a new notation for the character, as well as other abbreviations:

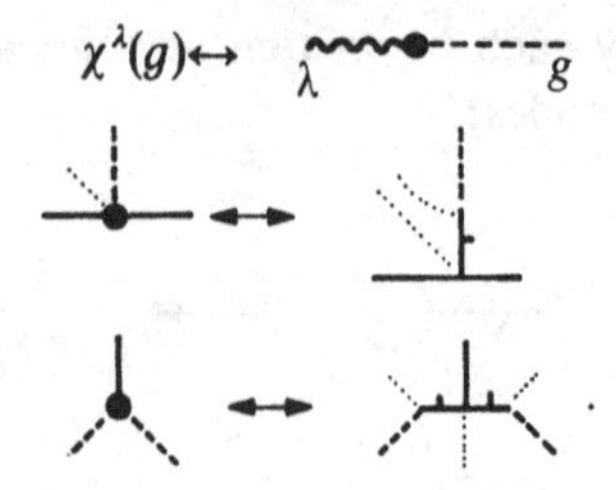

These may be regarded as the natural nodes to describe all possible class–irrep functions independent of component or elements within irrep or class.

Problem 3.16
Prove the class-sum-theoretic theorem: if any function $f(g)$ of the group element g is a class function, i.e. if it is invariant under group conjugacy:

$$f(hgh^{-1})=f(g)$$

then it is a linear sum of the characters.

Solution
This amounts to the theorem that all invariant diagrams of the form $N(1,0)$ are a linear combination of the characters $M(1,0)$.

Comment
In general a class sum renders one line an invariant, and so reduces m by one.

Problem 3.17
Find the coefficients (reduced matrix elements) of the expansion of the group operator (a node of type $N(1,2)$) in terms of the standard nodes $M(1,2)$.

Solution
The invariant node corresponding to the operator is reducible using the unitarity of the duality transformation with JLV3 – the adapted or generalised form of JLV3:

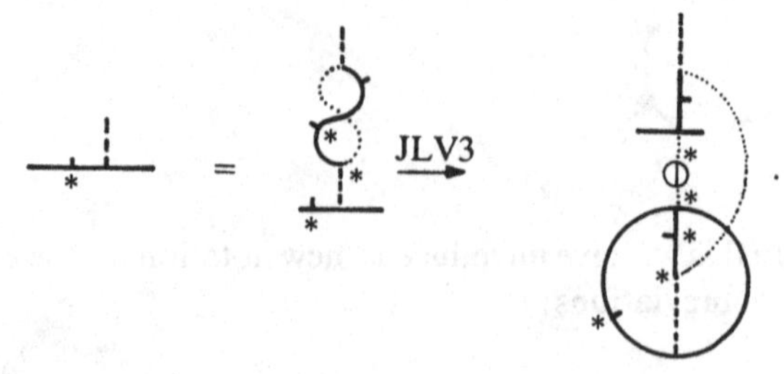

Comment

Hence the group operator may be used to get an irreducible tensor operator of the form:

3.8 Projection operators

As one example of the application of these techniques, consider the projection operator, whose properties may be summarised using a generalised great orthogonality theorem (Black and Stedman 1982) using the method of §3.7.

From Problem 3.8, the projection operator

$$\mathrm{P}^{\lambda} = |\lambda|/|\mathrm{G}|\Sigma_{g \in \mathrm{G}} \chi^{\lambda}(g)^{*} \mathrm{O}_{g} \leftrightarrow \quad \overset{\lambda}{\longleftarrow} \quad \xrightarrow{\text{GOT}}$$

clearly projects out the irrep λ. Its completeness and idempotence also follow from the Character Orthogonality Theorem and its dual (see Comment (a) of Problem 3.15; note that $|\mathrm{E}| = 1$):

$$\Sigma_{\lambda} \mathrm{P}^{\lambda} \leftrightarrow \quad \Sigma_{\lambda}$$

while

$$\mathrm{P}^{\lambda}\mathrm{P}^{\mu} \leftrightarrow$$

$$= \quad \delta_{\lambda\mu} \quad = \mathrm{P}^{\lambda}\delta_{\lambda\mu}.$$

Note in this derivation that the group element lines are being pinched after a fashion, using the group rearrangement theorem:

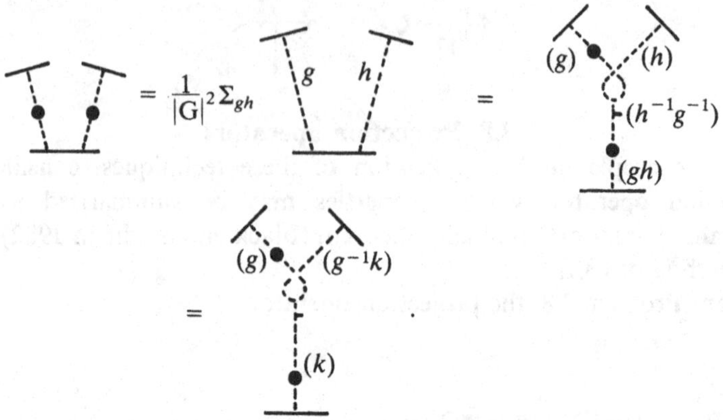

$$= \frac{1}{|G|^2}\Sigma_{gh} \quad g \quad h \quad = \quad (g) \quad (h) \quad (h^{-1}g^{-1}) \quad (gh)$$

$$= \quad (g) \quad (g^{-1}k) \quad (k)$$

After this is performed once, one group label which is actually fixed by two others is nominally summed over the group (the extra terms vanish through the definition of the $3Kk$ vertex), giving a cyclic symmetry which enables us to reverse the pinching theorem in a different orientation. In the last step, after the use of the Great Orthogonality Theorem, we obtain a loop denoting a character.

The relation between the projection operator and the transformation coefficients is

$$\langle \Lambda L|\mathrm{P}^{\lambda}|r\lambda l\rangle \equiv \langle \Lambda L|r\lambda l\rangle \leftrightarrow \quad \Lambda L \quad \lambda \quad r \quad \lambda l \quad = \quad \Lambda L \quad \lambda l \quad r,$$

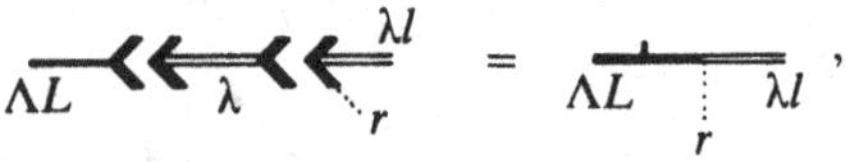

$$\langle \Lambda L|\mathrm{P}^{\lambda}|\Lambda L'\rangle = \Sigma_r \langle \Lambda L|r\lambda l\rangle\langle r\lambda l|\Lambda L'\rangle \leftrightarrow \quad \Lambda L \quad \lambda * \quad * \quad \Lambda L',$$

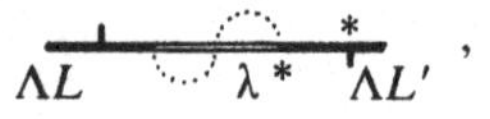

so that

$$\mathrm{Tr}_{\Lambda}\mathrm{P}^{\lambda} \leftrightarrow \quad * \quad = \quad \Lambda$$

$$\leftrightarrow |\Lambda|.$$

4

Symmetric and Unitary Groups

The most resolute speaker . . . picked up the chalk and began to draw lines on the board. 'Give us the formulae first,' said Landau. 'Just coming to it,' replied the speaker, again drawing . . . This dialogue was repeated several times. At the tenth repetition, Dau exclaimed, 'Oh, draw then, blast you!' . . . Landau of course was fully aware of the stratagem . . . known as the 'Pekar algorithm' after its 'inventor' . . .

Livanova (1980)

4.1 Introduction

This chapter is in the nature of an interlude; we review, again with a diagram notation, some combinatoric results from the theory of symmetric and unitary groups, including some interconnections. This not only illustrates the general approach but will be of use later in the development of Lie group theory.

We offer two warnings to the reader: it is particularly in the area of combinatoric problems with the symmetric group and its applications that diagram techniques have been used to give rules of thumb for establishing theorems whose proofs are nonexistent, and also in which they have been used with an inconsistent mixture of labelling conventions. This is one reason why our treatment is not as advanced as it might be; we concentrate on the foundations, and endeavour to clarify the significance of old results which are often cursorily treated.

4.2 Class theory and Young's symmetrisers

A thorough review of S_n, the group of permutations on n objects, will not be attempted here, and should be sought from Weyl (1931), Hamermesh (1962), Messiah (1962), Boerner (1963), and similar texts. A helpful introduction is given by Coleman (1968). However, some basic points will be stated and illustrated using the diagram notation for generalised Kronecker deltas (§4.3) following Mandula (unpublished Lecture notes,

1980), Cvitanovic (1984) etc., but with minor modifications to reflect our general rules (§1). For partitions, group labels, etc., we use the notation of Wybourne (1970), who gives a useful summary of definitions.

As a finite group, S_n has as many irreps as conjugacy classes (§3.7). The latter are the set of possible permutation cycle structures $(1^{\upsilon_1}2^{\upsilon_2}\ldots i^{\upsilon_i}\ldots n^{\upsilon_n})$ where υ_i is the number of i-cycles (equal to 1 if suppressed). Since $\Sigma_i \upsilon_i = n$, each cycle structure may be identified with a *partition* $[\lambda] \equiv [\lambda_1, \lambda_2, \ldots \lambda_n]$, defined by $\Sigma_i \lambda_i = n$, $\lambda_i \geq \lambda_{i+1}$, $\lambda_n > 0$. For example, with $n=6$ the partition (321) corresponds to (123)(45)(6), or a cycle structure (123) and so a partition [321].) The corresponding irrep is associated with a *Young diagram* (sometimes called a Young frame) of the form

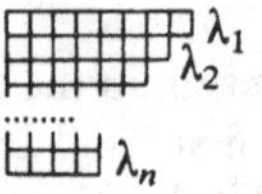

λ_1
λ_2
λ_n

A basis function of this irrep will have n labels and will be symmetric within each set of λ_i labels (rows) and antisymmetric within columns. More precisely, it will be projected out by means of a Young symmetriser, the product of the column antisymmetrisers and the row symmetriser operators. The various components of the basis for any irrep (or Young diagram) are labelled and enumerated by the different possible standard labellings of the Young diagram, that is the *standard Young tableau*, according to which numbers 1 through n are included in the boxes without repetition and in such a way that numbers strictly increase within rows and within columns. (Sometimes the name of standard Young diagram is reserved for this application, and the name standard Young tableau for the different standard numbering appropriate for unitary group irreps, as described later.) Examples will be given shortly.

Problem 4.1

Construct diagrams for the permutation operators $\hat{\pi}$ which are elements of S_2, and hence for the S_2 projection operators for the irreps.

Solution

The integer 2 has two partitions: [2] and $[1^2]$, corresponding to the permutation diagrams (following §1)

$$[2] \supset (1)(2) \leftrightarrow \begin{matrix} 2\ 1 \\ \times \\ 1\ 2 \end{matrix}, \quad [1^2] \supset (12) \leftrightarrow \begin{matrix} 2\ 1 \\ |\ | \\ 1\ 2 \end{matrix}$$

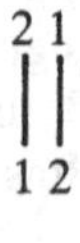

respectively. The group is abelian and the two classes have just one element each; the character table has the form

S_2 class:	(1^2)	(2)
irrep:		
[2] □□	1	1
$[1^2]$ ⊟	1	−1

The two irreps associated with the Young diagrams

$$[2] \leftrightarrow \square\square, \quad [1^2] \leftrightarrow ⊟$$

correspond to symmetric and antisymmetric combinations of two labels, i.e. the symmetriser and antisymmetriser of §1.3:

$$P^{\square\square} \leftrightarrow \quad , \qquad P^{⊟} \leftrightarrow$$

Comments

(a) We attach external labels in anticlockwise order for clarity in this example only; in general, a permutation will be designated by

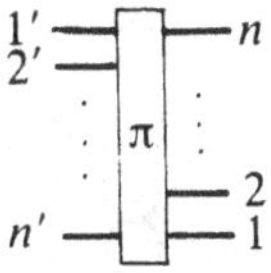

$$\leftrightarrow \Pi_{ii'} \delta_{i', \pi(i)}.$$

The distinction between π and its inverse is preserved in this notation merely by the orientation of the symbol π in the box: a 180° rotation would invert this symbol, and so would fail to be an invariance operation. The diagram is to be read as stating that the line linkage in the box takes the label i on the right side to the label i' on the left, where $\{i'\} = \pi(\{i\})$. As such, this notation is unambiguous, although the reader might prefer the addition of a dot on the side of the box to specify, independently of rotations or of labelling, which set of labels is the leading set.

(b) Our policy of choosing as standard a rotational order (anticlockwise) around each subdiagram necessarily involves crossing lines in a linear cascade of subdiagrams, which can be laborious here. For example, when compounding permutations we need to cross lines in between to ensure that labels match:

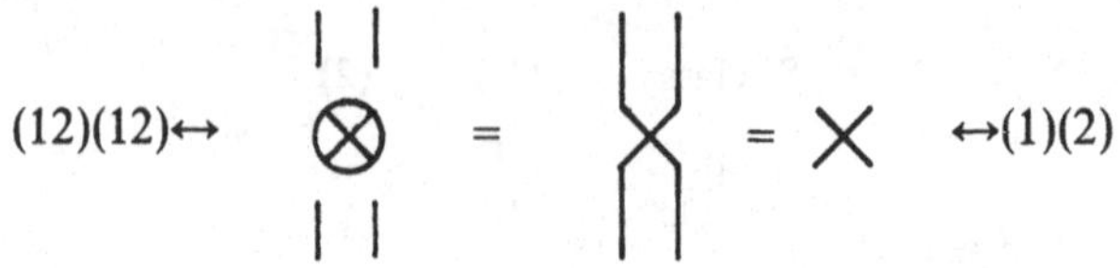

(12)(12)↔ ⊗ = = ↔(1)(2)

As a useful mnemonic, we include the multiplication symbol when combining permutations. This sort of thing was first done in Problem 1.13, though the crossing of lines under the multiplication was not mentioned there.

Other authors/artists are divided on this matter in this context. Certainly our technique adds labour, especially for permutations on a greater number of lines – lines must be crossed at every juxtaposition. Most authors would prefer to make the diagram space anisotropic and allow the labels to be ordered from left to right (or, alternatively, top to bottom) on all sides of all nodes. This would make the identity permutation the uncrossed, rather than the crossed, diagram. If this were the only consideration, we should unhesitatingly adopt this convention. However, once substantial nodes are combined with these projection operators, the advantages of crossing as a matter of course quickly resurface. The real problems, as hinted above, arise when one tries to have one's cake and eat it too, by mixing these conventions and thus introducing inconsistencies. In practice, as in §1, our policy of crossing lines becomes less laborious with practice: one may simply perform a given permutation by switching legs on one side of a diagram in the obvious manner, for example.

(c) Idempotence is readily checked:

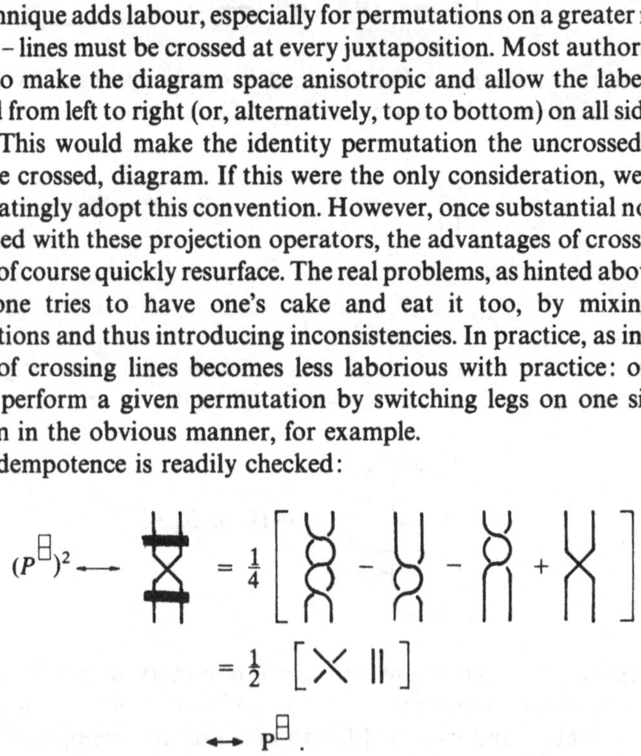

$(P^{\boxminus})^2$ ↔ $= \frac{1}{4}[\;-\;-\;+\;]$

$= \frac{1}{2}[\;\;]$

↔ $P^{\boxminus}$.

(d) The forms of the symmetric and antisymmetric projectors may be confirmed from the character table of S_2 by using the construction (§3.8) of a projection operator in terms of group elements:

$\mathbf{P}^{[\lambda]} = \frac{|\lambda|}{n!}\Sigma_\pi \chi^{[\lambda]}(\pi)\hat{\pi} \leftrightarrow\; = \frac{1}{2}[\;\pm\;]$ for $[\lambda]=[2], [1^2]$, respectively.

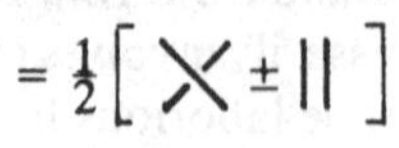

Problem 4.2

Find the diagrams representing elements of S_3 and classify them by conjugacy class; also the corresponding Young diagrams and projection operators for the irreps of S_3, given their characters.

Solution

The three classes are associated with the cycle structures (1^3); (12) or interchange; and (3) or cyclic permutations. There are three interchange and two cyclic permutations, and thus the elements of S_3 may be classified by

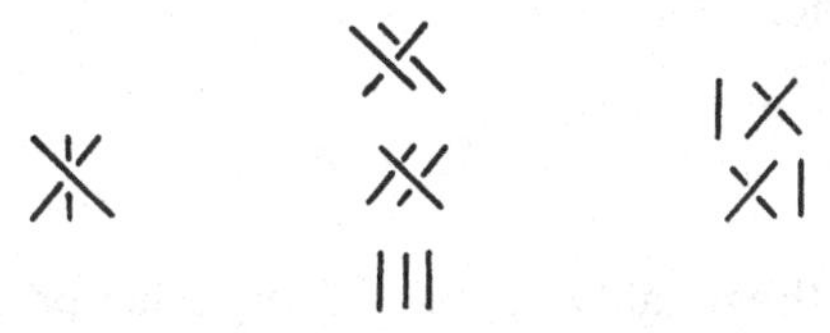

Invariant (1^3) Interchange (12) Cyclic (3) .

Correspondingly, we have three Young diagrams

for the symmetric, mixed and antisymmetric irreps of S_3 (corresponding to the partitions [3], [2 1], $[1^3]$, respectively). The character table has the form:

S_3 irrep: class:	(1^3)	(12)	(3)
[3]	1	1	1
[21]	2	0	−1
$[1^3]$	1	−1	1.

The irreps correspond to symmetric, mixed and antisymmetric projection operators. The first and last of these are obviously the symmetriser and antisymmetriser on three labels. These are also readily recognised as the combinations of the permutations above with the appropriate character (+1 for even permutations, and +1/−1 for the three interchange permutations for the symmetriser/antisymmetriser, respectively), according to the general construction of a projection operator:

$$P^{[3]} \leftrightarrow \quad = \frac{1}{6}\left[\ \ + (\ \ + \ \ + |||\ ) + \ \ + \ \ \right]$$

$$P^{[1^3]} \leftrightarrow \quad = \frac{1}{6}\left[\ \ - (\ \ + \ \ + |||\ ) + \ \ + \ \ \right].$$

For the projection operator for the mixed irrep, the general construction gives:

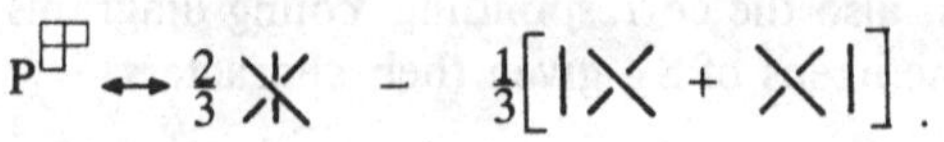

$$P^{\square\!\square\atop\square} \leftrightarrow \tfrac{2}{3}\,[\text{diagram}] - \tfrac{1}{3}\left[[\text{diagram}] + [\text{diagram}]\right].$$

Comments

(a) From the Character Orthogonality Theorem, this mixed symmetriser is clearly a uniquely orthogonal combination of the permutations (S_3 elements) to those in the symmetriser and antisymmetriser.

(b) Its idempotence

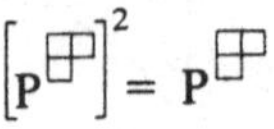

$$\left[P^{\square\!\square\atop\square}\right]^2 = P^{\square\!\square\atop\square}$$

and similarly the orthogonality of this and any other projection operator, e.g.

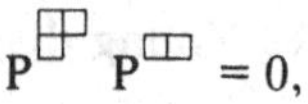

$$P^{\square\!\square\atop\square}\,P^{\square\square\square} = 0,$$

are useful elementary exercises in diagram manipulation; by this stage, they may be beneath the dignity of the reader.

(c) The symmetrisation and antisymmetrisation implicit in any Young diagram indicates that the relevant projection operator may be constructed from a suitable set of symmetriser and antisymmetriser operations. This may be done in a variety of ways. The most fundamental of these is the expansion of the projection operator for the irrep [21] in terms of essentially idempotent and orthogonal Young's symmetrisers for each component of the irrep, i.e. each standard Young tableau:

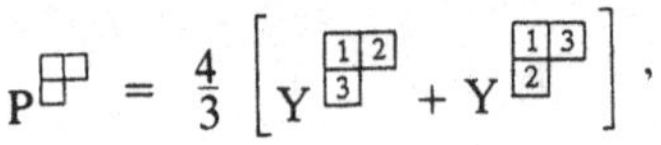

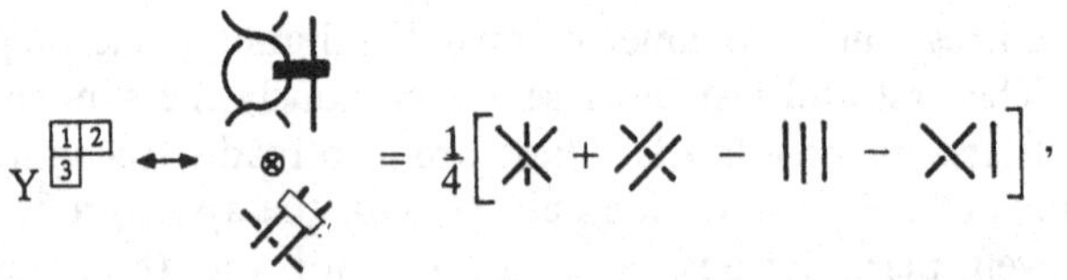

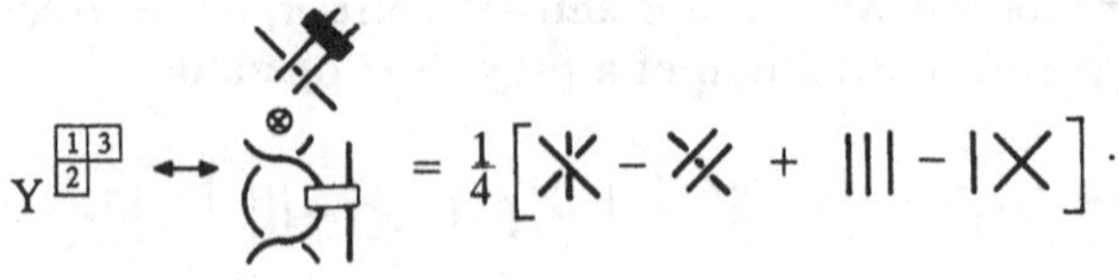

$$P^{\square\!\square\atop\square} = \tfrac{4}{3}\left[Y^{\boxed{1}\boxed{2}\atop\boxed{3}} + Y^{\boxed{1}\boxed{3}\atop\boxed{2}}\right],$$

$$Y^{\boxed{1}\boxed{2}\atop\boxed{3}} \leftrightarrow [\text{diagram}] \otimes [\text{diagram}] = \tfrac{1}{4}\left[[\text{diagram}] + [\text{diagram}] - |||\, - [\text{diagram}]\right],$$

$$Y^{\boxed{1}\boxed{3}\atop\boxed{2}} \leftrightarrow [\text{diagram}] \otimes [\text{diagram}] = \tfrac{1}{4}\left[[\text{diagram}] - [\text{diagram}] + |||\, - [\text{diagram}]\right].$$

(d) Another illustration of this combination of symmetrisation and antisymmetrisation operations is the expansion:

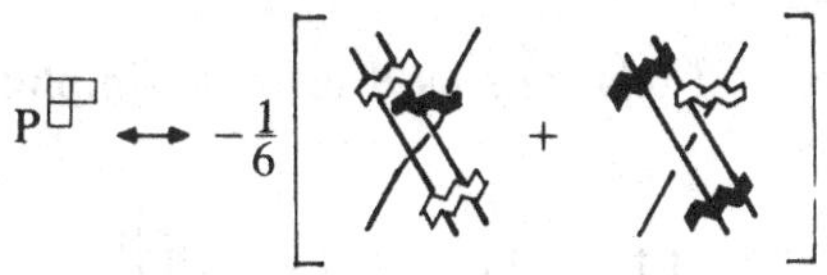

$$P^{\boxplus} \leftrightarrow -\frac{1}{6}\left[\quad + \quad\right]$$

as is most directly proved by direct expansion of the right side.

Problem 4.3

Prove this result by establishing that the right side satisfies (a) idempotence and (b) orthogonality to the symmetriser and the antisymmetriser on three labels.

Solution

(b) is obvious. For (a), the square of the right side becomes

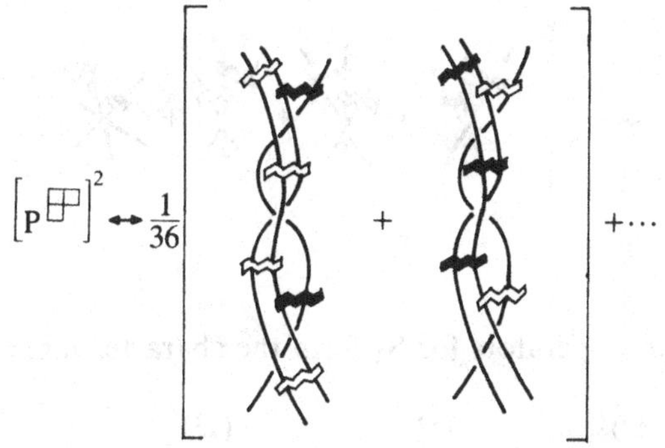

$$\left[P^{\boxplus}\right]^2 \leftrightarrow \frac{1}{36}\left[\quad + \quad\right] + \cdots$$

plus cross terms that vanish, so that, using the result (obtained by expansion)

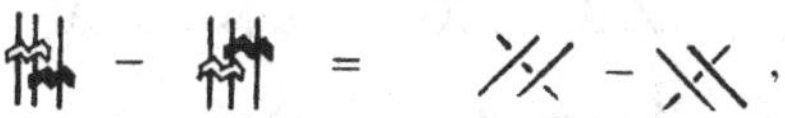

$$\quad - \quad = \quad - \quad ,$$

the above expression reduces to

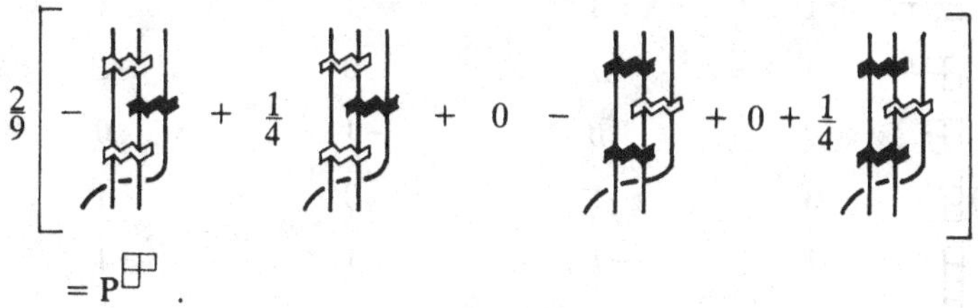

$$\frac{2}{9}\left[- \quad + \frac{1}{4} \quad + 0 - \quad + 0 + \frac{1}{4} \quad\right]$$

$$= P^{\boxplus} .$$

Comments

(a) Some fundamental relations used in these simplifications are

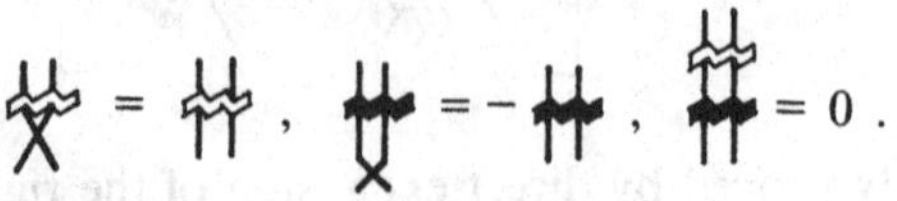

(b) Note similarly that symmetrisers dissolve into symmetrisers, and similarly for antisymmetrisers if intermediate labels are properly connected:

(c) Alternative representations are possible; for example

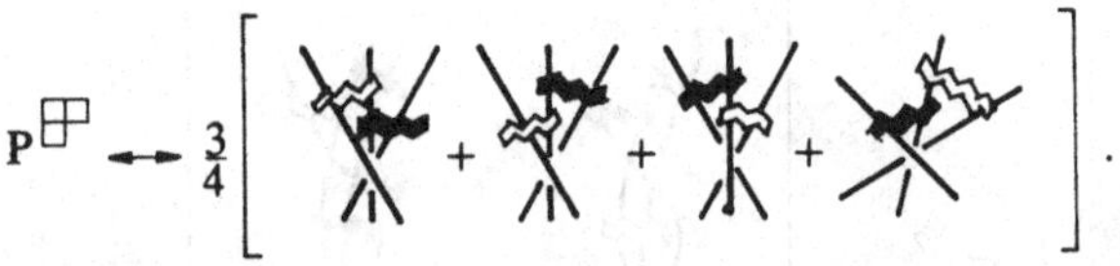

$$P^{[2^2]} \leftrightarrow \frac{3}{4}\left[\quad + \quad + \quad + \quad\right].$$

Problem 4.4

Derive projection operators for S_4 from the character table

S_4 class:	(1^4)	$(1^2 2)$	(13)	(4)	(2^2)
dimension	1	6	8	6	3
elements: 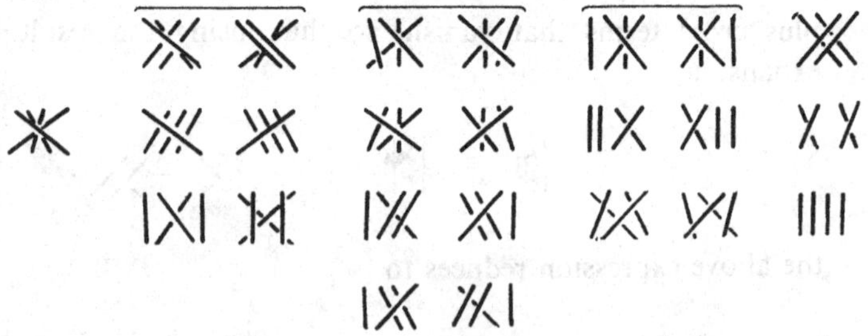					
irrep:					
[4]	1	1	1	1	1
[31]	3	1	0	−1	−1
$[2^2]$	2	0	−1	0	2
$[21^2]$	3	−1	0	1	−1
$[1^4]$	1	−1	1	−1	1

Solution

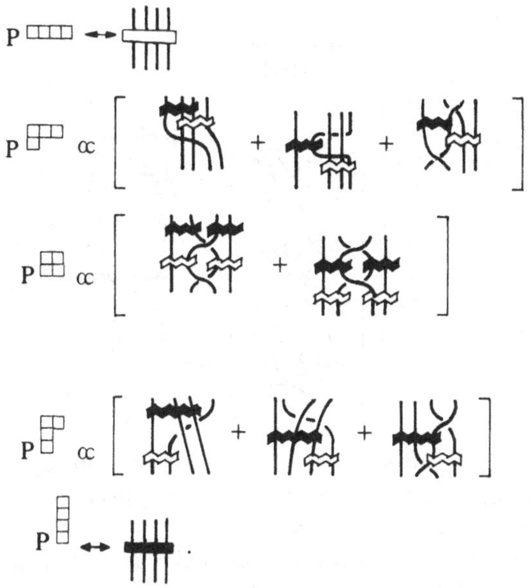

P ↔

P ∝ [+ +]

P ∝ [+]

P ∝ [+ +]

P ↔ .

Comments

(a) Normalisation factors have been omitted.

(b) A brute force method of obtaining the fundamental expansion

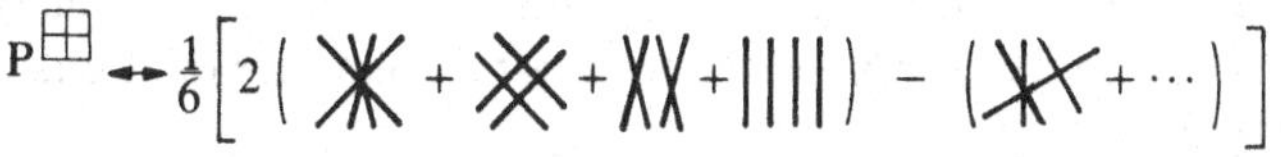

P ↔ $\frac{1}{6}$[2(+ + +) − (+ ⋯)]

derived from the character table is to compare the coefficients in this result with the general expansion (where $\alpha = +1, -1$ denotes a symmetriser, antisymmetriser, respectively):

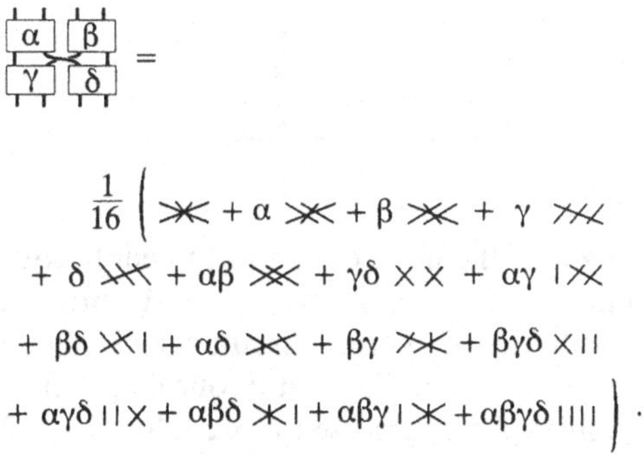

α β γ δ =

$\frac{1}{16}$(+ α + β + γ

+ δ + αβ + γδ + αγ

+ βδ + αδ + βγ + βγδ

+ αγδ + αβδ + αβγ + αβγδ) .

(c) More elegantly, we may determine the Young symmetrisers:

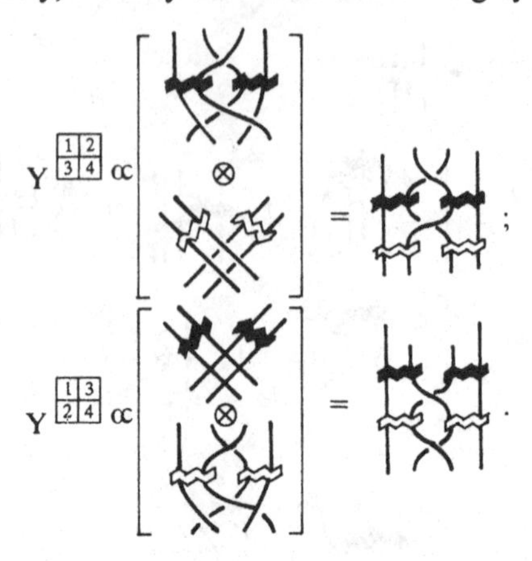

(these relations are proved simply by tracing lines through) to show that each term may be associated with a projection onto one component of the irrep; similarly for the other projectors.

(c) Penrose (1971) has already mentioned that such diagrams have the symmetries of the Riemann–Christoffel curvature tensor.

4.3 Generalised symmetrisers

We discuss briefly algebraic relations satisfied by symmetrisers and antisymmetrisers: their recursion relations and the relationship with the antisymmetric or alternating (Levi–Civita) symbol in the general case. Again this will not be complete; we concentrate on those relations of use in the following.

Our old notation (§1) of a black dot for the antisymmetric symbol ε assumes a cyclic symmetry which is not valid in general, since in an even number of dimensions the symbol changes in sign under a cyclic permutation. We revert to the bar notation of Penrose (1971) (see also Aitken 1958):

$$\varepsilon_{i_1 i_2 i_3 \ldots i_p} \leftrightarrow \quad i_p,\ \vdots,\ i_1 \quad (p) .$$

The arrows pointing into the node on the right denote covariant terminal labels, corresponding to the subscripting on the algebraic symbol. Arrows flowing out of a node denote contravariant labels. These arrow senses corresponds to the senses of the chevron in our diagram notation for bra and ket of §2.7.2, and will be discussed more fully in §§5.2, 5.3.

An antisymmetriser dissolves into this symbol:

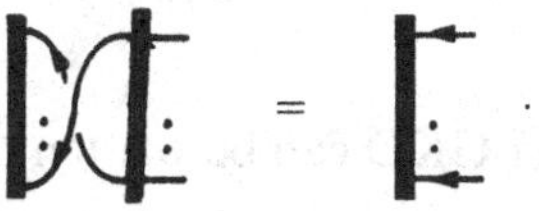

The contraction of such a symbol and its contravariant counterpart is a constant:

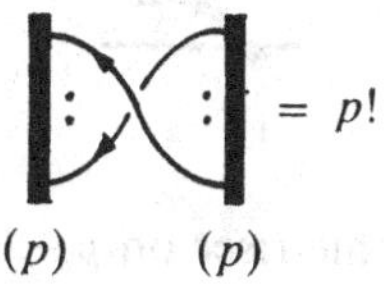

$$= p!$$

$(p) \quad (p)$

and the product of two is the antisymmetric generalised Kronecker delta (GKD) symbol:

This is illustrated in Problem 1.13.

The determinant of a matrix (Problem 1.15) can therefore be rewritten as

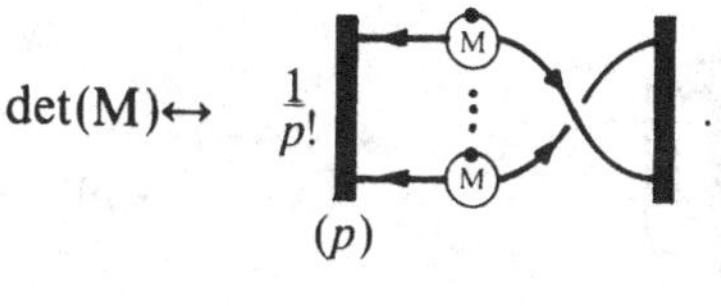

$$\det(M) \leftrightarrow \frac{1}{p!}$$

(p)

The symmetric and antisymmetric GKDs obey recursion relations (Cvitanovic, unpublished lecture notes, 1980):

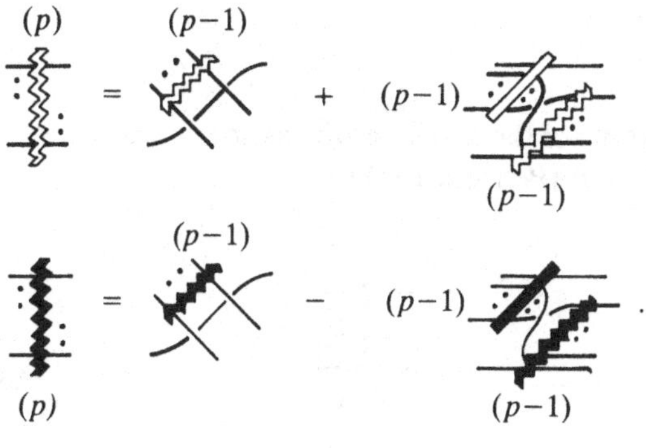

$(p) \quad (p-1)$

$$= \quad + \quad (p-1)$$

$(p-1)$

$(p-1)$

$$= \quad - \quad (p-1)$$

$(p) \quad (p-1)$

Problem 4.5

Prove the second of these as an example.

Solution

By definition, the p-order GKD can be written in terms of p $(p-1)$-order GKDs:

(p) = ... − ... + ... − · · · ·
(p−1)

Multiplying with an antisymmetriser on $p-1$ labels from the left gives a series in which $p-1$ terms are equal, since:

= − = .

Problem 4.6

Show that the qth order trace in a pth order antisymmetric GKD is related to the $(p-q)$th order GKD by the factor $(p+1-N)!/(p-q+1-N)!$, where N is the dimension of the index space $\{i_r\}$:

$$\underset{(p)}{\overset{(q)}{}} = \frac{(N+1-p)!}{(N+1-q-p)!} \underset{(p-q)}{} .$$

Solution

Take $q=p-1$. Use Problem 4.5. Then iterate.

Comment

This contains some of the above results as special cases, and also contains the general relation (Penrose 1971)

$$\underset{(N+1)}{} = 0 .$$

4.4 Schur's theorem for the reduction of unitary group irreps

We consider now more particularly the application of the symmetric group S_n to the irreducible decomposition of n-tensors of the unitary group. General definitions relating particularly to the latter will be given in §5; for the moment we give a skeleton review of vital points.

The set of $N \times N$ unitary matrices define the group $U(N)$, and also constitute its fundamental or defining irrep ω, with dimension N. It has as basis N independent and orthogonal complex vectors $\{\mathbf{b}_o | o = 1,2, \ldots, N\}$ with a component structure $\mathbf{b}_o \equiv \{b_{o\alpha} | \alpha = 1,2, \ldots, N\}$ which may be defined by $b_{o\alpha} = \delta_{o\alpha}$. In diagrams we may write these relations as

$b_{o\alpha} \leftrightarrow$ ωo → ···· α , $b^*_{o\alpha} \leftrightarrow$ ωo ← ···· α ;

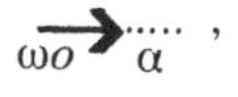

→···→ = – , ····→→····· = ····· ,

$O_g b_{o\alpha} \leftrightarrow$ →···(g)···· = (g)→····· .

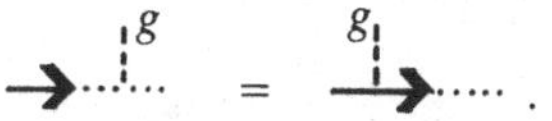

According to Schur's theorem (Coleman 1968) n-tensors formed from Kronecker products of these basis vectors:

$$\mathbf{b}^{\times n} \equiv \{\underbrace{\mathbf{b} \times \mathbf{b} \times \ldots \times \mathbf{b}}_{N}\} \leftrightarrow$$

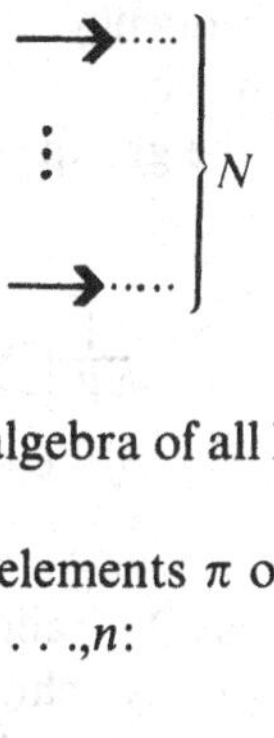

generate all irreps of $U(N)$; furthermore, the algebra of all linear operators which commute with $\mathbf{b}^{\times n}$ is given by S_n.

We introduce S_n by defining the action of elements π of the symmetric group S_n as permuting the labels o_j for $j = 1, \ldots, n$:

$\pi \mathbf{b}^{\times n} \leftrightarrow$

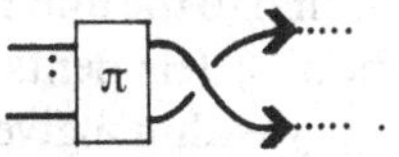

We wish to reduce the basis of $U(N)$ afforded by the set $\mathbf{b}^{\times n}$ of all n-tensors. The unitary basis transformations performing the irreducible block diagonalisation:

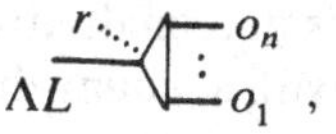

$\langle \mathbf{b}^{\times n} \{o_j\} | r\Lambda L \rangle \leftrightarrow$ ΛL r o_n o_1 , where Λ denotes an irrep of U(N), and L the corresponding component, have the following properties:

(a) They are unitary matrices, since the basis set $\mathbf{b}^{\times n}$ is orthonormal;

(b) they commute with the elements $\{g\}$ of U(N) (see Comment (b) on the definition of the Clebsch–Gordan coefficient (§3.3.1));

(c) they correspond to the matrix element of the projection operator on V_Λ (§3.8):

$$\langle \mathbf{b}^{\times n}\{o_j\}|\mathrm{P}^\Lambda|r\Lambda L\rangle \leftrightarrow$$

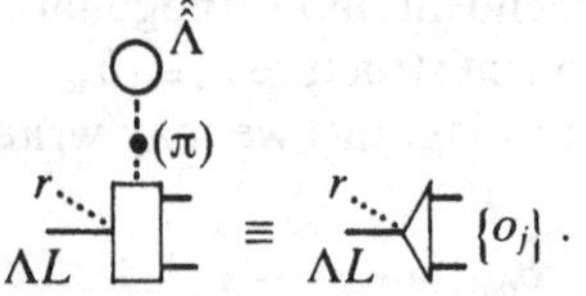

It follows that since in the irreducible basis, the trace of the projection operator is the dimension $|\Lambda|$ (§3.8), the trace in the original basis is related:

$$\mathrm{Tr}_{\times n}\mathrm{P}^\Lambda \leftrightarrow \quad = \quad = \quad = R|\Lambda|.$$

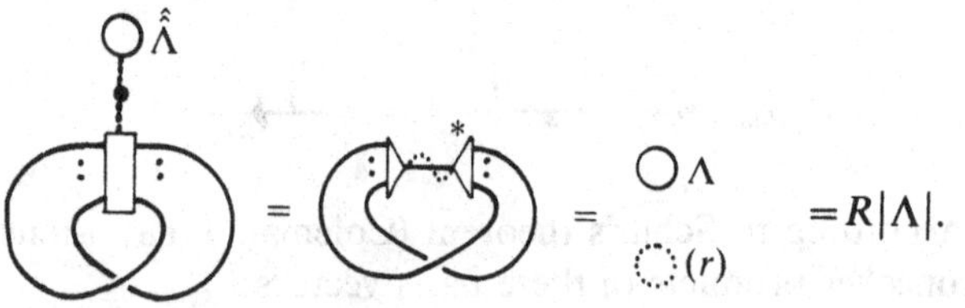

where R is the frequency, or total number of repetitions r, of Λ in the n-tensor representation.

The actions of the group elements $\{\pi\}$ of S_n and $\{g\}$ of U(N) commute:

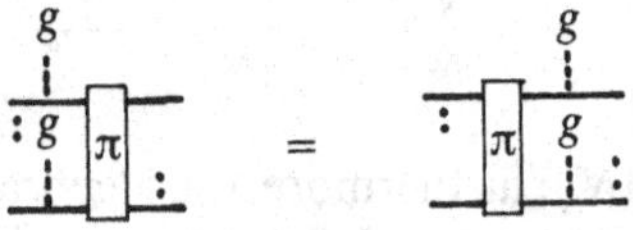

since the matrices O_g 'railroad' through the linkages implied by the permutation π. Since by Schur's lemma II, the various elements π of S_n have matrix elements proportional to the unit matrix within any irrep Λ of U(N), and conversely the various g (elements of U(N)) have matrix elements proportional to the identity within a given λ (irrep of S_n), these two sets of matrices may be simultaneously diagonalised.

If the Young's symmetrisers of S_n are applied to $\mathbf{b}^{\times n}$ to project out components of different symmetry, these are not mixed by U(N) operations, and so furnish a block diagonalisation. Schur's theorem (Coleman 1968) states that this diagonalisation is irreducible. Hence each irrep Λ of U(N) is associated with a Young diagram corresponding to a partition [λ]

of n, and the frequency R of Λ in $\mathbf{b}^{\times n}$ is the dimension $|\lambda|$, and conversely, the frequency of λ in $\mathbf{b}^{\times n}$ is $|\Lambda|$. r can be identified as a component label of λ, and l as a repetition label for the irrep Λ, in the reduction of $\mathbf{b}^{\times n}$:

Problem 4.7
Verify these results for $N=2$, $n=3$.

Solution
The projectors for the three irreps of S_3 have traces given by:

$$\mathrm{Tr}\,\mathrm{P}^{\square\square\square}$$

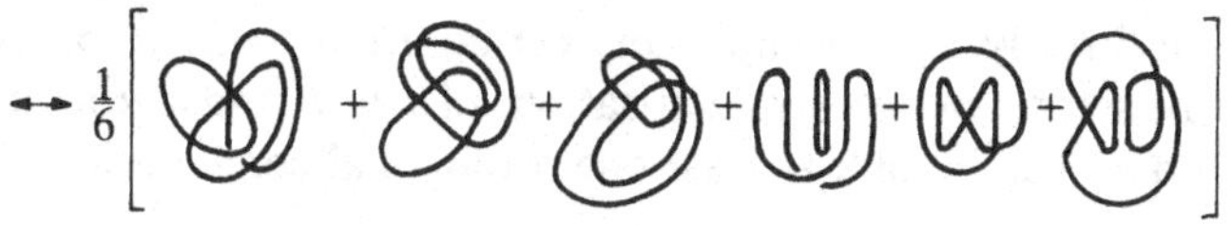

$$\leftrightarrow \tfrac{1}{6}[N^3+3N^2+2N]$$

$$=N(N+1)(N+2)/6;$$

$$\mathrm{Tr}\,\mathrm{P}^{\square\square,\square}$$

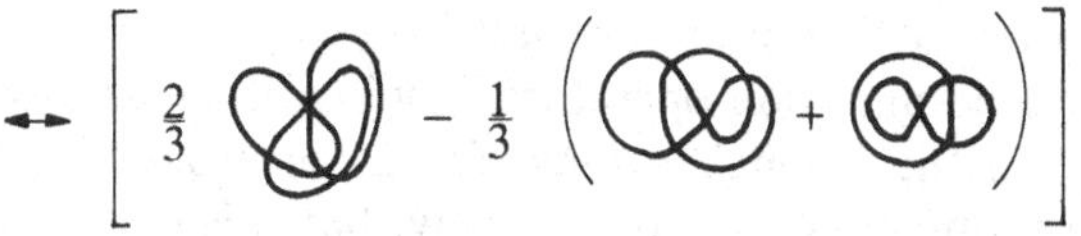

$$\leftrightarrow 2N(N^2-1)/3;$$

and similarly

$$\mathrm{Tr}\,\mathrm{P}^{[1^3]} = \tfrac{1}{6}[N^3-3N^2+2N]=N(N-1)(N-2)/6.$$

(A check is to note that $\Sigma_\lambda \mathrm{Tr}\mathrm{P}^{[\lambda]}=N^3(=\mathrm{Tr}\,1)$.) For $N=2$, these three traces have the values 4, 4 and 0, respectively. (This illustrates that more than two indices are needed to obtain a fully antisymmetric expression.) These correspond to the joint U(2)–S_3 components of $\mathbf{b}^{\times 3}$:

$\square\square\square$: $u^1u^1u^1$, $u^2u^2u^2$, $Y^{\boxed{1}\boxed{2}\boxed{3}}\, u^1u^1u^2$, $Y^{\boxed{1}\boxed{2}\boxed{3}}\, u^1u^2u^2$;

$\begin{smallmatrix}\square\square\\ \square\end{smallmatrix}$: $Y^{\begin{smallmatrix}1&2\\3\end{smallmatrix}}\, u^1u^1u^2$, $Y^{\begin{smallmatrix}1&3\\2\end{smallmatrix}}\, u^1u^1u^2$,

$Y^{\begin{smallmatrix}1&2\\3\end{smallmatrix}}\, u^2u^2u^1$, $Y^{\begin{smallmatrix}1&3\\2\end{smallmatrix}}\, u^2u^2u^1$.

In the second case, we note that:

(a) $u^1u^1u^2$ and $u^2u^2u^1$ span the unitary irrep in the mixed case when symmetrisation of two variables and antisymmetrisation with the third is required, and hence different *columns* give the *two* components of different occurrences of the corresponding irrep of U(3);

(b) rows give partners or components of S_3, corresponding to the two standard tableaux associated with the one Young diagram;

(c) the unit-matrix character of S_3 operators in U(2) and vice versa may be checked directly. For example, any permutation π has no off-diagonal elements between Young's symmetrisations of $u^1u^1u^2$ and $u^2u^2u^1$, since the scalar product will involve at least one orthogonal combination $\langle u^1|u^2\rangle$:

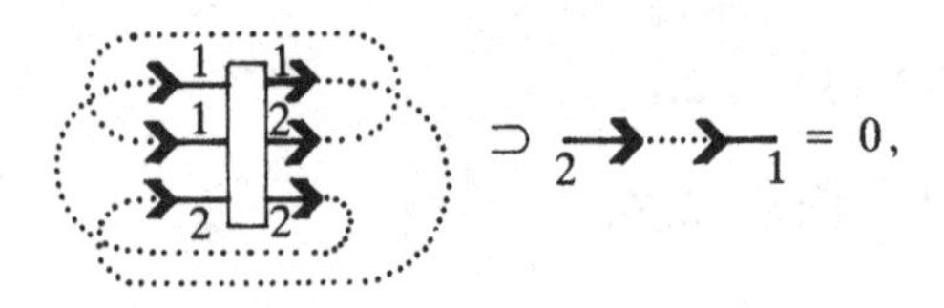

and the diagonal matrix elements will be the same, since they are related merely by a relabelling of dummy labels.

Indeed, the permutation operators may in particular cases have simple eigenvalues on the Young's symmetrisers, corresponding to symmetrisation or antisymmetrisation, respectively; but this is not universal.

5

Lie groups and Lie algebras

On ne trouvera pas de Figures dans cette Ouvrage . . . Les méthodes que j'y expose ne demandent ni constructions ni raisonnement géometriques ou mécaniques, mais seulement des opérations algébraiques . . .

J.-L. Lagrange (reprinted 1965)

Under the demoralizing influence of quantum mechanics, the knowledge required of a theoretical physicist was reduced to a rudimentary knowledge of the Latin and Greek alphabets . . .

attributed to R. Jost (Salam 1988)

But, as Alice inquired, what is the use of a book without pictures?

Ziman (1960)

5.1 Introduction

The linear matrix transformations $\{(g)\}$ of a complex basis $\{x_i | i=1, \ldots, N; x_i \in \mathbb{C}\}$

$$x_i \rightarrow g_i{}^j x_j$$

represent the various groups following under the restrictions stated (restrictions earlier in the list are implied throughout, unless explicitly changed):

$GL(N, \mathbb{C})$	$g_i{}^j \in \mathbb{C}$, $\det(g) \neq 0$
$SL(N, \mathbb{C})$	$\det(g) = 1$
$U(N)$	$(g)(g)^\dagger = I_N$, $\det(g) \neq 0$
$SU(N)$	$\det(g) = 1$
$O(N)$	$g_i{}^j \in \mathbb{R}$, so $(g)(g)^T = I_N$, $\det(g) \neq 0$
$SO(N)$	$\det(g) = 1$
$Sp(N)$, $N = 2k$	$g_i{}^j \in \mathbb{C}$, $(g)(g)^\dagger = I_N$, $(g)^T J(g) = J$,
	$J \equiv \begin{pmatrix} 0 & I_k \\ -I_k & 0 \end{pmatrix}$.

We shall concentrate on U(N) and its subgroups in this and subsequent chapters. The analysis assumes a standard Racah algebraic approach to group representation theory (Wybourne 1974, Sharp 1984) and therefore pertains principally to connected compact Lie groups; however, some results are capable of generalisation to noncompact groups (Sharp 1984). We shall merely assume this further generalisation of the representation algebra illustrated in §2 for SO(3) and in §3 for point groups, and exhibit the major results for Lie groups in general.

5.2 Defining and contragredient irrep

In all these classical group structures, we form the diagram notation for the group action in the obvious way according to the defining matrix relation (paralleling the material of §2.7.2, §3.6.1:

$$O_g b_o \leftrightarrow \quad \omega o \xrightarrow[g]{} .$$

Again we use the arrow-like notation for a ket-type basis function here. It interfaces with our earlier notation; with the introduction of the contragredient representation shortly in terms of the bra notation (reversed arrow) it also interfaces with and makes more precise and general the arrow notation of many authors (Cvitanovic, Kennedy, etc.) for these characteristic representations.

We call the associated irrep of the group the *defining* or *fundamental* irrep ω; it has dimension N. From now on we use the corresponding Latin letter o as a component label in place of i. We associate with the irrep the Young diagram

$$\{b_o\} \leftrightarrow \square .$$

If the matrices (g) form a representation of any group, the matrices $(\tilde{g}) \equiv ((g)^{\mathrm{T}})^{-1}$ also form a representation; the group laws are equally obeyed, and $(\tilde{g})(\tilde{h}) = (\widetilde{gh})$. We call this the *contragredient* representation ϖ. However, we must distinguish here, for later discussion in §5.3, the basis labels o, o' used in labelling of elements of the matrices $(\tilde{g})$ as well as (g), and the *standard* basis component labels $\bar{o}$ associated with ϖ. For example, in a unitary group, the matrices $(\tilde{g})$ are the complex conjugate matrices $(g)^*$ and are related by a unitary transformation to the standard basis for the complex conjugate irrep (e.g. the matrix T of §3.1; for the relation to the 2jm symbol, see Comment (d) of §3.3.2, and Comment (b) on Problem 2.1). The basis vectors on which the matrices $\{(\tilde{g})\}$ act will be denoted by a contravariant index in the ω basis thus: $\{b^o\}$; for unitary groups these basis functions will be the complex conjugates of the basis functions $\{b_o\}$. The change of basis function $b_o \to b^o$ without change of label (ωo) will be denoted diagrammatically by reversing the sense for the chevron denoting the basis function, as for the ket→bra transformation of §2.7.2. With this notation, we may write the action of the group operator on a contravariant basis function (in the defining irrep basis ωo) in the form

$$O_g b^o \leftrightarrow \quad \omega o \text{---}\overset{g^{-1}}{\perp}\!\!\text{<} \; .$$

Problem 5.1

Find the relation between the characters of the contragredient irrep and the defining irrep.

Solution

The character $\chi^{\bar{\omega}}(g) = \chi^{\omega}(g^{-1})$:

$$\overset{\tilde{\omega}}{\bigcirc}\text{---}g \;=\; \overset{\omega}{\bigcirc}\text{---}g^{-1}$$

$$= \overset{\omega}{\bigcirc^{*}}\text{---}g \; .$$

Comments

(a) The representation behaviour of this information of the contragredient basis transformation is readily checked in this notation:

$$O_g O_h b^o \leftrightarrow \quad o\text{---}\overset{h^{-1}\;g^{-1}}{\perp\,\perp}\!\!\text{<} \;=\; \text{---}\overset{h^{-1}g^{-1}}{\perp}\!\!\text{<} \;=\; \text{---}\overset{(gh)^{-1}}{\perp}\!\!\text{<} \; .$$

Note that the second operator O_g acts on the ket, not the external indices. The elements of the group representation matrices 'flow out' from the chevron with sequential operations.

(b) A useful mnemonic is to note that in either irrep transformation notation, if the arrowhead on the node is translated along the irrep lines, the arrow on the *leading* irrep line (leading, that is, with respect to anticlockwise rotation from the group element line) points in an *outwards* sense, away from the group element node.

(c) Comment (b) is our first hint of the arrow notation of Kennedy (1982) and Cvitanovic (1984), for example, to distinguish the two representations. However, the above notation, associating the arrow with the terminal rather than the line, has the important advantages of:

(i) associating arrows with terminals which are themselves the proper source of the representation character, and
(ii) satisfying the elementary rule of §1.1 that the significance of any notation be unchanged by gentle deformations of the diagram, including rotations. It is best to avoid associating an irrep label merely by means of the sense of an arrow on a line, since a π rotation changes its meaning. When the arrow is tied down to a basis function node there is no ambiguity, and diagrams may be freely deformed.

Nevertheless the link with the arrow-on-line notation of Cvitanovic and others is apparent as indicated above, once basis functions are present in the diagram. For complicated mixed tensors, the arrow-on-line notation will be more convenient (§5.3).

(d) The contrast between these basis functions (for a representation and its contragredient representation) is a generalisation of the distinction between left and right invariance made in earlier chapters. We say that (under O_g) a covariant basis function label is *left-transformable by element g* (the sense of the group operation being clockwise), while a contravariant basis function label is associated here with a contragredient transformation character, i.e. it is *right-transformable by element* g^{-1}, the sense being anticlockwise.

(e) The definitions of right- and left-transformable nodes in Comment (d) go over to right-invariant and left-invariant nodes (as in §2.5) for invariant tensors. In particular, the 2jm symbol discussed in §§2–3 corresponds to the contravariant form of a second-rank invariant tensor.

(f) The distinctions discussed here arise essentially since we are using the notation of the defining irrep for both irrep and component (o) labels in discussing the contragredient irrep, and so must find a different way to make a distinction between ω and the irrep ϖ (and between the respective choices of basis even when ω and ϖ are equivalent). Similarly we may use the standard diagram definition of the character when using ϖ labels, but must adapt the definition when using ω labels, as depicted in the above problem. A fuller discussion is given in §5.3.

Problem 5.2

Show that the contraction of two basis functions corresponding to a representation and its contragredient counterpart is a group invariant.

Solution

This is a direct application of the notation for the group operation in the two irreps ω, ϖ (see Problem 5.1):

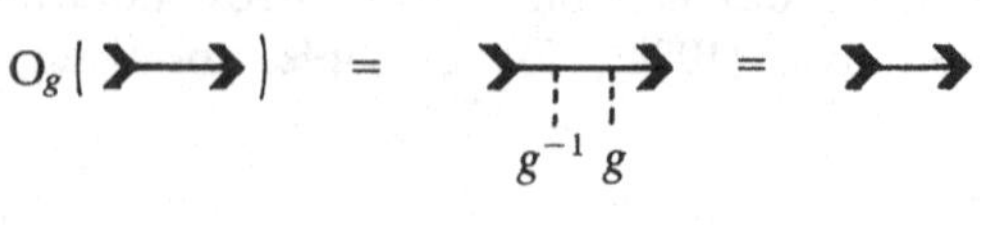

$$O_g(\succ\!\!\longrightarrow) = \succ\!\!\overset{}{\underset{g^{-1}\;g}{\longrightarrow}} = \succ\!\!\rightarrow$$

Comments

(a) This implies that the Kronecker product $\omega \times \varpi$ must contain the identity 0. This is a standard result which generalises Problem 3.2, and which also may be checked from the Character Orthogonality Theorem:

$$n(0 \in \omega \times \overline{\omega})$$

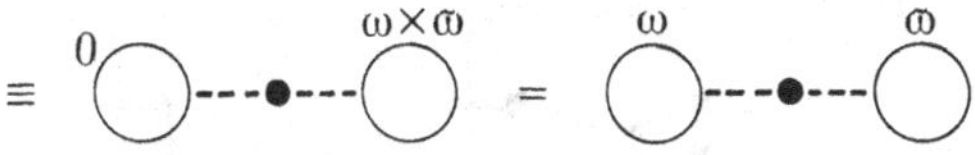

0 $\quad \omega\times\overline{\omega}$ $\quad \omega$ $\quad \overline{\omega}$

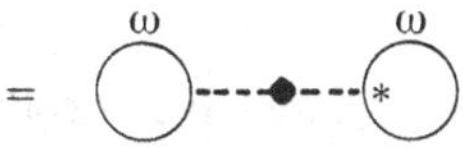

$\equiv$ … $=$ …

ω $\quad \omega$

$=$ … $* \rightarrow \delta_{\omega\omega} = 1.$

The first step follows since $\chi^0(g) = 1$, and $\chi^{\omega\times\varpi}(g) = \chi^{\omega}(g)\cdot\chi^{\varpi}(g)$. We then use Problem 5.1 and the Character Orthogonality Theorem (§3.4).

(b) The 2jm symbol associated with this triple $(\omega\varpi 0)$ performs the coupling to a group invariant:

$\omega \quad \overline{\omega}$

and is itself a (right) invariant tensor.

(c) Different classical Lie groups may be distinguished by their invariants, and hence by the relation of ω to ϖ. For example, the unitary groups have an invariant $b_o^* b_o$. Corresponding to this, the unitarity relation

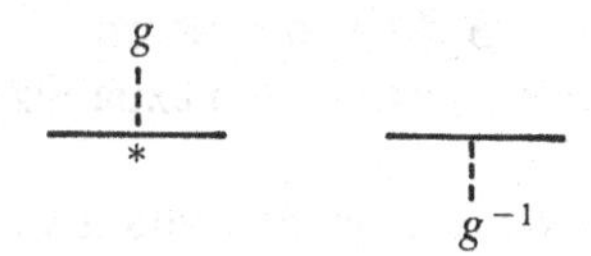

shows that if $b^o \equiv b_o^*$, b^o transforms contragrediently to b_o; in U(N), $\varpi = \omega^*$. As a consequence, in O(N), ω and ϖ are equivalent.

(d) From Comment (c) we might regard the 2jm symbol as an invariant *metric tensor* of U(N). This will be clarified in §5.3.

Problem 5.3

Show that the covariant alternating symbol on N indices is an invariant tensor with respect to the group SU(N):

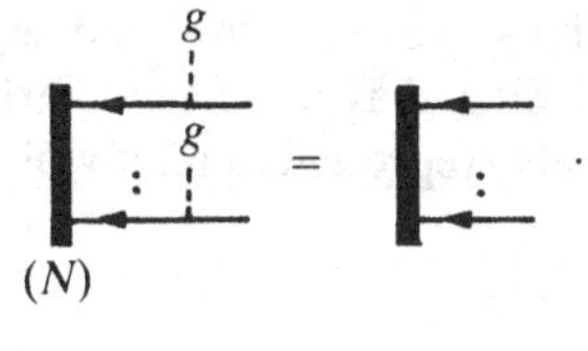

g $\quad g$ $\quad = \quad .$

(N)

Solution

Since det(g) is a constant for all g in SU(N):

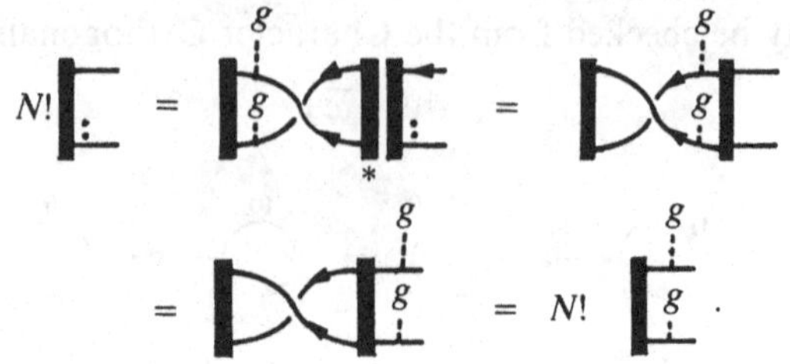

$N!$ = = = = $N!$.

Comments

(a) As detailed in §5.3 (see also §§4.3, 5.1), we have introduced arrows on lines here to indicate the covariance or contravariance of indices at a node. The contravariant tensor would be written:

$$\varepsilon^{i_1 i_2 \ldots i_N} \leftrightarrow$$

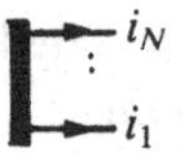

i.e. by outgoing arrows.

(b) The sense of the group operation is given by the rule (see Problem 5.1 Comment (d)), that a covariant label is left-transformable (by the group element), and a contravariant label right-transformable (by the inverse group element).

5.3 Mixed tensors

This is another housekeeping section, and expands our earlier notation in several directions.

First, we use the arrow notation of Cvitanovic (1984) and others to denote the transformation type of individual labels. We discuss this in the context of mixed tensors.

Mixed tensors correspond to commutative nodes in the sense of §3, the two classes of legs corresponding to covariant and contravariant indices, respectively. The property of cogredience or contragredience of transformation character is associated independently with each terminal. This could be denoted diagrammatically, for example by adding a parity label to each terminal, as described in Comment (f) on Rule 6 in §1. The logical but complicated and uncouth technique suggested by §5.1 would be to add dimples and warts at the root of each hair follicle, i.e. inward or outward protrusions at each terminal. Alternatively, we could follow Penrose (1971), who distinguishes two kinds of lines: 'arms' and 'legs'; this requires a mild violation of the rules of §1 and is not particularly convenient. Another popular notation, which also represents a mild violation of basic ideas, is to use arrows. In this notation, an arrow on a line pointing into a node will signify a covariant index for that node, etc.:

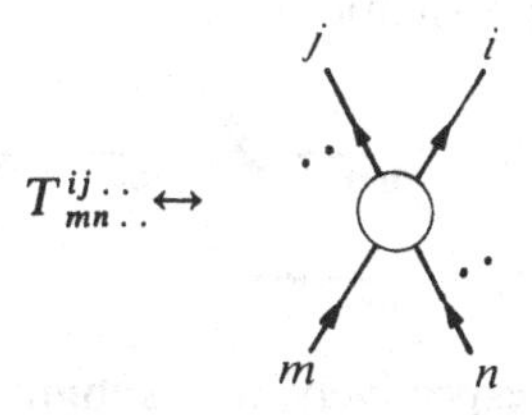

We emphasise that such arrows have meaning only in relation to a nearby terminal. They do not correspond to the 2jm symbol (for which we reserve the stub symbol), or to the metric tensor. One strong reason for adopting this notation is that, since one invariably wishes to combine a covariant with a contravariant node, a single arrow does service for both terminals linked by any internal line. To this we add that the reversal of the arrow corresponds to the exchange of contravariant and covariant label in the corresponding contraction or scalar product, which is an invariance operation.

Second, we may view the 2jm symbol for the defining irrep as being related to the components of a metric tensor in some basis. It may be used to raise and lower indices thus establishing a covariant/contravariant notation (Butler 1975). The 2jm symbols of the Racah algebra of Lie groups (for which see Sharp 1984, for example) may be compared with the metric of Lie group theory: the groups U(N) and O(N) preserve a sesquilinear and a bilinear metric, respectively. (Indeed, in §2.5 we have already used the 2jm symbol as a metric.) This identification derives from the preservation of the metric under unitary operators, since the latter amounts to the invariance or commutativity of the corresponding node in the manner defined in §2.5. This ensures the applicability of JLV2, and hence that the metric must be proportional to the pinched diagram consequent on applying that theorem. For an invariant node, this diagram will be a 2jm symbol; for a commutative node, the Kronecker delta. In U(N), the sesquilinear metric is essentially trivial, being a unit matrix 1_N corresponding to the invariant $b_o^* b_o$ where $\{b_o\}$ span the defining irrep, provided we work in a unitary basis (as we do throughout this book). It is then reasonable to make nontrivial use of the mixed tensor notation using the 2jm symbol as a metric $\boldsymbol{\eta}$.

Since the 2jm symbol acts on a cogredient basis to give a contragredient basis (i.e., being a right-invariant node, it maps a left-transformable to a right-transformable node) its terminals may be associated with outgoing arrows, and conversely for the complex conjugate node:

$$\text{—}\!\!\perp\!\!\text{—} \equiv \leftarrow\!\perp\!\rightarrow ,$$

$$\underset{*}{\text{—}\!\!\perp\!\!\text{—}} \equiv \rightarrow\!\underset{*}{\perp}\!\leftarrow .$$

There are now two ways of generating a set of contravariant basis functions $\{b^o\}$ from a covariant set $\{b_o\}$: complex conjugation, and application of a 2jm transformation:

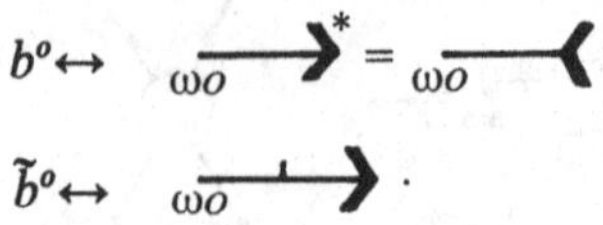

$$b^o \leftrightarrow \quad \omega o \quad * \quad = \quad \omega o$$

$$\tilde{b}^o \leftrightarrow \quad \omega o \quad .$$

We write these as b^o, $\tilde{b}^o$, respectively, to distinguish them in the following discussion.

Comments

(a) In the first case, we choose to make complex conjugation the definitive method (i.e. $b^o \equiv b_o^*$) as far as the covariant↔contravariant notation interchange $b_o \leftrightarrow b^o$ (and for that matter, the bra–ket notation change where applicable) is concerned, on the grounds that in the unitary basis the familiar notation $b^o b_o$ then corresponds to the fundamental invariant $b_o^* b_o$; we shall shortly illustrate this notation with the metric tensor.

(b) In the second case, since it is the external labels that should correspond in any definition (e.g. for there to be any prospect of a simple relation between the definitions) it is useful to define the construction of $\tilde{b}^o$ by applying the 2jm transformation to $b_{\bar{o}}$ rather than to b^o, i.e. in the contragredient basis.

(c) We then see that, as the notation suggests, b^o, $\tilde{b}^o$ are related to each other in a similar manner as the conjugate nodes b, $\tilde{b}$ of §2.5:

$$\tilde{b}_o = (\tilde{b}^o)^* = \eta_{o\bar{o}} b_{\bar{o}}^* \leftrightarrow \quad \equiv \quad * \quad *$$

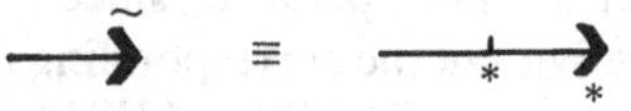

We call b_o, $\tilde{b}_o$ (or equally b^o, $\tilde{b}^o$) *conjugate bases.*

(d) For a self-conjugate basis, the tilde is unnecessary, as for §2.5.

(e) In general, however, we now have found distinct representations of bases related by these various transformations:

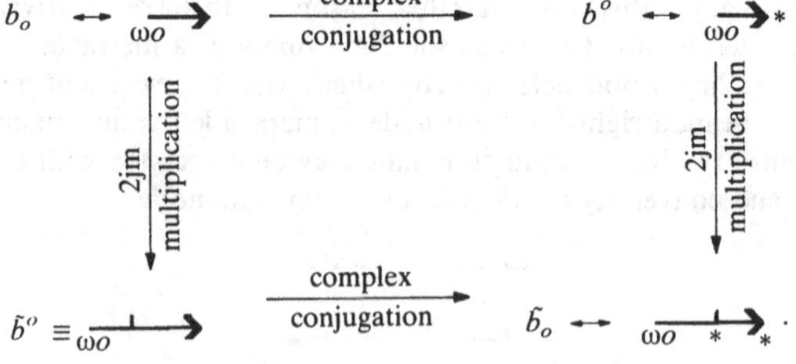

$b_o \leftrightarrow$ ωo —complex conjugation→ $b^o \leftrightarrow$ ωo *

↓ 2jm multiplication (left) ↓ 2jm multiplication (right)

$\tilde{b}^o \equiv$ ωo —complex conjugation→ $\tilde{b}_o \leftrightarrow$ ωo * * .

Two of those nodes are right-transformable nodes, and two left-transformable (the former by the *inverse* group operator). All these four nodes carry the same external labels. Another four related nodes are those corresponding to contragredient irrep labels ($b_{\bar{o}}$, $b^{\bar{o}} = b^*_{\bar{o}}$, and their conjugates $\bar{b}_{\bar{o}}$, $b^{\bar{o}}$).

The relation between the 2jm symbol and the associated metric tensor is then

$$\eta^{cd} = \begin{pmatrix} \gamma & \delta \\ c & d \end{pmatrix} \leftrightarrow \quad \gamma c \qquad \delta d ,$$

where γ, δ stand for either ω, ϖ or ϖ, ω, respectively. The relationship between algebra and diagram notation is specified as before, by making the first column of the matrix correspond to the leading leg of the diagram, taken in an anticlockwise sense from the stub, and by interpreting outward arrows from a node as contravariant labels. We include below our old notation in which a stub when unconjugated is assumed to be a contravariant (and right invariant) tensor, omitting arrows on the legs in that case.

The covariant metric is naturally chosen as the inverse matrix:

$$\eta_{cd} \leftrightarrow \quad c \overset{*}{\quad} d \quad \equiv \quad c \qquad d .$$

Comments

(a) Thanks to unitarity, this coincides with our above convention of making the definition of the contravariant index in terms of the complex conjugate node, provided that the indices are also exchanged: $\eta_{cd} \equiv (\eta^{dc})^*$. Hence in the above figure the first stub points downwards. Only this is consistent with the requirement $\eta_{bc}\eta^{cd} = \delta_b{}^d$.

(b) The second stub points upward since the node represents the same tensor $\boldsymbol{\eta}$ as does the previous diagram, the difference in the representation being sufficiently and logically denoted by the arrow senses. (Our earlier depiction of the complex conjugated 2jm symbol in this section with arrows on the terminals did not have this reversal of sense of the stub. However, it was then explicitly complex conjugated.) The corresponding mixed tensor is the Kronecker delta:

$$\eta^c{}_d = \delta^c{}_d \quad c \qquad d \quad = \quad \overset{*}{\quad} \quad = \quad \text{———} \quad = \quad \text{—◄—} .$$

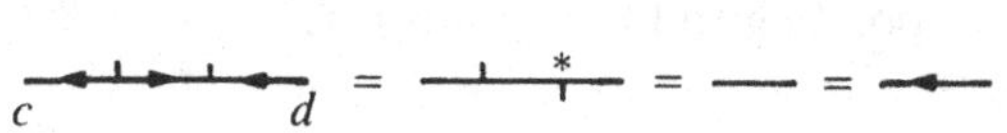

Hence the 2j phase could be written as:

$$\underset{*}{\quad} \quad = \quad \text{—◄—┴—►—┬—◄—} .$$

Likewise, the 2jm symbol for other irreps may be given a metric interpretation. We show in §5.6 the relationship between the 2jm symbol for the adjoint irrep and the Killing form for the Lie algebra.

For example, we return to the vector notation of §1.2 and its elaboration in §§2.5, 3.6. In view of the above distinctions, we might write the covariant vector components e_o for b_o, etc.:

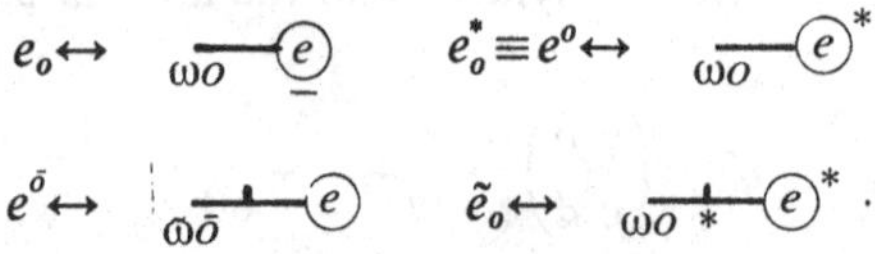

$e_o \leftrightarrow$ ωo (e) $e_o^* \equiv e^o \leftrightarrow$ ωo (e)*

$e^{\bar{o}} \leftrightarrow$ $\varpi\bar{o}$ (e) $\tilde{e}_o \leftrightarrow$ ωo * (e)* .

The distinction between the bases $\{\omega o\}$ and $\{\varpi\bar{o}\}$ is then maintained in principle even if as this example $\omega = \varpi$ (the $j=1$ irrep of SO(3)). The distinction is important in this case for a spherical (jm), basis, in which the 2jm symbol is nontrivial.

5.4 Tensor products of the defining irrep

Tensor products of the defining irrep generate all other irreps of these groups. The irreducible basis functions themselves need projection from the reducible representations derived by forming straight Kronecker products. We shall concentrate on results for U(N) and its subgroups.

An essential step in this projection is that of using Young's symmetrisers and antisymmetrisers to project out basis functions of S_n in any nth rank tensor product. As discussed in §4.3, these commute with the group operator and preserve the transformation character of any basis. In particular, the fully symmetrised and the fully antisymmetrised tensors transforming as $[n]$ and as $[1^n]$, respectively, under S_n are also irreducible under GL(N) and its subgroups:

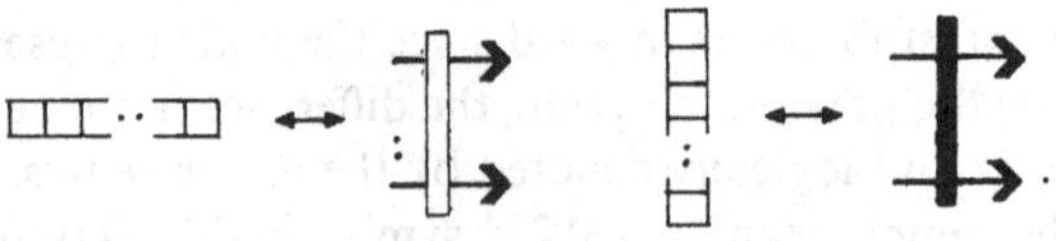

Problem 5.4

Show that within U(N), for example, the ($N-1$)th rank antisymmetrised tensor corresponding to $[1^{N-1}]$ of S_{N-1}:

$(N-1)$

transforms contragrediently to [1] in an appropriate basis.

Solution

Choose the Nth order contravariant antisymmetric symbol to provide the Young symmetrisation and also the choice of basis labelling:

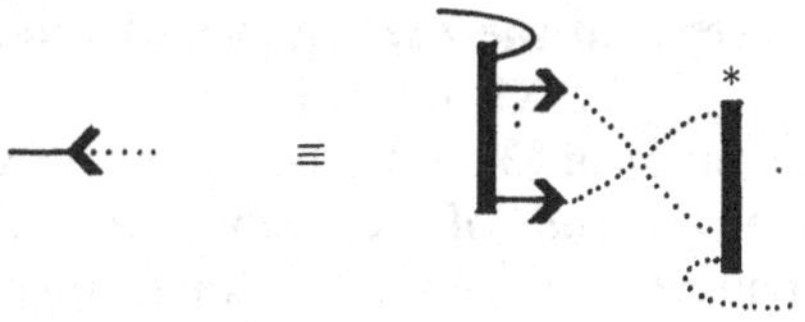

The representation $[1^{N-1}]$ has dimension N, since all labels must be different except one, which is chosen in N ways; this label is the terminal label on the left in the above notation. Similarly for the basis component labels: the basis functions of $[1^{N-1}]$ may be given the same labelling as that of $[1]$.

Under a group element, the various terminals transform in the standard manner:

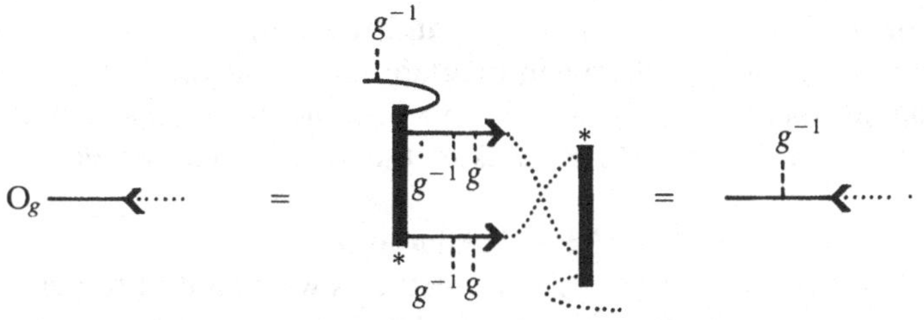

Since the group elements cancel on internal lines, the whole figure has the standard transformation of the contragredient irrep ω.

Comments

(a) Similarly, the irrep $[1^N]$ transforms as the identity 0. This confirms that $\omega \times \varpi$ contains 0, since adding a box to $[1^{N-1}]$ certainly can produce $[1^N]$.

(b) We know from Problem 5.4 that on forming the tensor product $V_\omega \times V_\varpi$ of the two representation spaces, we generate the invariant from the contraction. In other words the matrix transformation of basis associated with the reduction of the product representation into irreducible parts includes this projection of the trace term:

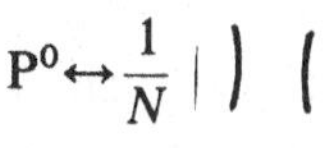

$$P^0 \leftrightarrow \frac{1}{N} \mid \,) \; ($$

Note that the normalisation factor ensures idempotency: $(\mathrm{P}^0)^2=\mathrm{P}^0$. On multiplying two such diagrams, the intermediate terminals link to form a loop and thus a factor N.

5.5 Adjoint irrep and the Lie algebra of group generators

5.5.1 Introduction

We may use the machinery of §3 on the theory of finite groups to lead quite directly to many central results of Lie algebras. As Cvitanovic (1984) has pointed out, the fundamental commutator expression for the Lie algebra of the group and the Jacobi relation between Lie brackets are both essentially statements of the invariance of the 3jm symbols of the Lie group, a theme well studied in earlier chapters. They may therefore be derived as exercises, once an adequate foundation has been laid.

When finite group elements are used, invariance and commutativity etc. may be defined exactly as in earlier sections, and the JLVn theorems may then be applied in essentially the same form. This underscores the essential unity of the development for continuous and for finite groups. This unity is not obvious, however, from the work of Cvitanovic (1984). Invariance of a node under infinitesimal operations and invariance under finite group elements look rather different in diagram form, and the JLVn theorems were not appealed to by Cvitanovic (1984), although links were made with Schur's lemma and the Wigner–Eckart theorem in that approach.

5.5.2 Adjoint irrep

In §5.4 we noted that if Young's symmetrisers were applied to the tensors formed from the basis functions of the defining irrep [1], then irreducible basis functions of other irreps of U(N) were obtained. In particular, $[1^N]$ was identified as the identity irrep 0, while $[1^{N-1}]$ is identified as the irrep contragredient to the defining irrep [1]. The Kronecker product rule for S_N irreps gives $[1]\times[1^{N-1}]=0+\alpha$, where

$$\square\times\begin{array}{c}\square\\ \vdots\\ \square\end{array}_{(N-1)}\supset\begin{array}{c}\square\\ \vdots\\ \square\end{array}_{(N)}.$$

We consider the character of mixed second-rank tensors formed from one defining and one contragredient basis function. We denote this tensor space as $\mathrm{V}_\omega\times\mathrm{V}_\varpi$ for brevity of reference. A physical application in SU(3) is to products of quark and antiquark, say.

The other term in the Kronecker product will have the extra box in a second column:

$$\square\times\begin{array}{c}\square\\ \vdots\\ \square\end{array}_{(N-1)}\supset\begin{array}{l}\square\square\\ \vdots\\ \square\end{array}_{(N-1)}.$$

This irrep is the *adjoint* irrep, and we shall denote it by α (and its components by a, a') rather than the fuller notation $[2\,1^{N-2}]$. In short, $[1]\times[1^{N-1}]=0+\alpha$.

Since the irrep 0 has dimension 1, and each of the original irreps the dimension N, the adjoint irrep will have dimension N^2-1 in U(N). In general, the dimension n of the adjoint irrep is called the *dimension* of the group. By a fundamental theorem of Lie group theory (Cornwell 1984), this is the dimension of the corresponding Lie algebra, and the number of generators of the group.

Since the irrep 0 corresponds to the contraction or trade of the basis functions of the two original irreps $[1]$, $[1^{N-1}]$ when expressed in covariant and contravariant notation, respectively, the basis functions of the adjoint irrep α will correspond to the symmetric traceless and the antisymmetric components of the $N\times N$ Kronecker product of basis functions. The appropriate projection operator is therefore the operator removing the trace:

$$\mathrm{P}^{\alpha}\leftrightarrow \;\times\; - \frac{1}{N}\;)(\;.$$

We have included the arrows here to indicate the connection with the formalism of Cvitanovic; as explained in §5.3 they are not to be understood as discriminating between lines, but are properly associated with terminals. In this case, they indicate that this operator is understood to act on the basis

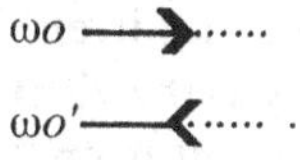

$$\omega o \longrightarrow \cdots\cdots$$
$$\omega o' \longleftarrow \cdots\cdots\;.$$

Indeed both sets of irrep labels are those of the *defining* irrep ω; otherwise, it would be impossible to equate them through the linear coupling in P^{α}.

We may use the Clebsch–Gordan series to write

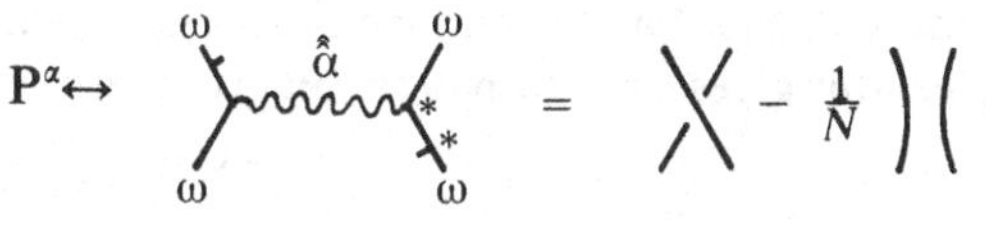

$$\mathrm{P}^{\alpha}\leftrightarrow \;\overset{\omega\qquad\omega}{\underset{\omega\qquad\omega}{\succ\!\!\sim\!\hat{\alpha}\!\sim\!\!\prec}}\; = \;\times\; - \frac{1}{N}\;)(\;.$$

Note that no repetition index is necessary for the triple $(\omega\varpi\alpha)$. A transposition of stubs via unitarity gives a corresponding relation for mixed $(\omega,\ \varpi)$ labels:

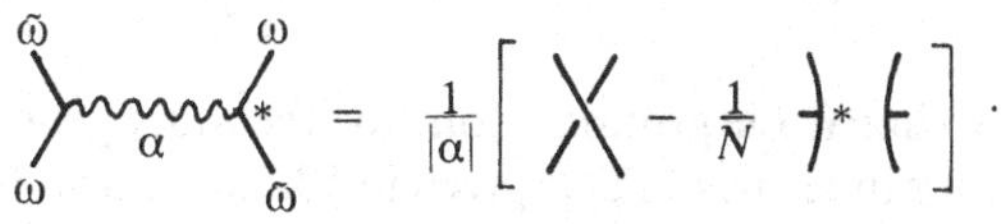

$$\overset{\tilde{\omega}\qquad\omega}{\underset{\omega\qquad\tilde{\omega}}{\succ\!\!\sim\!\alpha\!\sim\!\!\prec}} = \frac{1}{|\alpha|}\left[\;\times\; - \frac{1}{N}\;\succ\!\!\prec\;\right].$$

We leave it as an exercise for the reader to check that P^{α} is idempotent, that it is orthogonal to P^0 and that P^0+P^{α} is the unit operator.

In the monograph of Cvitanovic (1984), a general strategy for creating such projection operators for a wide variety of irrep and of tensor spaces is given detailed illustration in many different groups.

5.5.3 Infinitesimal generators and Lie algebras

A fundamental theorem of linear Lie group theory (Gilmore 1974, Wybourne 1974, Cornwell 1984) allows us, near the identity element, to associate a group element g with n real coordinates $\mathbf{x}=\{x_a|a=1,2,\ldots|\alpha|=n\}$ so that the matrix elements of O_g are analytic functions of $\mathbf{x}$. This allows the definition of the *infinitesimal Lie generators*

$$\left\{X_a \equiv \frac{\partial O_{g(\mathbf{x})}}{\partial x_a}\bigg|_{\mathbf{x}=0} \bigg| a=1,2,\ldots,n\right\}.$$

These span the Lie algebra (see Problem 5.8) associated with the group.

Problem 5.5

Show that the generators $\{X_a\}$ transform as irreducible tensor operators (ITOs) for the adjoint irrep α.

Solution

The left side is the first derivative with respect to x_a of a product of group operators, and hence of a group operator. As a result, the action of this operator on any component of a basis for the defining irrep will generate a superposition of components of that irrep. The operator couples ω to ω, and can transform therefore only as the identity, which is excluded in taking the first derivative, or as the adjoint irrep. Fundamental theorems ensure that these operators span α; this is seen from the above to be consistent with the obvious dimensional considerations.

Comments

(a) For a standard account and proof, see Cornwell (1984, vol. 2, p. 417).

(b) In diagrams, this gives a relation between group elements and ITOs:

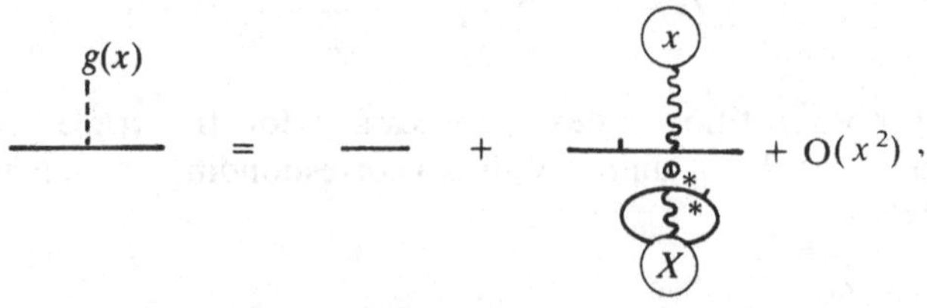

which is special to linear Lie groups. (Contrast the more general expansion of a group operator in terms of ITOs given in §3.7.) We now focus on these linear terms in the expansion, in effect on the adjoint irrep of the Lie algebra (see Comment (d) on Problem 5.8).

(c) The necessity of carrying a stub as well as the adjoint irrep line to represent each infinitesimal group operation can be avoided if desired by using a Clebsch–Gordan node in place of the 3jm node with stub. We find it as convenient to use a 3jm node, because of its greater rotational symmetry.

Problem 5.6

Define the nature of the constraints associated with invariance and with commutativity of a diagram node (as defined in §2.5 for finite group elements) in the infinitesimal case.

Solution

For infinitesimal group elements, the above expansion generates from the finite form of invariance a series of terms, one for every index, with differing signs for class 1 and class 2 legs on a commutative node, i.e. for covariant (left-transformable) and contravariant (right-transformable) indices, which sums to zero.

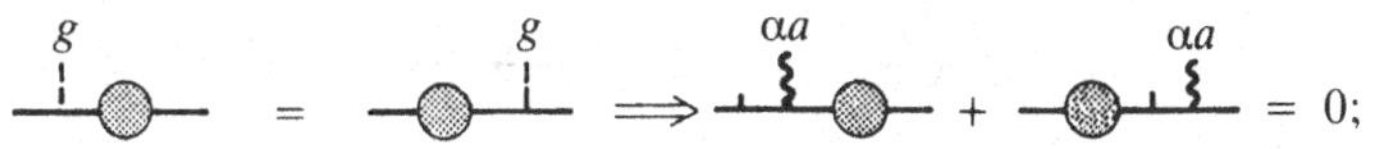

g g αa αa = $\Rightarrow$ + = 0;

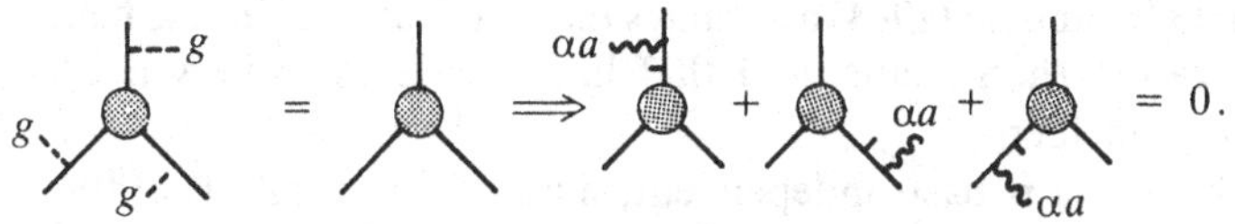

g g g = αa $\Rightarrow$ + αa + αa = 0.

Comment

This forms the link between the diagram definitions of invariance for Lie groups (e.g. Cvitanovic 1984) and for finite groups (e.g. Stedman 1975).

Problem 5.7

Prove that the product of 3j phases $\{\alpha\omega\varpi\}\{0\omega\varpi\} \equiv \{\alpha\omega\varpi\}\{\omega\} = -1$.

Solution

Use the group invariance of the metric:

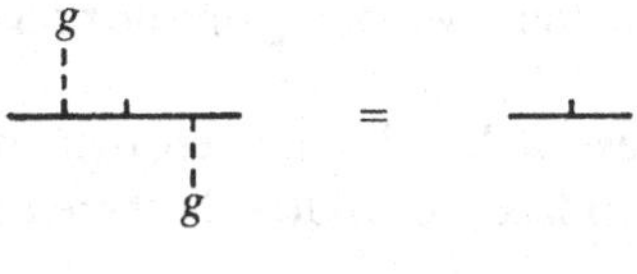

g = g

in infinitesimal form:

and rearrange:

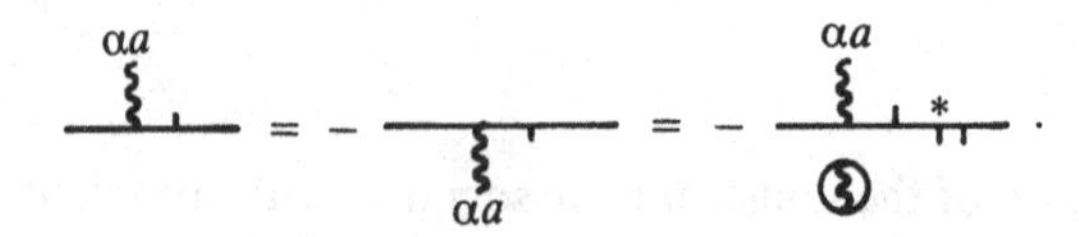

Comments

(a) We use a wiggly line to denote the adjoint irrep/component labels αa. This follows Kennedy (1982) for example, and is suggested by the standard Feynman diagram notation for the (adjoint) photon propagator in QED, and similarly for gluons in QCD.

(b) This corresponds to and generalises Cvitanovic's (1984) result:

Note that Cvitanovic (1984) associates the arrow at least in this figure with a node, rather than a line, and that his node is essentially a Clebsch–Gordon coefficient.

(c) This result is basis-independent, since all 2jm symbols of the group are invariant. It therefore holds not only in the defining irrep, but in any irrep when coupled with the correspondingly contragredient irrep via the corresponding 2jm symbol.

For example, we may use this in the adjoint irrep. We note there that the adjoint irrep is self-contragredient; inverting the corresponding Young diagram and adding this as in Kronecker multiplication of S_n irreps creates an invariant:

Hence $\alpha \times \alpha \supset 0$. Also note that by moving one box off the bottom to the top right we find $\alpha \times \alpha \supset \alpha$.

(d) However, in the case of the adjoint irrep, there may be more than one repetition of α in $\alpha \times \alpha$. Indeed, for $SU(l+1)$ there are certainly two such occurrences (Cornwell 1984, p. 631). This means that we should adapt the above problem solution to take the form:

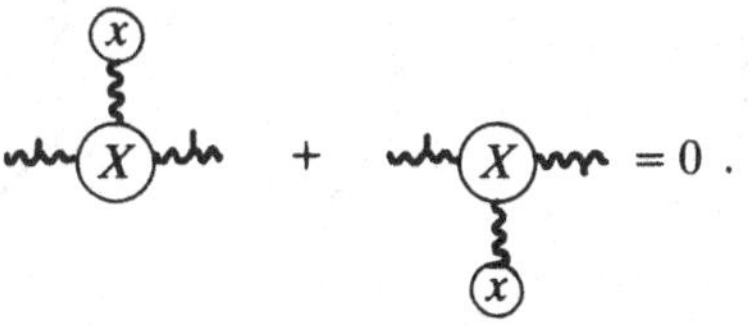

Since this holds for all $\{x_a\}$, we may contract with another 3jm symbol to give

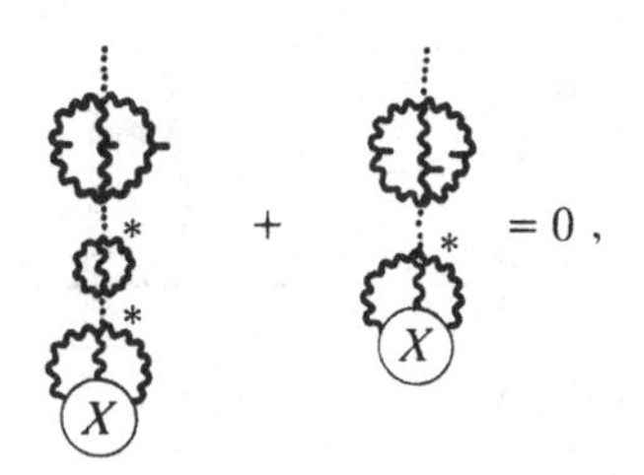

which means that, for any repetition label s, either the reduced matrix element associated with that repetition vanishes, and so that repetition is not associated with the group action, or else (if it is associated with the group action: $s \equiv r$, say)

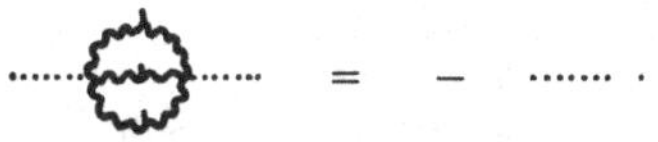

We shall for simplicity include only one such repetition in the following, and label it by r.

Problem 5.8

Derive the fundamental equation for the *Lie algebra* $\mathscr{A}$ corresponding to the group G:

$$[X_a, X_b] = f^c_{ab} X_c,$$

where the *Lie bracket* [. . .] is the commutator, $\{X_a\}$ are the generators and the constants $\{f^c_{ab}\}$ are the *structure constants.*

Solution

Use the group invariance of the 3jm symbol for the triple $\{\omega\varpi\alpha\}$:

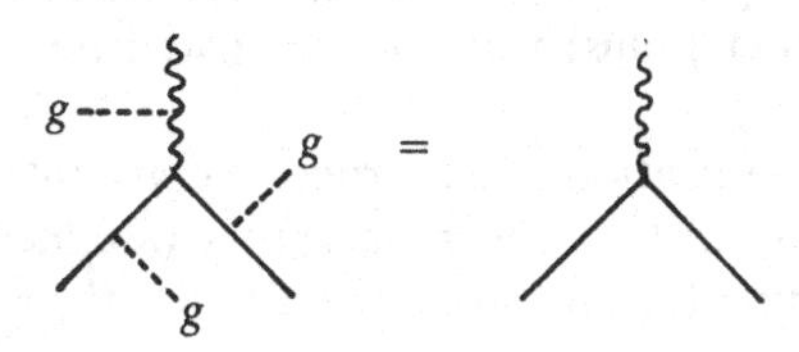

in the infinitesimal form:

αa + + αa = 0 .

aα

We use Problem 5.7 in the second diagram, add a stub to one external line in each term, and rearrange:

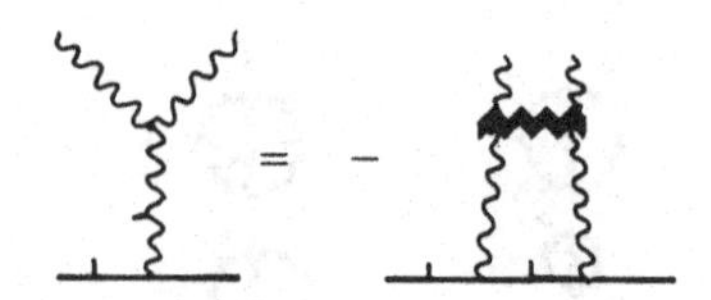

This is the required result, with the scalings:

αa αa

$<X^{\alpha}_{a}>_{\omega}$ ↔ X ↔ $<\omega||X^{\alpha}||\omega>$

αa αa

f^{c}_{ab} ↔ αc X αb ↔ $<\alpha||X^{\alpha}||\alpha>^{r\,|}$ αc αb .

r

Comments

(a) The structure constants f^{c}_{ab} contain the 3jm symbol for the triple $(\alpha\alpha r)$. The above solution shows that the structure constants are antisymmetric in a, b ($f^{c}_{ba} = -f^{c}_{ab}$). Hence $\{\alpha\alpha r\} = -1$; the 3jm symbol for this triple is totally antisymmetric.

(b) Combining this with Problem 5.7, Comments (c) and (d), we have $\{\alpha\} = 1$, i.e. the (self-contragredient) adjoint irrep is *real* rather than symplectic.

This may also be seen by using quasiambivalence (§3.3.2) – the product of all 2j phases at any *n*jm node equalling unity – for the triples $(\omega\varpi0)$, $(\omega\varpi\alpha)$ together, or for the triple $(\alpha\alpha r)$. Indeed, all operators (such as $\mathbf{X}^{\alpha}$) within a manifold (such as $\{\omega o\}$) must transform as true irreps (Comment (f) on Problem 3.12).

(c) The above identification of the structure constants of the Lie algebra with matrix elements of the tensor operators obtained from the group generators in the adjoint irrep shows directly that the structure constants give a matrix representation of the adjoint irrep of the Lie algebra $\mathscr{A}$ (Cornwell 1984). For example, the SO(3) irreducible tensor operators **L**, the generators of angular momentum, have matrix elements in a cartesian vector basis proportional to the SO(3) structure constants $f^k_{ij} = \varepsilon_{ijk}$, which in turn are proportional to the 3jm symbols

$$\begin{pmatrix} 1 & 1 & 1 \\ i & j & k \end{pmatrix} = \varepsilon_{ijk}/\sqrt{6}$$

(Problem 2.7, Comment (d)).

(d) A general member X of the Lie algebra may be expanded in the basis $\{a\}$ afforded by the chosen component basis of the adjoint irrep α: $X = \Sigma_a x_a X_a$, and will have matrix elements in this irrep given by

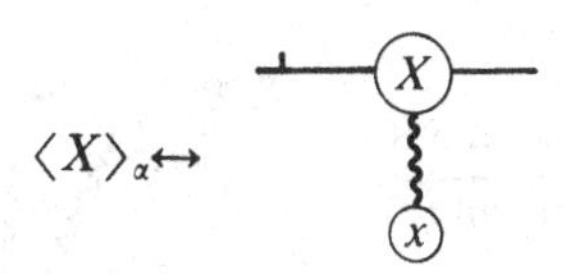

The expansion coefficients $\{x_a\}$ are real/complex for a real/complex Lie algebra, respectively.

(e) Again, the derivation of the algebraic consequence of the invariance of the 3jm symbol may be repeated for any representation; any 3jm symbol of a group is left invariant under the group operators.

In particular, the triple $(\alpha\alpha\alpha r)$ may be used in the above analysis. The infinitesimal form of the associated invariance relation is:

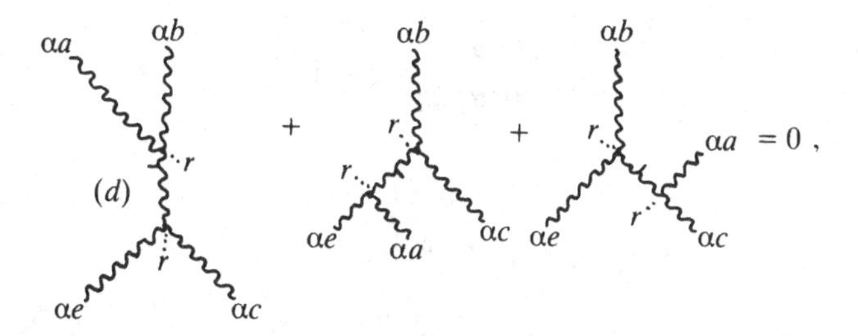

which transliterates to the algebraic identity

$$f^d_{ab} f^e_{dc} + f^d_{bc} f^e_{da} + f^d_{ca} f^e_{db} = 0,$$

in turn equivalent to the *Jacobi relation* between the Lie brackets:

$$[[X_a, X_b], X_c] + [[X_b, X_c], X_a] + [[X_c, X_a], X_b] = 0.$$

Since this is a homogeneous relation, no normalisation factors are needed. The Jacobi relation, like the fundamental equation of the Lie algebra, is equivalent to the group invariance of group 3jm symbols.

Problem 5.9

Prove the general relations

(a) $\begin{Bmatrix} \alpha\,\alpha\,\alpha \\ \alpha\,\alpha\,\alpha \end{Bmatrix}_{rrrr} = \dfrac{1}{2|\alpha|}$,

(b) $\begin{Bmatrix} \alpha\;\omega\;\varpi \\ \varpi\;\alpha\;\alpha \end{Bmatrix}_r + \{\omega\}\begin{Bmatrix} \alpha\,\omega\,\varpi \\ \alpha\,\omega\,\omega \end{Bmatrix} = -\dfrac{1}{|\omega|}$.

Solution

(a)

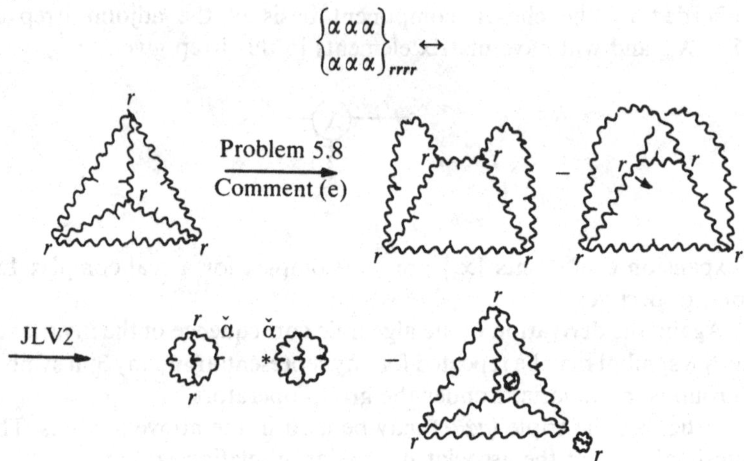

Hence, collecting terms,

$$2\begin{Bmatrix} \alpha\,\alpha\,\alpha \\ \alpha\,\alpha\,\alpha \end{Bmatrix}_{rrrr} = 1/|\alpha|.$$

(b)

$$\begin{Bmatrix} \alpha\;\omega\;\varpi \\ \varpi\;\alpha\;\alpha \end{Bmatrix}_r \leftrightarrow$$

Problem 5.8

JLV2

$$\leftrightarrow -\frac{1}{|\omega|} - \{\omega\}\begin{Bmatrix} \alpha\,\omega\,\varpi \\ \alpha\,\omega\,\omega \end{Bmatrix}.$$

Comments

(a) Many fundamental properties of groups may be associated with, and hence in a sense (which is not compelling) derived from, the properties of its 6j symbols. For example, it may be argued that $G_2 \supset SO(3)$ 'essentially' because a certain 6j symbol of G_2 vanishes.

(b) Nevertheless, such 6j symbol values and relations are not easy to confirm from the literature (see for example Haase and Butler 1984, 1986).

5.6 Casimir operators and Killing form

Casimir operators, or commuting combinations of group generators, may be constructed on a general prescription (Problem 5.14 and Comments). By Schur's lemma they are proportional to the unit matrix within an irrep. Their eigenvalues (the proportionality constants) appear often in the reduction of a general expression, and are essentially products of appropriate *n*j symbols and reduced matrix elements.

In diagram manipulations the eigenvalues arise as closed diagrams, namely the reduced matrix elements and *n*j symbols which decouple from a network on invoking a JLV*n* theorem, at least for a compact group; a complicated network will reduce to a minimal coupling diagram multiplied by these trace terms. In this way the group theoretic content of a diagram in some field theory satisfying a formal symmetry may be determined rapidly.

Given the viewpoint already developed in this book, such reduction is merely an exercise in Racah algebra by diagram techniques. As a consequence, we give only a brief survey which de-emphasises and hopefully reduces the mystique of the reduction, avoiding the conventional ancillary definitions of various 'Casimirs' – these being nothing but products of reduced matrix elements and *n*j symbols.

It is granted that this emphasis has its practical limitations for present explicit calculations, 6j symbols even of SU(3) not being immediately available in texts. Nevertheless, and especially for a book like this which concentrates on formal structure, we believe it is time to take the wider view.

Problem 5.10

Show that $\mathrm{Tr}_\omega X_a^\alpha = 0$.

Solution

The left side is proportional to

which vanishes from JLV1: $\alpha \neq 0$.

Comment

This may be regarded as the near-ultimate generalisation of, for example, Problems 1.2, 1.7 etc.

Problem 5.11

Show that $\mathrm{Tr}_\omega[X^{\alpha}_{a} X^{\alpha}_{a'}]$ is proportional (by a constant A, say) to the 2jm symbol $\begin{pmatrix} \alpha & \alpha \\ a & a' \end{pmatrix}^*$.

Solution

Comments

(a) Note that, as first remarked in Problem 1.15, the coupling of operators in a trace involves *no* crossings of those lines corresponding to summed labels.

(b) The constant A together with the reduced matrix element of X may be regarded as the *normalisation constant* for the generators $\{\mathbf{X}^\alpha\}$.

(c) Such a trace arises when applying JLV2 to the field theoretic self-energies of particles in both the defining ('quark') and adjoint ('gluon') irrep:

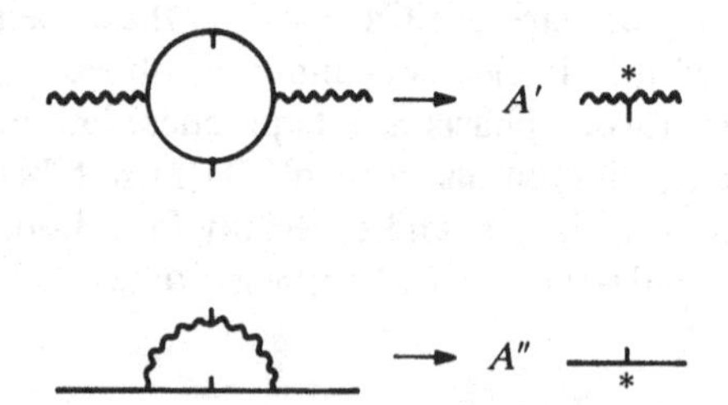

Hence, even with the ambiguities of normalisation choice, we have one nontrivial relationship between these group theoretic factors, viz:

$$\frac{A''}{|\alpha|} = \frac{A'}{|\omega|} \leftrightarrow \check{\omega}$$

Problem 5.12

Find some general constraints on the group theoretic factors associated with the third order field theoretic vertex corrections:

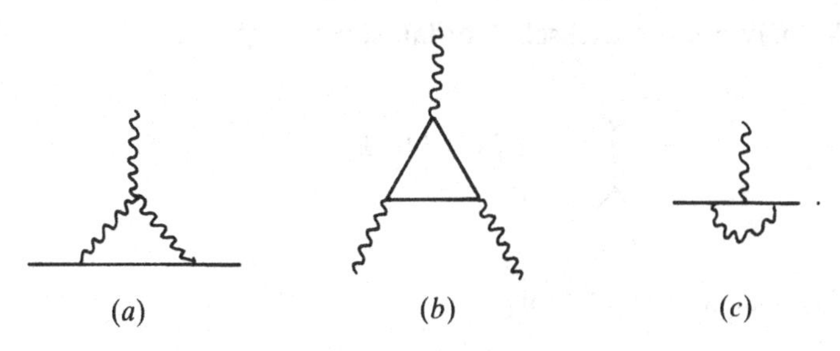

Solution

Field theoretic diagrams always correspond to group theoretic diagrams which are fully invariant/commutative, e.g. which carry stubs on all internal lines (§7). Apart from reduced matrix elements, then,

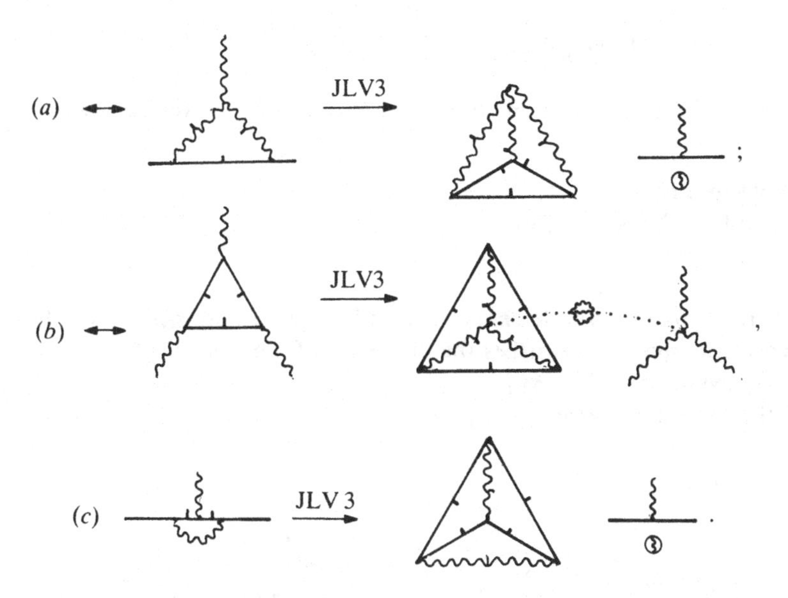

i.e. the group theoretic factors for (*a*), (c) are same to within *n*j phases, dimensional factors and reduced matrix elements.

Comments

(a) Problem 9(b) gives a relation between the 6j symbols appearing above, and so a relation between the field theoretic diagrams of the general form:

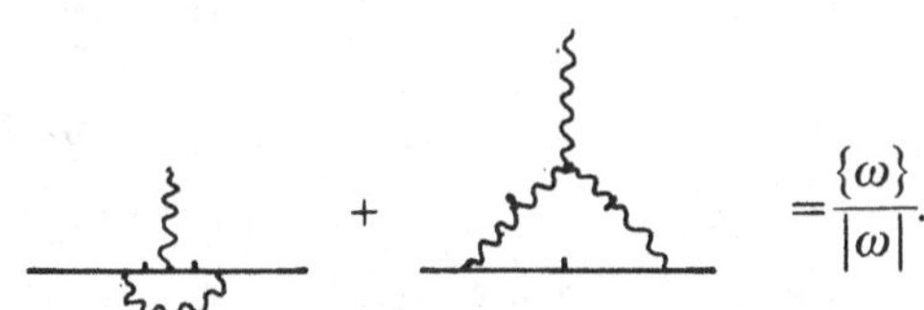

(b) We may use the Clebsch–Gordan series of §5.5.2:

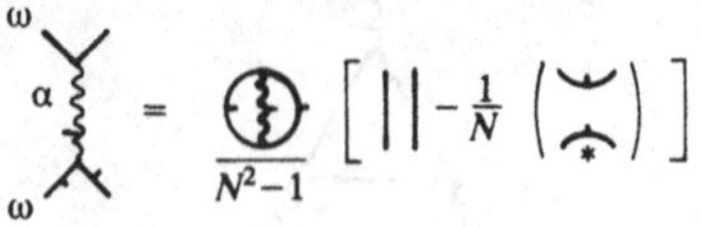

to simplify case (c), for example:

$\left\{ \begin{matrix} \alpha\,\omega\,\varpi \\ \alpha\,\omega\,\omega \end{matrix} \right\} \leftrightarrow$ 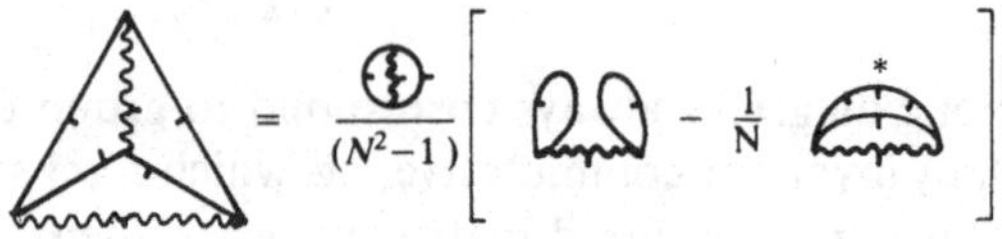

$$\leftrightarrow -\{\omega\varpi\alpha\}/[N(N^2-1)].$$

We use JLV1 to cancel one term, and Problem 5.7, in the last step.

Problem 5.13

Show that the *Killing form*

$$g_{XY} \equiv \mathrm{Tr}[\alpha(X)\alpha(Y)],$$

where X, Y are two elements of the Lie algebra and $\alpha(X)$, the adjoint representation of X, satisfies the properties (for example) of

(a) symmetry: $g_{XY} = g_{YX}$;

(b) form-invariance: $g_{[X,Y],Z} = g_{X,[Y,Z]}$.

Solutions

(a)

$g_{XY} \leftrightarrow$ 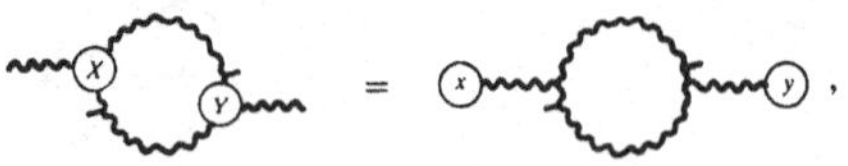

which has an obvious rotational symmetry.

(b)

$g_{[X,Y],Z} \leftrightarrow$ 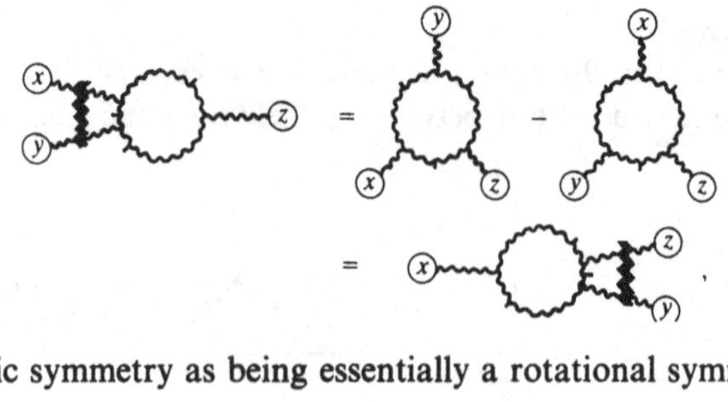

showing the cyclic symmetry as being essentially a rotational symmetry.

Comments

(a) Other properties of the Killing form, such as bilinearity, follow almost trivially.

(b) In any representation Λ (not necessarily irreducible), we can define a related form by:

$$h_{XY} \equiv \mathrm{Tr}[\Lambda(X)\Lambda(Y)] \leftrightarrow$$

The symmetry ($h_{XY} = h_{YX}$) and form-invariance properties follow as before.

(c) One may see these properties in the original problem also through application of JLVn and using the results of previous problems for interchange phases.

Symmetry, for example, may be checked by the manipulation

JLV2

where the factor corresponding to the closed diagram is $\{\alpha\omega\varpi\}\{\omega\}/|\alpha|$ and is equal to $-1/|\alpha|$ from Problem 5.7. (a) then follows since $\{\alpha\} = 1$, i.e. since the direction of the stub does not affect the answer.

For form-invariance, the manipulations

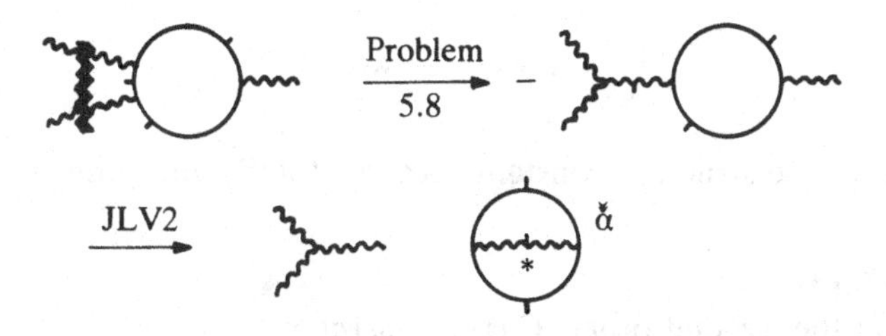

lead to an expression obviously possessing cyclic symmetry.

(d) Within restrictions (Cornwell 1984: a simple complex Lie algebra) we may extend this to the more general definition of form, and show that any form-invariant symmetric bilinear form must reduce to the Killing form up to a factor. The above definition of form-invariance is related to the definition of invariance of §§2.5, 3 in such cases. The application of JLV2 to the diagram for the form associated with representation Λ (Comment (b)), which is invariant in the original sense of §§2.5, 3, immediately demonstrates a relationship to the Killing form (complicated perhaps by sums over repetition labels and reduced matrix elements in the irreducible decomposition of Λ, which we suppress). Each form amounts to the 2jm symbol of irrep α times a basis-independent factor akin to a reduced matrix element:

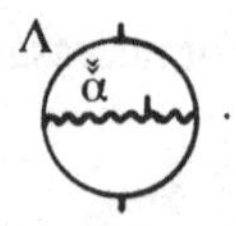

The proportionality constant linking the two forms is known as the *Dynkin index* of the representation Λ. We recognise this for irreps Λ as being essentially a 3j phase with normalisation factors.

(e) According to a theorem of Weyl (we shall not discuss this, the associated definitions or proofs; see Cornwell 1984) a connected semi-simple linear Lie group is *compact* if and only if its corresponding Lie algebra is also compact, or equivalently if and only if its Killing form is negative definite. This enables a test for compactness based on study of the Lie algebra. In view of the sign found above when relating the Killing form to the 2jm symbol (i.e. the negative sign of the closed diagram factor in Comment (b)) this means that for a compact group

$$(x)\sim\sim(x) \geqslant 0 .$$

Hence contraction of two members of the Lie algebra on a 2jm symbol satisfies the axioms of an *inner product*, and it becomes possible to choose a basis of α such that the 2jm symbol is trivial:

$$\underset{*}{\sim\sim} = \sim\sim .$$

In this case, the structure constants become totally antisymmetric.

Problem 5.14

Show that the (second order) *Casimir operator*

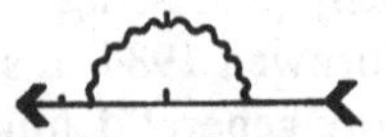

commutes with all elements of the Lie algebra.

Solution

A general element of the Lie algebra written in the operator form

has the commutator

which, using the fundamental relation of the Lie algebra (Problem 5.8), leads to cancelling terms:

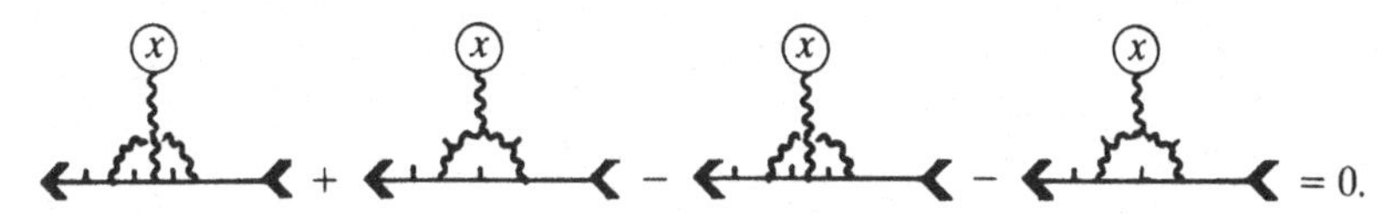

Comments

(a) The trace of the Casimir operator, which is its eigenvalue apart from normalisation (see the original form of JLV2), is clearly the *n*j symbol found for various self-energy diagrams in Problem 5.12.

(b) The Dynkin index and the eigenvalue of a Casimir operator are related, each to an *n*j phase:

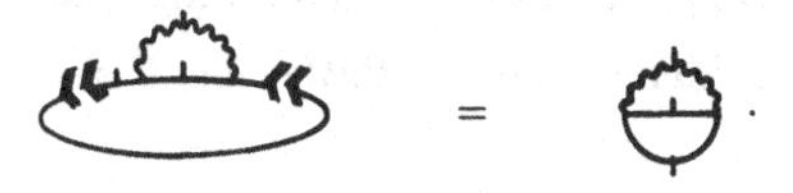

(c) Higher order Casimir operators may be constructed:

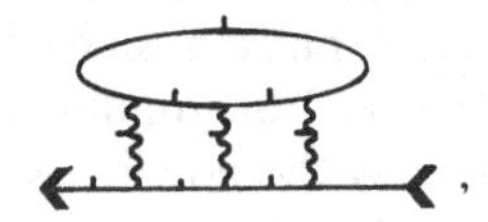

corresponding to a trace over more than two operators in the adjoint irrep. The proof of their commutativity may be given by extending the above argument. The crucial point is diagrammatically obvious: the structure is an invariant diagram. It is not even necessary that the loop be in the adjoint irrep, and Gruber and O'Raifeartaigh (1964) have exploited this to generate a more complete and useful set of Casimir operators.

The application of the material of this chapter to gauge theories such as QCD is direct, at least when the symmetry group is compact. We shall not discuss this in detail in this book. Full and specialist accounts include those by Cvitanovic (1976, 1983, 1984; unpublished lecture notes, University of Oxford, 1980), Canning (1978), Cvitanovic, Lautrup and Pearson (1978), Butera, Cicuta and Enriotti (1980), Cvitanovic, Lauwers and Scharbach (1981), Cvitanovic and Kennedy (1982). Some aspects and illustrations of this work will be mentioned in §8.6.

6

Polarisation dependence of multiphoton processes

For some time now I have been worrying about what new developments in theoretical chemistry are likely to be most important in the next 10–15 years. One has to hazard a guess. 20 years ago we opted for group theory, and turned out to be absolutely right. . . . Today something new must be encouraged. Rightly or wrongly we have decided on diagram methods . . .

C.A. Coulson, reported in Paldus and Čížek (1975).

6.1 Introduction

6.1.1 Radiation–matter interaction

Group theory is important in physics since it separates questions of dynamics from geometry. We need the fundamental laws of physics – the dynamical equations – to specify the possible kinds of couplings between excitations of the quantum-fields. Once these are determined and their symmetry specified, the consequences for a physical observable or matrix element in a particular situation often has a strong geometric element, dependent only on the choice of irrep partner or component in the symmetry space, or equivalently the 'geometric' orientation of state vector and operator in the abstract space associated with the symmetry. Group theory, based on the symmetry properties of the dynamical laws, suffices to determine the dependence of the matrix element on choice of geometry.

Nowhere is this more clearly illustrated than in the theory of the polarisation dependence of the interaction of light with matter. The space of symmetry operations is real space and the point group symmetry of the matter (say a molecule or a crystal), as well as that of a photon beam, is readily visualised. The geometrical clarity of the problem has influenced many presentations, with a variety of theoretical techniques (see Buckingham, Graham and Raab 1971, McClain 1971, Barron and Buckingham 1972, 1975, Buckingham and Raab 1975, McClain and Harris 1977, 1983,

Buckingham and Shatwell 1978, 1980, Jerphagnon, Chelma and Bonneville, 1978, Andrews 1980, de Figueiredo and Raab 1980, 1981, Graham 1980, Barron 1982, Craig and Thirunamachandran 1983). Similar analyses exist of polarisation dependence in relativistic collision theory (Craigie, Hidaka, Jacob and Renard 1983, Goldstein and Moravcsik 1984a,b, Conzett, Goldstein and Moravcsik 1985).

In the field of quantum optics, a vulgarised form of Feynman diagram has been adopted as standard to assist in bookkeeping within perturbation theory (Cohen-Tannoudji et al. 1977, Shen 1984, Friedberg and Hartmann 1988). Diagram techniques in group theory have also been applied in detail, drawing on such links with field theoretic diagrams. Originally (Yuratich and Hanna 1976a,b, 1977, Luypaert and van Craen 1977, 1979, Nienhuis 1984) this was confined to atomic physics where atomic states had irreducible character under SO(3) and the method of §2 sufficed to analyse the symmetry information. This has been extended to molecular and crystal systems (Stedman 1985); this section is essentially a summary and (in part) extension of this work. The methods of §3 are relevant with G being the point group for the molecule or crystal in interaction with the light.

It is these methods which will be illustrated here for some representative optical processes: optical absorption (including refractive index, birefringence, dichroism and optical activity, also field-dependent processes such as the Faraday effect) and also Rayleigh and Raman scattering.

The *interaction* between radiation and matter contains the term

$$H_{\text{int}} = \Sigma_i (q/2m)\mathbf{A}(\mathbf{r}_i,t)\cdot\mathbf{p}_i,$$

where $\mathbf{p}_i$ is the canonical momentum operator for electron i, and in second quantisation the vector potential has the form (Sakurai 1967)

$$\mathbf{A}(\mathbf{r},t) = \Sigma_{\mathbf{k}j}(\hbar/2N\omega_{\mathbf{k}})^{\frac{1}{2}}\mathbf{e}_{\mathbf{k}j}c_{\mathbf{k}j}\exp\mathrm{i}(\omega_{\mathbf{k}}t - \mathbf{k}\cdot\mathbf{r}) + \text{h.c.}$$

In this expression $\mathbf{e}_{\mathbf{k}j}$ is the polarisation vector for the photon (radiation) mode with wavevector $\mathbf{k}$ and polarisation j, and $c_{\mathbf{k}j}$ is the corresponding boson annihilation operator. h.c. refers to the hermitian conjugate.

A quadratic term, proportional to $\mathbf{A}^2$, should also be included in the interaction. The gauge dependence of the formalism may be exploited to transform the $\mathbf{A}\cdot\mathbf{p}$ form into an $\mathbf{E}\cdot\mathbf{r}$ form. The wavefunctions used in the analysis should be consistent with the gauge choice made when describing the interaction (see, for example, Lamb, Schlicher and Scully 1987, Reid 1988).

The photon frequency and wavevector are linked by the dispersion relation $\omega_{\mathbf{k}} = c|\mathbf{k}|$. Transversality gives $\mathbf{e}_{\mathbf{k}j}\cdot\mathbf{k} = 0$. In the particle physicists' convention, a right circularly polarised beam has particles of positive helicity, i.e. with angular momentum parallel to $+\mathbf{k}$; if $\mathbf{k}$ is parallel to the $+z$ axis, the corresponding polarisation vector may be taken as $(1,\mathrm{i},0)/\sqrt{2}$.

The complex conjugate vector describes left circularly polarised light. We note that their vector product is proportional to the wavevector:

$$i\mathbf{e}\times\mathbf{e}^* = S\mathbf{k} \leftrightarrow \sqrt{6}\,i$$

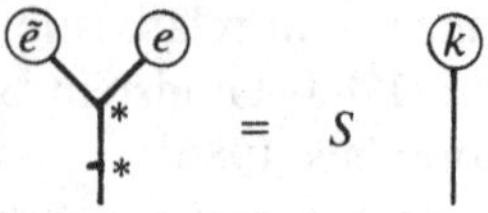

where the Stokes parameter $S = +1(-1)$ for right (left) circularly polarised light.

Comments

(a) We have incorporated the finesses discussed in §§2.5, 3.6, 5.3, according to which we have defined the conjugated vector node $\tilde{e}$ by

The reader with a healthy contempt for all but the most essential phases can read e^* for $\tilde{e}$ and ignore all stubs in this and the following diagrams.

(b) The imaginary unit i was included in the above expression to make the result real. Its otherwise imaginary value may be verified either by noting that the relevant 3j phase $\{111\} = -1$, or, as proved in §2.5, that the coupled node is anticonjugate. The expansion of the exponential will often provide this imaginary factor in our practical applications.

(c) The radiation–matter interaction is a rotational invariant: $\mathbf{A}\cdot\mathbf{p}$ is a scalar. This makes SO(3) the natural group for the discussion of symmetry effects. We use an electronic–radiation product basis for the unperturbed kets (eigenstates of the uncoupled systems) in the perturbation calculations. The electronic operator includes not only the vector $\mathbf{p}$, but also in general the position vectors $\mathbf{r}$ in the multipole expansion of the vector potential in powers of $|\mathbf{k}|$.

6.1.2 Lowest order multipole expansion terms

Expanding the exponential in the vector potential, we obtain terms in the interaction of the form (dropping the indices $\mathbf{k}j$ and the factor $(1/n!)$ for simplicity)

$$\mathbf{e}\cdot\mathbf{p}(i\mathbf{k}\cdot\mathbf{r})^n \leftrightarrow$$

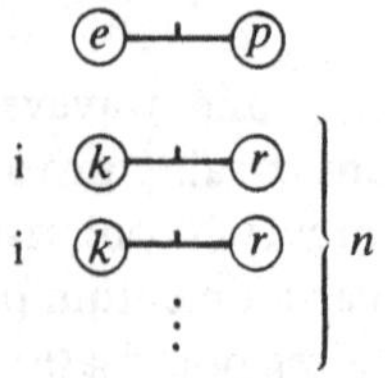

The first term is the *electric dipole* (E1) coupling term, with operator form **e·p** in this gauge.

We now re-express the second of these terms ($n=1$) in the form of purely electronic tensor operators coupled to purely radiation tensor operators, and classify these by their coupling symmetry.

We couple the electronic vector operators **p**, **r** as in the quantum theory of angular momentum (§§1–2) to give the form

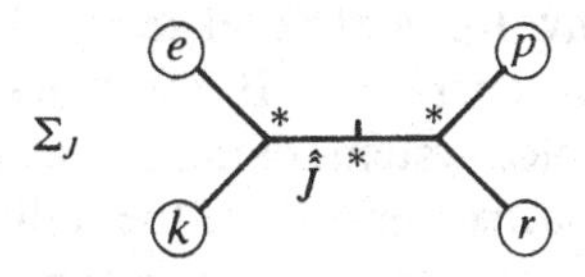

When the coupling angular momentum J is zero, the contribution contains the factor **e·p** which vanishes from transversality. $J=1$ gives an interaction term which may be variously written as proportional to

$$\tfrac{1}{2}\mathrm{i}(\mathbf{p}\times\mathbf{r})\cdot(\mathbf{e}\times\mathbf{k})\leftrightarrow$$

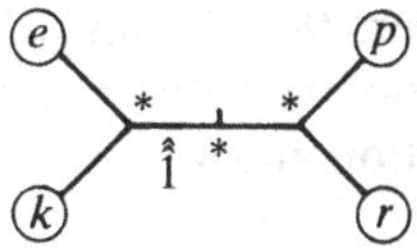

$$=\tfrac{1}{2}\mathbf{L}\cdot\mathbf{m}=\tfrac{1}{2}\mathrm{i}[(\mathbf{p}\cdot\mathbf{e})(\mathbf{k}\cdot\mathbf{r})-(\mathbf{p}\cdot\mathbf{k})(\mathbf{e}\cdot\mathbf{r})]\leftrightarrow\ \mathrm{i}$$

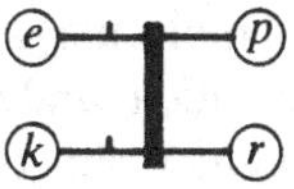

This is the *magnetic dipole* (M1) interaction. Note that the auxiliary vector

$$\mathbf{m}\equiv \mathrm{i}\mathbf{k}\times\mathbf{e}\leftrightarrow \quad = \quad \sqrt{6}\,\mathrm{i}$$

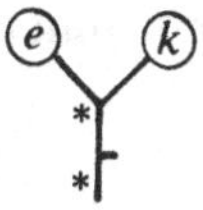

gives the polarisation of the magnetic part of the electromagnetic radiation field, and that **L** is the angular momentum operator **r** × **p**.

Similarly the term $J=2$ gives the *electric quadrupole* (E2) interaction:

$$\mathrm{i}\mathbf{k}\cdot\mathbf{Q}\cdot\mathbf{e}\leftrightarrow$$

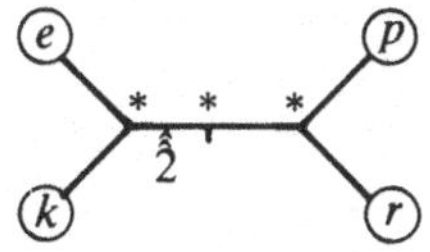

where the E2 tensor $\mathbf{Q}\equiv(\mathbf{pr}+\mathbf{rp})$, whose matrix elements in the electronic system describe the quadrupole moment of the electronic charge distribution.

6.1.3 Method

According to the Golden Rule of quantum mechanics, the interaction rate is given by the square modulus of the matrix element of the interaction between initial and final states, summed over isoenergetic final states and thermally averaged when appropriate over initial states. The matrix elements of H_{int} will involve matrix elements both of the electronic operators and radiation operators. If we work in a product basis, appropriate to the uncoupled systems of radiation and matter, these matrix elements factorise. The matrix elements of the radiation second quantised operators $c_{\mathbf{k}j}^{(\dagger)}$ are simple to calculate. For a symmetry analysis, we may concentrate on the electronic matrix elements, which will have a tensorial symmetry conjugate to that of the radiation field, since $\mathbf{A}(\mathbf{r})\cdot\mathbf{p}$ is an invariant. Further, the consequences of this tensorial character for the interacting radiation states are conspicuous within the electronic calculation, not only from the coupling of the electronic operators $\mathbf{p}$, $\mathbf{r}$, but also from the polarisation and wavevector $\mathbf{e}_{\mathbf{k}j}$, $\mathbf{k}$. These also appear within the electronic calculation and with the conjugate tensorial character.

6.2 Electric dipole absorption

6.2.1 Introduction

Consider a molecule whose symmetry is described by a point group G ($G \subset O(3)$) which undergoes a one-photon interaction with a linearly polarised light beam through E1 electric dipole) coupling. We find an expression for this coupling, taking full account of the symmetry.

We write the electronic matrix elements as an invariant diagram. In this case the appropriate electronic interaction is, apart from constants, $\mathbf{e}_{\mathbf{k}j}\cdot\mathbf{p}$. Its square modulus within electronic states has the form

$$\mathscr{A}v_i \Sigma_f |\langle i|\mathbf{e}_{\mathbf{k}j}\cdot\mathbf{p}|f\rangle|^2 \delta(E_f - E_i) \leftrightarrow$$

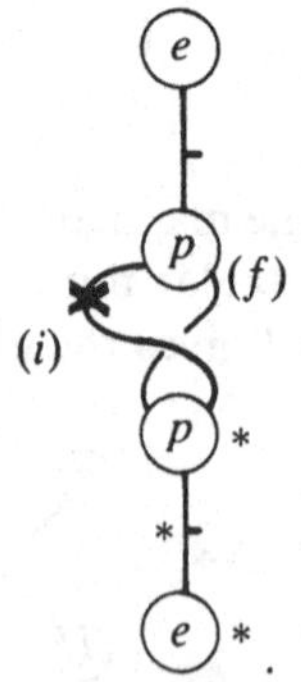

e
p (f)
(i)
p *
*
e *.

The lower part of the diagram represents the conjugate amplitude. The linkages between upper and lower parts of the diagram include thermally weighted summations over levels, and, within these, summations over degenerate states in any level. The summation over degenerate states amounts to the group-invariant simple sums over components of an irrep, since the various energy levels of a Hamiltonian are labelled by irreps of the appropriate symmetry group. The sum over irreps appearing in the sum over initial state label i has a nontrivial (thermal) weighting factor, $\exp(-\beta E_i)$, represented by a cross above; since the energy is the same for all degenerate states, such factors (along with perturbation-theoretic energy denominators, §7) do not affect symmetry considerations deriving from summations over components of an irrep. The whole diagram for an intensity contribution thus has an invariant internal structure so far as the initial and final electronic states are concerned. Its transformation character is determined by the external lines.

We extend this invariant character by adding to the external lines the unitary basis transformations reducing the polar vector representation $j^\pi = 1^-$ (appropriate to the operator $\mathbf{p}$) to irreps of the point group in question:

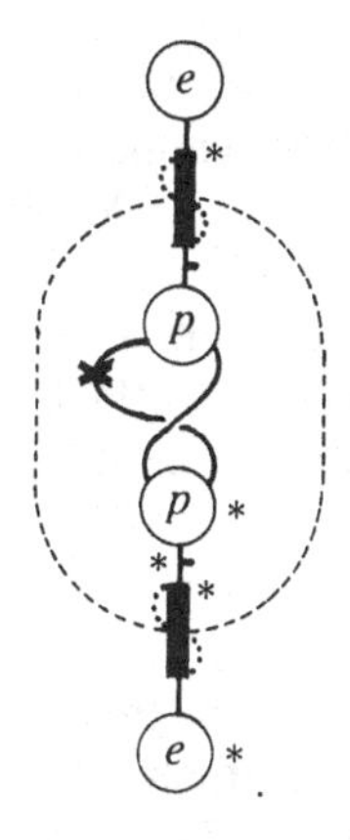

The dotted region then encloses a commutative diagram, to which JLV2 may be applied:

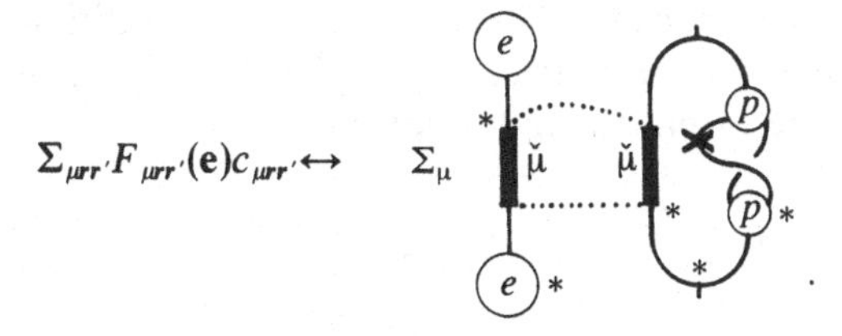

The closed diagram $c_{\mu r r'}$ on the right amounts to a physical or *dynamical constant*: it summarises in a geometrically invariant form the physical coupling strengths appropriate for this interaction and these systems.

The *geometric factor*,

$F_{\mu r r'}(\mathbf{e}) \leftrightarrow$ [diagram: e, r, $*$, $\check{\mu}$, r', $e*$] $=$ [diagram: e, $*$, r, $\check{\mu}*$, $*$, r', $\tilde{e}$],

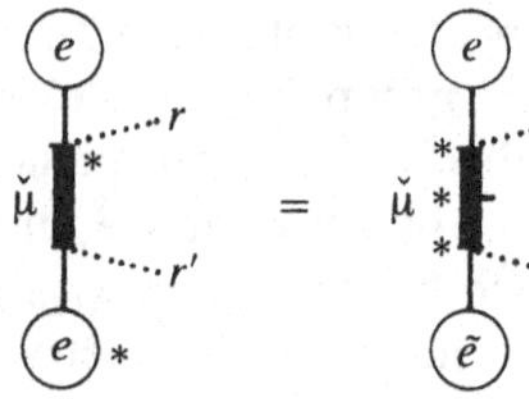

enshrines the essentials of the geometry of the absorption process, independent of the coupling strengths. A given term, with definite coupling irrep μ and repetition labels r, r', corresponds to a particular dependence of the whole process on the choice of photon polarisation vectors, governing how the intensity of the absorption varies as the orientation or polarisation of the light beam is changed. There are, say, N such terms, i.e. N sets of labels $(\mu r r')$. We may expect precisely N different types of geometrical dependence.

A very similar manipulation may be done when we consider fluids of identical absorbers (of arbitrary symmetry) which interact incoherently. The average over orientation of the molecules involves an SO(3) average over a joint rotation of all the labels connecting the material system to the radiation vectors (rotating from the arbitrarily oriented frame of the molecule to the lab frame). This is just the form required to validate JLVn for SO(3), and the diagram factorises as though the molecule had full SO(3) symmetry.

6.2.2 Examples

Octahedral or higher cubic or rotational symmetry

Consider the groups O, O_h, T, T_d, T_h, K, K_h, or SO(3). In these the O(3) irrep $J^{\pi}=1^{-}$ (where π denotes the parity, or inversion signature) – according to which the polar vector **p** transforms – reduces to a single triply degenerate irrep $\mu=1$. We are using Butler's notation; for conversion tables see Table A1 of the Appendix. There is no need for repetition labels r, r' in this case, and indeed no need to reduce to the point group. $N=1$ and this lone geometric factor has the SO(3) form

$$(e)\!\!-\!\!-\!\!(e)^{*} .$$

In this case the system appears isotropic to the light beam. All directions in a cubic crystal are equivalent, as far as optical absorption is concerned. In measuring one spectrum we measure all possible absorption spectra.

A tetragonal crystal, with symmetry D_{4h} say

In this case $J^{\pi}=1^{-}$ reduces to two irreps $\tilde{0}$ and 1, of which one is doubly degenerate. If the major symmetry axis is chosen as the z axis, then the corresponding geometric factors are

$$F_{\tilde{0}}(\mathbf{e})=e_z^2,$$
$$F_1(\mathbf{e})=e_x^2+e_y^2.$$

We need then two spectra to characterise the absorption of such a system in linear polarisation; x and z polarisation measurements would suffice. An absorption spectrum taken for any other choice of geometry would consist of a linear superposition of these two spectra, in proportions themselves determined also by geometry.

A system with a single self-inverse symmetry element

If the point group of the system is C_2 or C_s, the O(3) vector irrep 1^{-} reduces to three repetitions ($r=1,2,3$) of the single one-dimensional irrep $\mu=1$ (Table A3 of the Appendix). In this case we have in principle nine terms, as r and r' each vary in the range 1 to 3; however, the symmetry of the geometric factor with respect to exchange of r, r' reduces this to six terms: $(rr')=(11)$, (22),(33),(12),(23),(31).

Comments

(a) The first part of this section illustrates for the first time our general strategy in all the problems discussed in this chapter. One forms a coupling diagram appropriate to the intensity of the optical process from the matrix elements of all relevant electronic operators. It turns out to be an invariant diagram with p external legs coupling to radiation vectors only. Application of JLVp for the symmetry group of the problem separates the geometric factors and dynamical constants within each of N terms corresponding to all possible geometrical dependences.

(b) Ringing the changes on the geometry has strictly limited effectiveness; N geometrically independent measurements suffice to gain as much information as is possible by any such technique.

(c) It is perhaps surprising – and only recognised recently – that at least in the lowest symmetry case, so many measurements are needed. One might have guessed that three measurements with the electric field of the photon parallel to the three axes x, y, z would have been adequate, the spectrum for any other orientation being determined by a superposition of the spectra associated with each field component. However, this expectation fails, since, despite the geometrical independence (in some sense) of the amplitude contributions associated with different axial geometries, these contributions interfere in determining the intensity response.

6.3 Natural optical activity

6.3.1 The need for interference

In this and the following two subsections, we build up towards a succinct statement of selection rules for optical processes, stated generally in §6.3.4.

Optical activity, together with the associated absorption process of circular dichroism, is a one-photon interaction which demands consideration of circularly polarised modes, for which the polarisation vector **e** is necessarily complex. Further, it demands through the Kramers–Kronig (or causality) relations a difference in the refractive index or absorption of light of opposite circular polarisation. Hence it necessarily involves a geometric factor or factors which *reverse* sign under the replacement $\mathbf{e} \to \mathbf{e}^*$.

We call such effects, whose intensity contributions reverse under conjugation of the optical polarisation vectors, *chiral effects*. A fuller explanation and justification of the term will be deferred to §6.3.4.

Now any intensity contribution will be bilinear in these vectors, whatever other radiation vectors (such as **k**, from higher order multipole expansion terms) may be present as well. We introduce a coupling (or if necessary recoupling), called for brevity the *conjugate pair coupling*, of the radiation vectors such that **e** and **e*** are coupled prior to any other linkages:

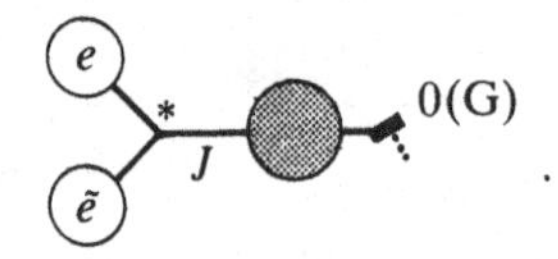

In such a coupling we can identify a possible contribution to optical activity by setting $J=1$. This is because $[1 \times 1]_- = 1$, i.e. among the relevant 3j phases, only $J=1$ gives $\{11J\}$ equal to -1, and only then does the appropriate 3jm symbol of SO(3) have the antisymmetry needed to give the sign change on conjugation (see for example Comment (e) on the diagram definition of $\varepsilon_{\alpha\beta\gamma}$ in §1.3, p. 8).

The E1 contribution discussed in §6.2.1 will not furnish such a term. At first sight we might recouple in that problem, and isolate the term corresponding to natural optical activity, as follows:

$I(\mathbf{e})|_{\text{NOA}} \leftrightarrow$

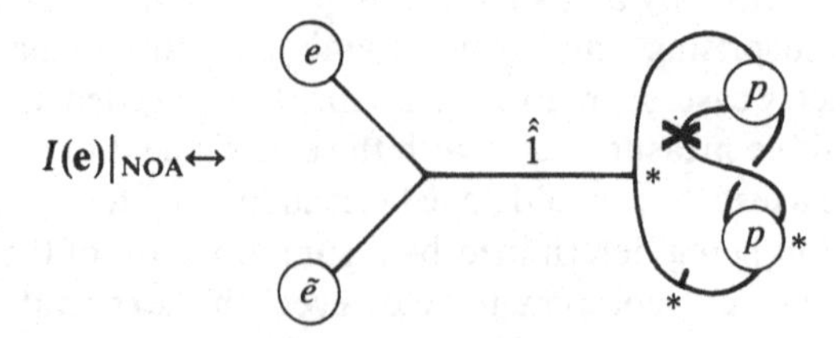

using the anticonjugacy (or sign change, equivalent to a 2jm transformation as specified by the generalised conjugation lemma of §3.5) of the rank-1 coupling of **e** and its conjugate node (§2.5). However, the above term often cancels to zero.

In a fluid this cancellation is obvious from rotational symmetry. The component of intensity relevant to optical activity involves the coupled tensor $[\mathbf{e}\tilde{\mathbf{e}}]^{j\pi}$ of spin-parity $j^\pi = 1^+$ in O(3), which reduces to the vector ($j=1$) irrep rather than the identity irrep ($j=0$) of the group SO(3) appropriate for orientational averaging of the constituents of a fluid.

One might expect a contribution when $j^\pi = 1^+$ does branch to the identity irrep of the point group in question. This includes (see Table A3 of the Appendix) $G = C_n$, C_{nh}, C_{ni}, C_s, C_i. However, even in these groups the contribution cancels if the system is time-even.

For example, consider $G = C_n$, where the z component of $j=1$ is an invariant, z being the n-fold axis. Using the antisymmetry of the triple (111) this implies that the right side of the above diagram has the combination of terms

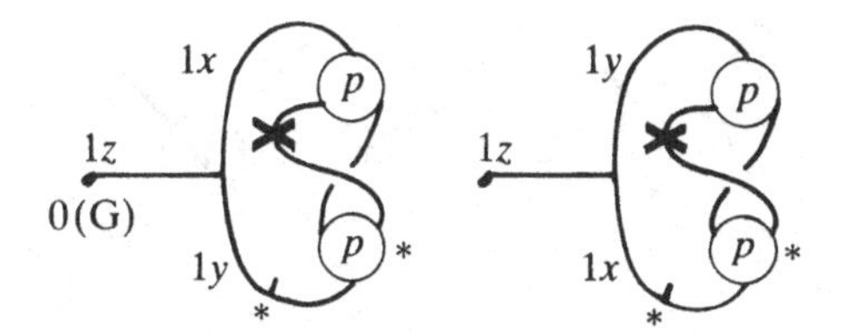

and so the combination of electronic matrix elements

$$C = \mathscr{A}v_i \Sigma_f C_{if};$$

$$C_{if} \equiv \langle i|p_x|f\rangle\langle i|p_y|f\rangle^* - \langle i|p_y|f\rangle\langle i|p_x|f\rangle^*.$$

Using the fundamental relation $\langle a|O|b\rangle^* = \langle \bar{a}|\bar{O}|\bar{b}\rangle$ for the conjugation of matrix elements under time reversal (§2.7.4), noting that the time-reversal phases of the two momentum operators in each term cancel, we find that the diagrams are antisymmetrically related under conjugation: $C_{if} = -C_{\bar{i}\bar{f}}$. We have implicitly chosen a time-even material system. By contrast, the symmetry of a system with a local magnetic field would properly be described by a magnetic point group. It is useful to separate external fields from the system in any case, so that for example linear and nonlinear effects in such fields may be distinguished. Under time reversal, then, the states i, f will be transformed into states $\bar{i}, \bar{f}$, each of which are degenerate with the original. Therefore the conjugation of labels $i \leftrightarrow \bar{i}$ has no effect on the averaging and summation operation; the energy-dependent factors (such as the thermal factor) are unchanged by time-reversal conjugation of the states. Since the summed quantity C_{if} reverses under conjugation, the whole contribution cancels.

This cancellation may be anticipated by noting the effects of the requirement that any contribution to the intensity must be real. If complex conjugation of the polarisation vectors alone reverses the sign of the contribution (as is required for optical activity), then that part of the geometric factor is imaginary, and the remaining terms must also be imaginary. However in E1 coupling the remaining terms comprise the dynamical constants, which themselves take the form of a summed square modulus of two conjugation-symmetric amplitudes, and which are therefore real.

In short, the generalised conjugation lemma (§3.5) as applied to $n$jm vertices (§3.3) and to matrix elements of physical operators (§3.6.4) shows that under complex conjugation, the nett effect is to (a) time reverse all labels; (b) add stubs to all external lines (these are inconsequential on identity irrep legs, as in our present example); and (c) add an overall phase τ, the product of the time-reversal phases of all operators in the physical constant:

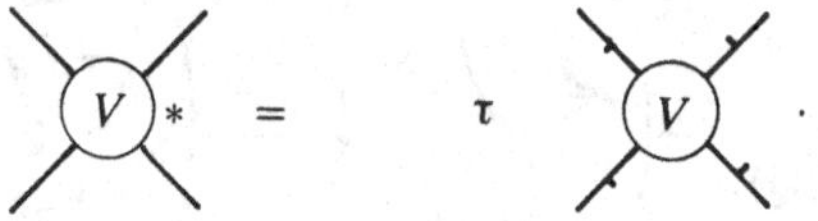

$$(V)^* = \tau \, (V) .$$

We conclude that the geometric factor is self-conjugate or -anticonjugate as the product of time-reversal phases τ is equal to $+1$, -1, respectively. Hence, for the above example, the physical constant is real; (a) and (b) are inconsequential, and the time-reversal phases are those of an even number of time-odd (momentum) operators, and cancel. An imaginary geometric factor combined with a real dynamical constant cannot induce a real intensity contribution.

6.3.2 Interference contributions

From the above discussion, the only way of inducing natural (external-field-independent) optical activity, in which a conjugation–antisymmetric amplitude must be formed, is through interference between multipoles of different order in the photon wavevector in the k-expansion of $\mathbf{A}(\mathbf{r})\cdot\mathbf{p}$. Since terms in the expansion of $\exp(i\mathbf{k}\cdot\mathbf{r})$ are alternately real and imaginary, the interference term will be imaginary as required if the orders n, n' of the associated terms in the expansion sum to an odd number, or equivalently if the interference term in the intensity is of odd order in photon wavevector.

This indicates that, for example, E1–M1 interference, when the polarisation vector $\mathbf{e}$ and the single photon wavevector $\mathbf{k}$ are coupled to $J=1$, furnishes a mechanism for optical activity:

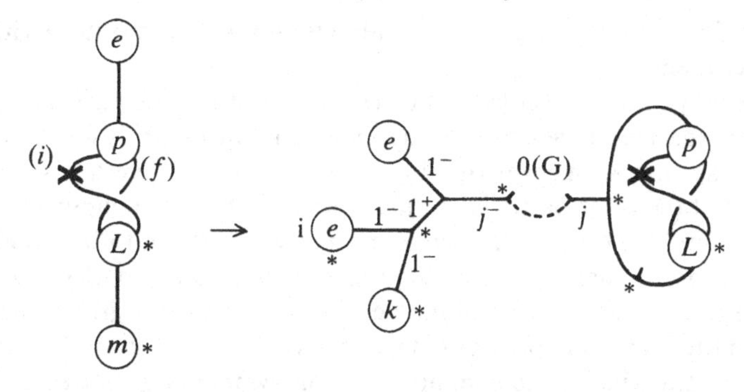

as does also the E1–E2 interference term in which the coupling is to $J=2$:

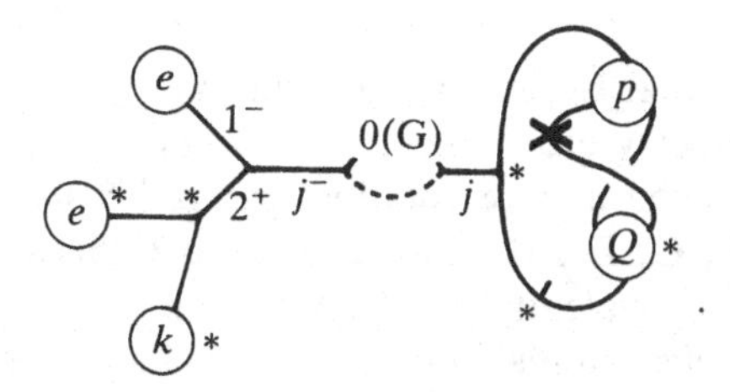

The geometric factors above should be recoupled internally so that **e** and **ẽ** are coupled to rank 1, thus ensuring anticonjugacy and optical activity. Using JLV3, a 6j may be factored off in this recoupling:

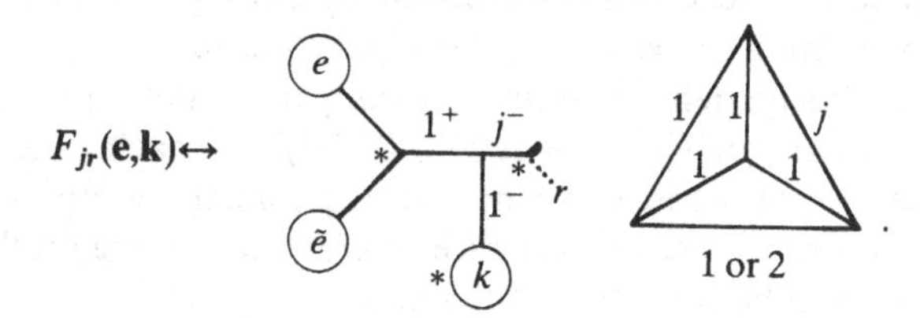

The odd number of extra polar (**k**) vectors in the geometric factor, and of the associated **r** operators in the dynamical constant, introduces an overall change of parity in the tensorial classification of both radiation and electronic operators; the interaction $\mathbf{A}(\mathbf{r})\cdot\mathbf{p}$ is an O(3) as well as an SO(3) invariant. The consequent selection rule may be analysed in either the radiation (geometric factor) or electronic (dynamical constant) context.

To take the latter first: the parities of the electronic operators in each amplitude are opposite. For example, in E1–M1 mixing, while **r** is a polar vector, **L** is an axial vector. Since such interference terms involve electronic operators with a negative net parity, the molecular absorber should not be inversion-symmetric. Otherwise both the initial and final electronic states would have definite parity, and it would be impossible for each of two electronic operators (one for each amplitude) with opposite parities to connect them.

This may also be seen from the viewpoint of the geometric factor. We recouple the vectors as above using O(3) spin-parity labels for the irreps, noting that the Kronecker rule for parity at every 3jm vertex is that the product of the three parity labels must be $+1$. The final coupling angular momentum j for the interference terms under discussion to be reduced to the point group identity irrep is associated with negative parity. For j^- to branch to the identity irrep of the point group, the point group must not contain the inversion operator (which sends $\mathbf{r} \rightarrow -\mathbf{r}$ etc.) as an element. We call a system which does contain inversion symmetry a *centrosymmetric* system, and one which does not so contain the inversion, a *noncentrosymmetric* system. Others use such nomenclature as centric and acentric systems.

Further, if j is equal to zero, as for rotationally averaged systems for example, the symmetry group must not contain any rotation–reflection operation; G must be a pure rotation group. This includes the point groups: O, K, T, $\mathrm{T_d}$, C_n, D_n, where $n = 1,2, \ldots, \infty$. Otherwise, any operation containing either (equivalently) a reflection or an inversion with a rotation will have the same effect as plain inversion, and change the sign of the term. We call a system whose symmetry group comprises pure rotations only a *chiral* system, as opposed to any *achiral* system which (in contrast) possesses a rotation–reflection symmetry operation (perhaps the inversion). We call a system a *rotation–reflection* system if its symmetry group possesses at least one rotation–reflection operation, but not inversion. Some of these terms have been used in roughly these senses in the literature, but the above precise definitions are especially important in the following. Note in particular (for later discussion, §6.3.4) that each of a chiral, an achiral, a centrosymmetric, and a noncentrosymmetric atom or molecule may or may not show a chiral *effect* (see §6.3.4), depending on the type of interaction with light and external fields.

We conclude that natural optical activity is forbidden for centrosymmetric systems in general, and for achiral systems (in particular, for rotation–reflection systems) in fluids.

6.3.3 Examples

Spherical symmetry

For a spherical absorber, or under orientational averaging of a fluid of absorbers, the geometric factor has the form:

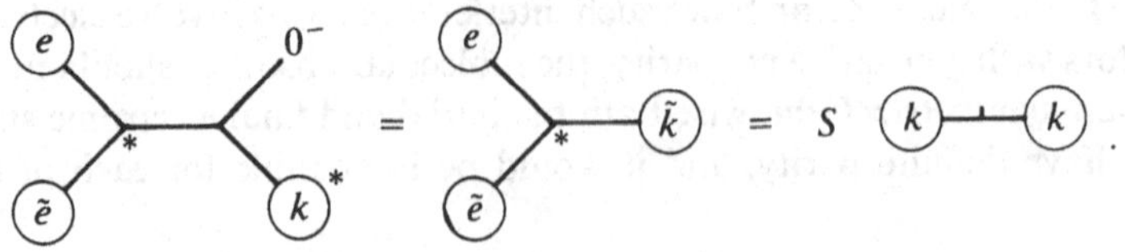

$$\left[e,\ \tilde{e},\ 0^-,\ k \right] = \left[e,\ \tilde{e},\ \tilde{k} \right] = S\ (k)\!-\!(k) .$$

This gives a trivial (i.e. an isotropic) geometric factor for E1–M1 of the form $S\mathbf{k}\cdot\mathbf{k}$. For E1–E2, the same coupling arises in principle; however, the associated 6j symbol depicted above (§6.3.2) cancels through triangle constraints on the triads. Alternatively one may see this through the application of JLV2 for SO(3) to the physical constant; the spherical tensor ranks of the operators contributing to each multipole must be identical for systems with rotational symmetry. Only E1–M1 mixing gives optical activity in such cases, and its strength is independent of geometry; one measurement suffices to determine it for all geometries.

Rotation–reflection groups

The final spin-parity labels $j^\pi=0^-$ in the geometric factors is disallowed as explained above, but the final labels $j^\pi=2^-$, for example, may give a nonvanishing contribution to the natural optical activity in rotation–reflection groups.

For example, let us consider a system of symmetry C_{2v}, such as a water molecule (or a set of such molecules adsorbed on a surface with identical orientation).

Method 1

The geometric factors may be defined by the coupling diagram

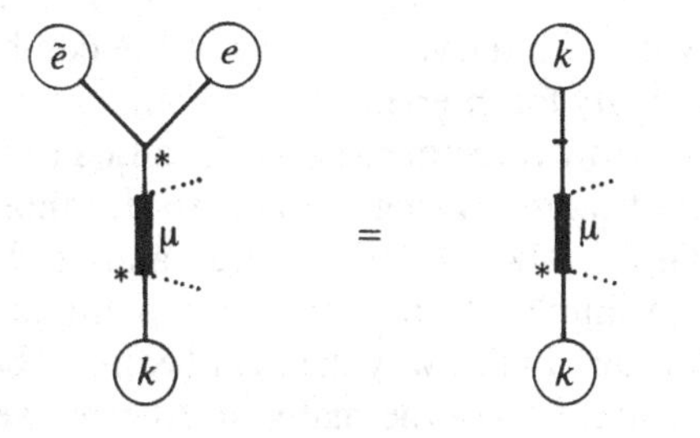

and as such may be found by coupling $j^\pi=1^+$ (from $\mathbf{e}\times\mathbf{e}^*$) and $j^\pi=1^-$ (from $\mathbf{k}$) to a component m of irrep μ of C_{2v}. The basis transformations required to evaluate this are the coefficients in the appropriately symmetrised kets given in Table A4 of the Appendix. In this case the basis is defined by the group branching $O(3)\supset O_h\supset C_{4v}\supset C_{2v}\supset C_s$, and is one in which the mirror reflection is σ_z and the rotation axis C_{2y}. The relevant kets (in a $C_{2v}\supset C_s$ basis (μm) on the left, and an $O(3)\supset SO(2)$ basis $J^\pi M$ on the right) are:

μ	m	$\{\lvert\mu m\rangle\}$
1	0	$i\lvert 1^+z\rangle,\ -i\lvert 1^-x\rangle$
$\tilde{1}$	1	$-i\lvert 1^-z\rangle,\ -i\lvert 1^+x\rangle$
0	0	$-\lvert 1^-y\rangle$
$\tilde{0}$	1	$\lvert 1^+y\rangle$

Since the irrep μ must be common to terms of each parity, one from each multipole, we find that the only possible geometric factors involve the basis transformation matrix elements

$$\langle \mu=1,m=0|j^\pi=1^-,m=z\rangle\langle \mu=1,m=0|j^\pi=1^+,m=x\rangle^*$$
$$\langle \mu=\tilde{1},m=1|j^\pi=1^+,m=z\rangle\langle \mu=\tilde{1},m=1|j^\pi=1^-,m=x\rangle^*$$

and hence induce a geometric factor of the form

$$F_1(\mathbf{e})\propto F_{\tilde{1}}(\mathbf{e})\propto k_X k_Z.$$

Method 2

In the original form, we couple 1^+ and 1^- within O(3) to J^- and reduce this to the identity of the point group:

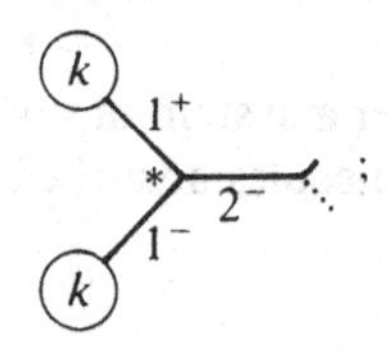

this time using the G-invariant ket derived from $J^\pi=2^-$. Again from Table A4 of the Appendix, this can only be $|2^-(1^+z,1^-x)\rangle$; this shows again that the k-components which are relevant appear in the combination $k_X k_Z$.

Physically, the necessity for interference between multipoles of odd and even rank in the photon wavevector may be visualised as the necessity for including the effects of finite wavelength and so distance-induced phase shift in the sample. This involves the mixing of an amplitude for absorption at one point, and an amplitude relevant to a point further along the direction of the light beam. In this way the light beam probes for a possible tendency to a helical structure of the material. Alternatively, we may say that any tendency towards helical structure favours one handedness of polarisation over the other. The electrons in the material tend to act as though they were confined to a helical waveguide, and set up currents in response to the applied optical field which have different strengths for the different handedness. In these ways the system acts differently for light beams of opposite handedness.

When the material is in the form of a fluid, it is impossible for such a helical tendency to be manifest unless each molecule lacks any mirror reflection symmetry, and even a symmetry operation which combines a rotation with a reflection. Hence the restriction discussed in §6.3.2 to pure rotation groups for optical activity in fluids. However, for an oriented absorber, even a water molecule can show optical activity if the light beam is sufficiently oblique. A water molecule has a reflection plane, along say the xy plane, and a two-fold axis, say the y axis; in this basis choice (which conforms with that used in our example) a light beam with a component of momentum in both the x and the z directions (so that the geometrical factor $k_z k_x$ is nonvanishing) sees the molecule as part of a spiral:

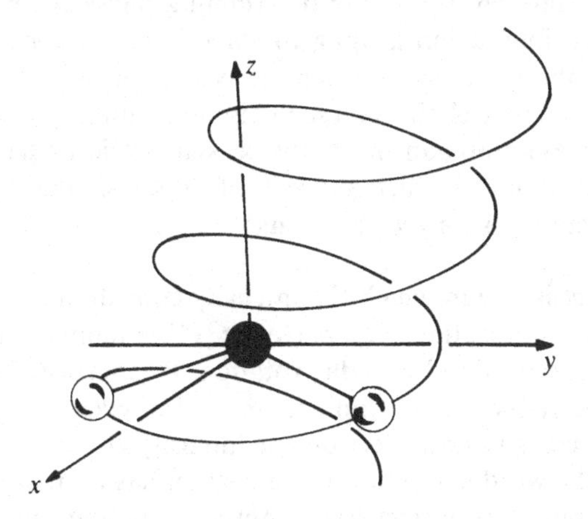

and in this configuration may have its plane of polarisation rotated.

6.3.4 Discrete symmetries and selection rules for chiral and achiral effects

The requirement, analysed in the last subsection, that the dynamical constant be imaginary is equivalent to its being odd under time reversal, and so, through the conjugation lemma, to its being an anticonjugate node. Just as an invariant network of 3jm symbols is related to its complex conjugate by the addition of a stub to each leg (§2.5), and the same addition to a ket diagram in a full $O(3) \supset G$ basis gives its time-reversal conjugate (§2.7.4), or an operator O may be similarly transformed under time reversal with the addition of a phase τ_O, a geometric factor is self-conjugate or -anticonjugate according as the product of the time-reversal phases of all the relevant physical operators is $+1$ or -1.

Since the interaction $\mathbf{A}\cdot\mathbf{p}$ fundamental to the above theory is linear in the time-odd electronic operator $\mathbf{p}$, there is a universal and so inconsequential sign factor of $(-1)^2$ from the time reversal of $\mathbf{p}$ in each amplitude. In addition, the imaginary factor appearing in the expansion of the exponential $\exp(i\mathbf{k}\cdot\mathbf{r})$ will give a minus sign on conjugation. The corresponding overall sign factor is $+1$ when the same multipoles participate in each amplitude, and -1 if the multipoles are different, and if the corresponding orders of perturbation (in $|\mathbf{k}|$) differ by an odd number (as required by the above argument for optical activity).

The time-odd character of the dynamical factor for contributions to optical activity may be justified in an even more pictorial manner, in which the time reversal or mirror reflection involved may apply only to one subsystem, radiation or matter (Stedman 1983a). At stake is the interplay between various discrete symmetry operations, notably parity (P) and particularly of time reversal (T) in determining physical constraints. If a notional optical interaction leading to some intensity contribution may have that contribution reversed when the system, or part of the system, is otherwise restored to itself through such discrete symmetry operators, then the interaction Hamiltonian must not be symmetric under the discrete symmetry. The more abstract analysis of §6.3.2 similarly depends on judicious transformations by parity and time reversal (complex conjugation) etc.

A *chiral effect* is one in which the intensity contribution is *reversed* by conjugating the polarisation vectors ($\{\mathbf{e}\} \rightleftarrows \{\tilde{\mathbf{e}}\}$). Examples of this include natural optical activity and Faraday rotation. By contrast, E1 absorption by a time-even system for example is an *achiral* effect; conjugating the polarisation vectors has no effect on the intensity contribution.

Our use of the word chiral in this connection has been argued (Barron 1986, 1987) to reveal, or at least risk, confusion. The issue may be seen in a simple example. Natural optical activity is distinguished from Faraday rotation since the former requires a chiral *system*, but both natural optical activity and Faraday rotation are chiral *effects* on our definition. We are quite deliberately using the term 'chiral' in this altered connection (i.e. in conjunction with the word 'effect') for an overall process, to emphasise the operational similarity of all kinds of optical activity. Such an emphasis is generally felt to be legitimate: what we call a chiral effect is called by others a *gyrotropic effect* or a *chiroptical process* (Silverman 1986, Craig and Thirunamachandran 1987).

There are indeed two reasons for introducing such a special term. First, the chirality of an effect as defined above, or its absence, is contingent on each of the system, the optical process, and the choice of interaction for each amplitude; all need to be specified before the chirality of the overall process can be specified. The alternative nomenclature advocated by Barron (1986) is inadequate at this point. Second, all specific contributions to the intensity at any order of perturbation are either (wholly) chiral or (wholly) achiral in the above sense, as is shown from the rules mentioned below.

If n is the joint order of the amplitudes in the multipole expansion of $\exp(i\mathbf{k}\cdot\mathbf{r})$, viz. the number of $\mathbf{k}$ vectors appearing in the geometric factor, we define the associated *perturbation order phase* $\chi \equiv (-1)^n$.

With these preliminary definitions, we can define two *general selection rules* that hold for all such, including higher-order and field-induced, multiphoton processes:

(a) *For a chiral* (alternatively achiral) *effect, the product* τ *of the time-reversal phases of the electronic operators and the perturbation order phase* χ *is negative* (positive, respectively):

$$\tau\chi = -1 \text{ (chiral)}$$
$$= +1 \text{ (achiral).}$$

Proof

The dynamical constant in any intensity contribution corresponds purely to sums over products of matrix elements of electronic operators derived from an even number of time-odd photon interactions, each in any multipole, and possibly some external field coupling. Its only external lines are trivial, being labelled by the identity irrep of some symmetry group. Hence, given that any time-odd character is associated with external fields, we may prove as in §6.3.2 from the conjugation lemma of §3.5 (see also §3.6) that the dynamical constant is real or imaginary as any field interaction is time-even or time-odd; all other phases will cancel:

We now consider the associated geometric factor. First we recouple it where necessary via the real 6j symbols of SO(3) in the conjugate pair coupling (in which conjugate pairs of polarisation vectors are coupled first, as in §6.3.1) to SO(3) spin labels $j, j', \ldots$ say. We then couple to any wavevectors arising from higher multipoles and to external fields, before final reduction onto the point group:

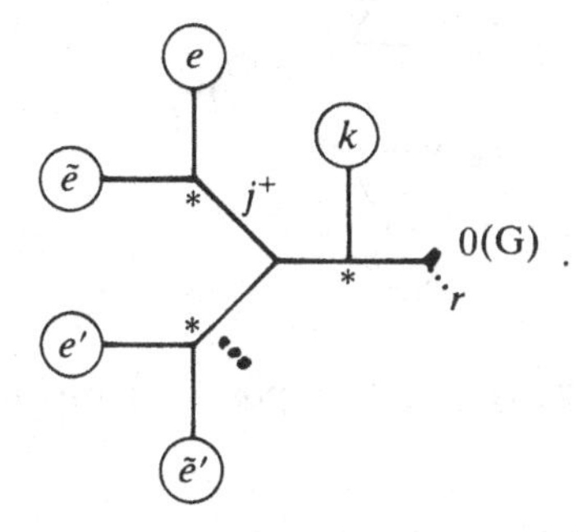

They will also carry a phase i^n arising from the imaginary constant in the expansion of $\exp(i\mathbf{k}\cdot\mathbf{r})$; as above, n is the total number (from both amplitudes) of wavevectors in the geometric factor.

Now let us complex conjugate the geometric factor. A phase $(-1)^j$ will arise from the interchange symmetry, or the 3j phase, appropriate when restoring the original form, from each coupled conjugate pair of polarisation vectors. The total phase $(-1)^{\Sigma j}$ from this source is by definition $+1$ for an achiral effect, and -1 for a chiral effect. In addition we will generate a phase $(-1)^n$ (from complex conjugation of the factor i^n) which by definition is equal to the perturbation order phase χ. Since the intensity contribution is an observable and so must be real to be finite, the conjugation phases of geometric factor and dynamical constant must be equal. This proves the rule.

We define a combined *field parity* Π as the product over interactions of the parity of the interaction of any external field $\mathbf{Q}$. Π is $+1$ for $\mathbf{Q}=\mathbf{B}$, -1 for $\mathbf{Q}=\mathbf{E}$, and $+1$ in the absence of any field. We now have:

(b) *Any centrosymmetic system will give a nonzero contribution to a given process only if*

$$\Pi\chi=1.$$

For fluids, this is also a restriction on the possibility of a contribution from any achiral system.

Proof

This may be seen as a straightforward consequence of combining the parities of all operators in the dynamical constant. Alternatively, we may look at the geometric factor. Every line in the diagram may be assigned a parity, and the terminal line, corresponding to the identity irrep label of the system point group, must have positive parity:

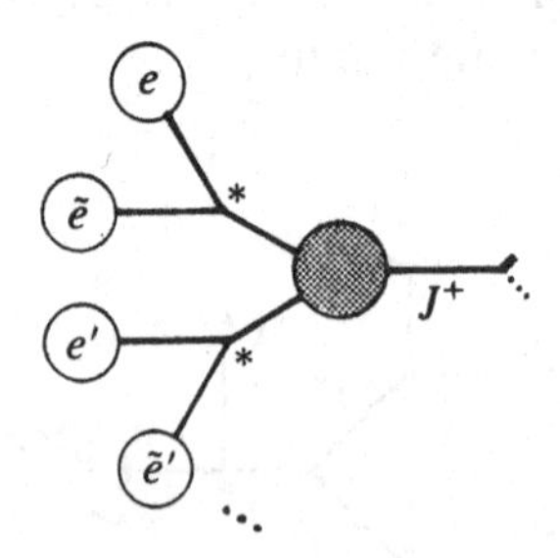

since the group identity irrep itself has positive parity in a centrosymmetric system.

6.4 Field-induced absorption

6.4.1 MCD (magnetic circular dichroism) or Faraday effect

We add a magnetic field $\mathbf{B}$, and look for that part of the response of the system which is linear in the field.

Within E1 coupling, we have the following selection rules from §6.3. Since the product $\tau\chi=(-1)(+1)=-1$, the resulting effect is chiral, and reverses on conjugating the optical polarisation vectors **e**. Hence **e** and **e*** are necessarily coupled to unit spin. Since $\Pi\chi=(+1)(+1)=1$, centrosymmetric (and, for fluids, achiral) systems are not prevented from participation.

The associated geometric factors have the form:

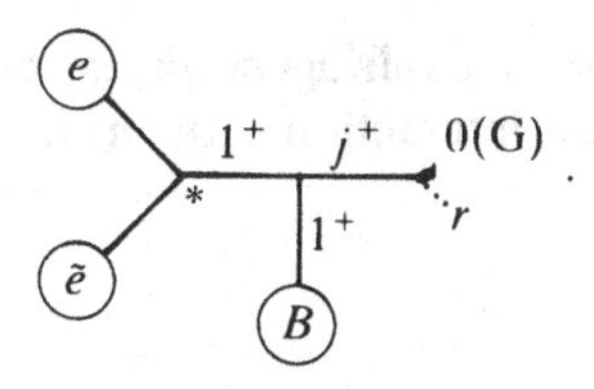

Since the choice $j^\pi=0^+$ always reduces to the invariant of any symmetry group, one geometric factor, **k·B**, exists for any system. This is the celebrated Faraday effect: a magnetic field induces optical activity and circular dichroism (hence the term MCD, Piepho and Schatz 1983) in any system, such as a fluid.

For systems of lower than spherical symmetry, the cases $j^\pi=1^+, 2^+$ may lead to further geometric factors to the extent that these spin–parity labels branch to the invariant of the group in question. As a result one easily deduces (Churcher and Stedman 1982, Stedman 1985) the result of Manson, Newman and Wong (1977) that MCD is isotropic for cubic systems (where 1^+, 2^+ cannot branch to 0(G) so that only the isotropic Faraday term is allowed), but axial in lower symmetries (as in Briat 1981, for example), etc.

6.4.2 Magnetic linear dichroism (MLD)

Linear effects in a magnetic field under E1–M1/E2 mixing will be associated with the geometric factors:

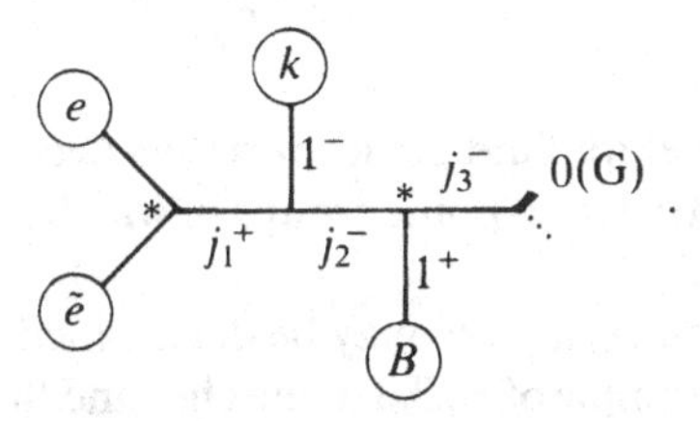

Both M1 and E2 multipoles are covered by this geometric factor. These terms necessarily pertain to an achiral process ($\tau\chi=1$), so that the coupling rank j_1 of the polarisation vectors must be even. Since $\Pi\chi=-1$, there is now a restriction to noncentrosymmetric systems.

The variety of possible geometrical dependences is now very extensive, and their experimental exploration is still in its infancy. The topic as a whole has received only partial theoretical coverage from a wide variety of viewpoints, each somewhat restricted, witness the plethora of names: transverse Zeeman effect, MLD, the Cotton–Mouton effect, or magneto-chiral birefringence and dichroism, in the literature for what is essentially the same phenomenon.

6.4.3 Electric birefringence/linear dichroism

Within E1 coupling, linear interaction with an electric field would give the geometric factors

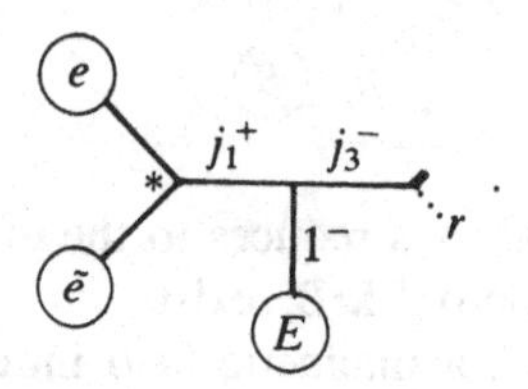

We must couple the polarisation vectors to even spin j_1 since $\tau\chi = 1$, and the process is necessarily achiral. Since $\Pi = -1$, we are restricted to a chiral system. The birefringent or dichroic tendency of many molecules and crystals is of course well known.

6.4.4 Electric circular dichroism/birefringence

When E1–M1/E2 mixing is considered, we have the corresponding chiral process, in principle for all systems, with the geometric factors:

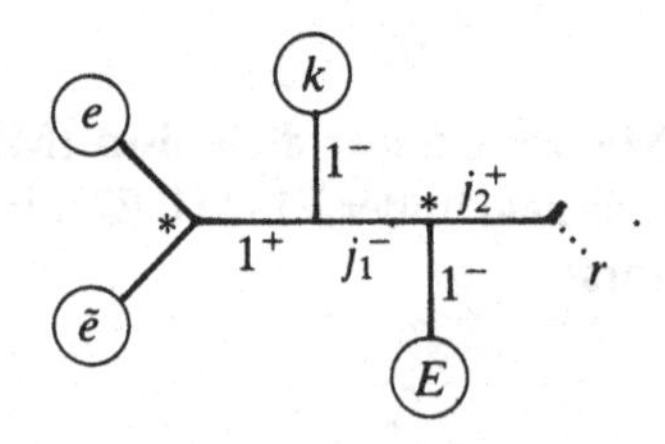

However, this vanishes for fluids, since j_1 is then necessarily unity and since $\mathbf{k} \times \mathbf{k} = \mathbf{0}$. We are restricted to point groups in which j^+ for $j = 1, 2$ branch to the group invariant.

The detailed geometric factors may be determined from Table A4 of the Appendix as in the example of §6.3.3 (Churcher and Stedman 1983). Results for the above optical processes are given in Stedman (1985).

6.4.5 Joint electric and magnetic field interaction

As a final example, we consider effects bilinear in a magnetic and an electric field. Since this combination corresponds to $\tau = \Pi = -1$, we conclude that

$\chi=1$ (e.g. E1 coupling) will induce a chiral effect in noncentrosymmetric systems, while $\chi=-1$ (e.g. E1–M1/E2 interference terms) will induce achiral effects in centrosymmetric systems.

In E1 coupling, this leads to an electromagnetic optical activity of the form

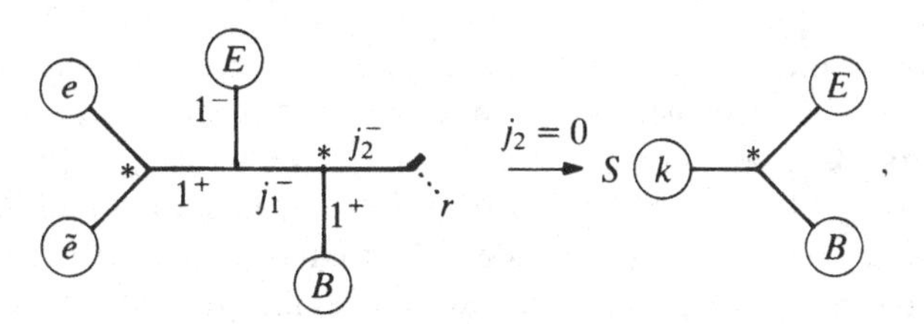

which for fluids will involve a geometric factor of the form **E** × **B**·**k** in chiral systems.

In E1–M1 mixing, we obtain a fields-dependent birefringence of the form

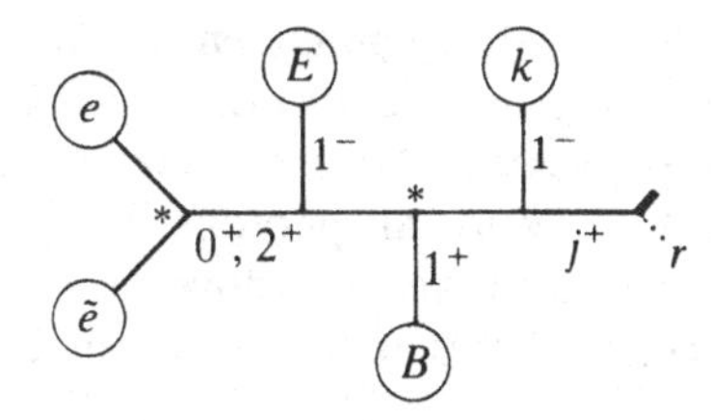

involving terms such as (**E**·**e**)(**B** × **k**·**e***) in the geometric factor.

Both these effects imply a geometry in which the fields are perpendicular to each other and to the direction of propagation of light.

Such results may be extended to higher orders in all interactions in an obvious manner (Ross, Sherborne and Stedman 1989).

6.5 Reciprocity principles

6.5.1 Reciprocity and conjugation theorems

The procedures listed above are readily extended to higher order processes; we shall not work through the details. For example, one may consider selection rules in hyper-Raman scattering from degenerate levels (Churcher 1982). A few such applications, including a fuller analysis of the material presented earlier in this chapter, are given in Stedman (1985), though many still await analysis (Bloembergen 1982).

In particular, the case of Raman scattering shows several new and interesting features. We may ask fundamental questions about the necessity for using scattering geometries other than 90°, and polarisations other than linear, for a full analysis of any system using Raman scattering (Borcherds and Alfrey 1975, Stedman 1985).

We shall concentrate on just one topic for this section. The involvement of two photons, one absorbed and one emitted, suggests some novel applications of our earlier formalism. For example, the possibility of the time-reversed process being identical to the original process suggests the study of reciprocity theorems via diagrammatic group theory.

Consider an optical process in which any number M of photons, with polarisation vectors $\{\mathbf{e}_p|p=1,2,\ldots,M\}$ and wavevectors $\{\mathbf{k}_p\}$ are absorbed, and N photons with polarisation vectors $\{\mathbf{e'}_q|q=1,2,\ldots,N\}$ and wavevectors $\{\mathbf{k'}_q\}$ are created. For brevity, we write $\gamma_p=\{\mathbf{e}_p,\mathbf{k}_p\}$ and similarly for γ_q'. The electronic system is initially in state i, and finally in state f. (We will later want to include a thermal average over i, and a sum over f compatible with energy conservation.) Let this process have a Golden Rule probability $I_{if}(\{\gamma_p,\gamma_q'\})$. The *reciprocity theorem* states:

$$I_{if}(\{\gamma_p,\gamma_q'\})=I_{fi}(\{\gamma_q',\gamma_p\}).$$

In brief, the intensity is unchanged if initial and final electronic radiation states are interchanged.

The reciprocity theorem is thus reminiscent of the principle of detailed balance. However, there is *no* requirement that any states, for example the optical states, be time-reversed. Time reversal would also entail the transformations $\mathbf{e}\rightarrow\tilde{\mathbf{e}}$, $\mathbf{k}\rightarrow-\mathbf{k}$ for all photons; we call these collectively $\gamma\rightarrow\tilde{\gamma}$. Such a transformation in fact introduces a nontrivial phase τ, the product of the time-reversal phases of the various multipole operators, into this relation:

$$I_{if}(\{\gamma_p,\gamma_q'\}) =\tau I_{\tilde{i}\tilde{f}}(\{\tilde{\gamma}_p',\tilde{\gamma}_q'\}).$$

This is an immediate consequence of the conjugation lemma, and we refer to it as the *conjugation theorem* to emphasise the distinction from the reciprocity theorem. In the literature on Raman scattering for example the reciprocity theorem and the conjugation theorem are almost universally confused; only the latter has scope for introducing phase changes.

Combining both theorems, we have the result:

$$I_{if}(\{\gamma_p,\gamma_q'\})=\tau I_{\tilde{f}\tilde{i}}(\{\tilde{\gamma}_q',\tilde{\gamma}_p\}),$$

which for convenience we call the *reversal theorem*: both time reversal and in$\leftrightarrow$out interchange are applied to all states.

To prove the reciprocity theorem, and hence the reversal theorem, we may apply hermiticity ($\langle a|\mathrm{O}|b\rangle=\langle b|\mathrm{O}|a\rangle^*$) to all matrix elements internal to the Golden Rule expression for the intensity contribution, and note that the intensity contribution must be real.

We have for simplicity glossed at one point. Strictly, the amplitude (on one multipole P, say) which is combined with a complex conjugate amplitude (for a multipole Q, say) in forming any interference term, is conjugated by this process; hence any such interference term of the form

PQ* is related by this theorem to the conjugate term QP* (rather than PQ*) for the interchange configuration. Indeed the above development may be expanded by considering the individual amplitudes.

Take a particular amplitude $A_{if}(\{\gamma_p,\gamma'_q\})$, which corresponds to a chain of electronic matrix elements in which state i and f are coupled via the annihilation or creation of photons with parameters $\boldsymbol{\gamma}$ or $\boldsymbol{\gamma}'$, respectively. The combination of time reversal and hermiticity, $\langle a|\mathrm{O}|b\rangle = \langle \bar{b}|\bar{\mathrm{O}}|\bar{a}\rangle$, applied to all matrix elements in the chain, leads to the relation

$$A_{if}(\{\gamma_p,\gamma'_q\}) = \tau A_{\bar{f}\bar{i}}(\{\tilde{\gamma}'_q,\gamma'_p\}) \leftrightarrow$$

τ being the product of the time-reversal signatures of all coupling operators in the amplitude. When this result is applied to both amplitudes, we obtain the reversal theorem. (In the diagram above, we have ignored for simplicity the possible k-dependence of the optical vectors, as if for E1 coupling only.)

6.5.2 Stronger selection rules for Rayleigh-like scattering

In the case of field-independent Rayleigh scattering, without restriction to any particular multipole coupling, an interchange of initial and final electronic state, $i \leftrightarrow f$, and a time reversal, $i \rightarrow \bar{i}$, are each inconsequential; each state is summed over the states in a degenerate and time-even level. Hence the reversal theorem of §6.5.1 relates contributions between processes which differ only by the time ordering of absorption and emission:

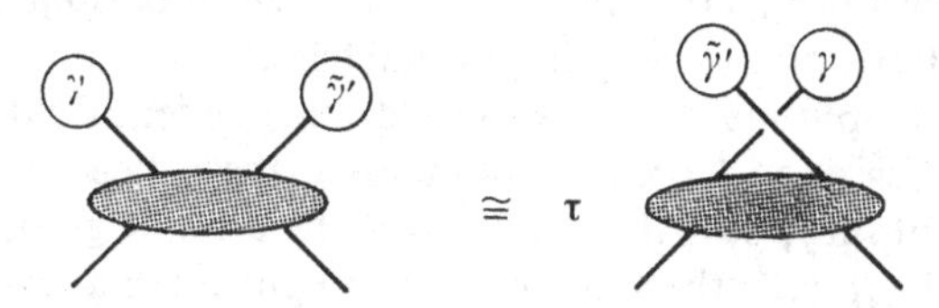

and $\tau = +1$ when fields are absent. If in forming the geometrical factors, we first couple the two polarisation vector sets $\boldsymbol{\gamma}, \boldsymbol{\gamma}'$ say to angular momentum j_1, j_2 in the two amplitudes (the suffices distinguishing the amplitudes), then the interchange 3j phase of their coupling will require the selection rule that j_1 and j_2 be *both even*, or *both odd*. In short, each pair of amplitudes giving a nonzero contribution to the intensity must be either symmetric or antisymmetric under the exchange $\{\gamma\}\leftrightarrow\{\tilde{\gamma}'\}$. Mixed symmetries never contribute. This result, and others to follow, may be generalised to nonresonant phonon Raman scattering, within the (very reasonable) approximation that phonon energies make negligible contributions to the perturbation energy denominators, themselves dominated by electronic energy differences.

(Note that this selection rule is distinct from those discussed in §6.3.4. Since field-independent Rayleigh scattering is time-even, those earlier rules gave for example that $(-1)^{\Sigma j}=\chi$ where the various j, labelled j and j', were the coupling angular momenta of each polarisation vector (**e**, **e**$'$, respectively) and its complex conjugate. For example, in E1 coupling Rayleigh scattering is achiral, so that j and j' are both even or both odd. However, for a chiral process in which one photon couples in one amplitude by M1 coupling, $j+j'$ is odd, but j_1+j_2 is still even.)

For fluids, for example, this rule is inevitably obeyed from rotational symmetry: j_1 and j_2 are equal because of JLV2:

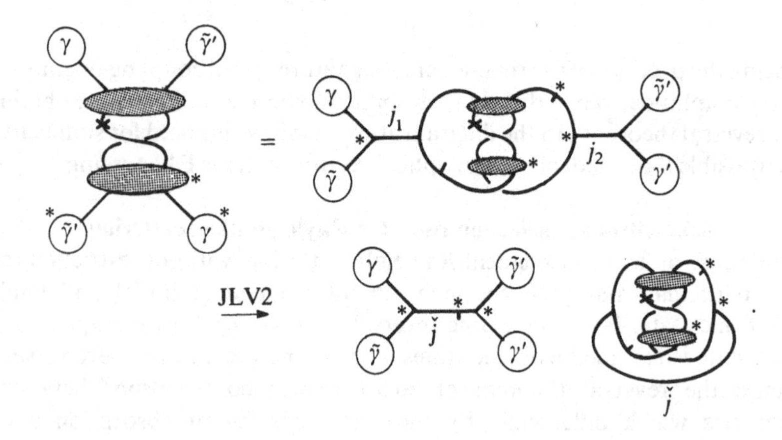

However, in lower symmetries this selection rule is important since it gives decisive limitations on the coupling symmetries.

Consider an amplitude $A_{if}(\boldsymbol{\gamma},\boldsymbol{\gamma}')$ for field-independent Rayleigh scattering. We couple (using a Clebsch–Gordan series) the optical operators to an angular momentum j_1 say, then reduce this to a point group irrep μ by unitary: $j(\mathrm{SO(3)})\downarrow\mu_r(\mathrm{G})$; the suffix r will distinguish the appropriate term should μ occur more than once in the reduction of j_1. The chain of matrix elements in the amplitude couple two levels, i and f, which are both in the time-even level λ. In §3.6.4 we derived in similar circumstances a selection rule of the form $\underline{\mu_r} \in [\lambda\times\lambda]_{\tau\tau_O}$ for nonvanishing matrix elements $M_{ll'}\equiv\langle\overline{\lambda l}|\mathrm{O}^{\mu}{}_{m}|\lambda l'\rangle$ of a *hermitian* operator O^{μ} coupling states within a level of symmetry λ; τ is the phase of the state $|\lambda l\rangle$ under double time conjugation, τ_O is the time reversal symmetry of the operator – in this case

$+1$ – and r labels the particular occurrence of μ in $\lambda \times \lambda$. This is based on the relation (obtained by combining hermitian and time-reversal conjugation) $M_{ll'} = \tau\tau_O M_{l'l}$ and needs some adaptation. There are two hermitian operators involved, for photon annihilation and creation; any term in perturbation theory will have these in some definite order. The product of two optical interaction operators is not hermitian in general. Reworking the argument for the present application shows that in addition to the phases $\tau\tau_O$, we should include the interchange symmetry of the coupling of these operators at the SO(3) level, i.e. $\mu_r \in [\lambda \times \lambda]_\theta$, and so $\{\lambda\lambda\mu r\} = \theta$, where $\theta \equiv \tau\tau_O(-1)^{j_1}$; similarly for the other amplitude.

Since, from our earlier rule, $j_1 + j_2$ must be even, we conclude that the 3j phases of the optical couplings within the electronic level must be the same in each amplitude.

On combining two amplitudes to form an intensity contribution, JLV2 applied within the point group also requires that μ be the common symmetry of the two optical operators (§6.2.1). The phase rule that we have just proved shows further that the particular repetitions of μ in $\lambda \times \lambda$ that combine are those which are either symmetric or antisymmetric (in both amplitudes), in conformity with the reciprocity theorem itself. Again this may be extended to phonon Raman scattering with the alteration that now $\mu_r \in [\lambda \times \lambda]_\theta \times \kappa$, where κ is the point group symmetry of the phonon. A consistent use of these rules explained several puzzling features of the phonon Raman spectra of inorganic crystals (Churcher and Stedman 1981a,b).

7

Quantum field theoretic diagram techniques for atomic systems

Armed with the theorist's standard panoply of Green's functions, I set off intrepidly into the complex frequency plane.

Leggett (1982)

7.1 Introduction

7.1.1 Summary of strategy

In this chapter we derive the links between the diagram techniques already discussed in this work and those familiar from field theory. In the latter we have in mind Feynman diagrams in relativistic gauge field theories, the closely related nonrelativistic diagram techniques of quantum statistical physics in many-body systems, and all associated techniques, including the adaptations to atomic and molecular physics, including situations where Maxwell–Boltzmann rather than boson or fermion statistics are relevant. We concentrate on the last of these because of its comparative novelty as well as its particular relevance to the applications of group theory emphasised in this book.

The important conclusion, developed in the next chapter, is that at least for nonrelativistic applications the field theoretic diagram may be thought of as a group theoretic diagram; indeed the two diagram techniques may be combined to make a coherent calculational tool superior to either tool on its own. Since the field theoretic interaction vertices correspond to matrix elements, any symmetries of the underlying Hamiltonians will be reflected in the possibility of an irreducible group theoretic classification of states and operators, and thus to a connection between the field theoretic vertex and our group theoretic diagram. The connection presumes the validity of the Wigner–Eckart theorem, at least within a compact subgroup, and the proportionality constant between the field theoretic and group theoretic diagram is just the appropriate reduced matrix element. The corresponding relation for the relativistic field theoretic diagram is not so straightforward, but can still be useful.

As indicated above, we will specialise our proof of diagram techniques in field theory. Many excellent dedicated works expound these thoroughly, especially in the (highly technical) relativistic gauge theory regime. Since it is the nonrelativistic applications of diagrammatic group theory which have so far been most usefully integrated with field theory, we concentrate on these. In order to exhibit the links, for example between the material of §6 and a fully fledged many-body approach to photon absorption and to Raman scattering in solids, we review the full development of the latter to the point of validating the Golden Rule expression for the interaction which formed the starting point of the analysis of §6.

The application of field theoretic techniques of the sort pioneered by Feynman in relativistic gauge theory and by Matsubara and others in many-body theory (see Abrikosov, Gor'kov and Dzyaloshinskii 1963, Schultz 1964, Taylor 1970, Fetter and Walecka 1971, Doniach and Sondheimer 1974, for example) and so in atomic, ionic and molecular physics involves in particular an adaptation of the standard quantum field theoretic formalism to describe states with Maxwell–Boltzmann rather than fermionic statistics. Of many possible techniques, we choose an adaptation due to Abrikosov (1965), and used by him and many others in various areas of many-body physics such as the Kondo problem for the description of the levels of an impurity with spin in a metal (e.g. Barnes 1985, 1986). It is precisely this many-body technique, and not its rival methods (such as the use of drone fermions, etc.) which permits a direct link with our diagram approach to group theory. Every fermionic propagator is labelled by a physical state of the atom, ion or molecule, and so by any appropriate symmetry label for that state. The effect of the many-body projection necessary to reduce from fermionic to Maxwell–Boltzmann statistics, at least as far as the numerator of the field theoretic expansion is concerned, is essentially to remove certain nonphysical diagrams, i.e. those involving multiple or zero occupancies of the levels of any ion. Many applications of diagram techniques in condensed matter ignore this complication, and as a consequence are rather less than systematic.

It seems doubly appropriate to offer a moderately extended account of this theory in this book, then, despite the lack of an appearance of group theory in at least the early part of the many-body theoretic development, since no other self-contained introduction or summary of this theory is available for the atomic, molecular and condensed matter physics applications on which this work concentrates. Even in the primary literature it is difficult to find a detailed explanation of the connection between the discontinuity of the relevant Green's function and the spectral density for absorption or light scattering experiments. We hope this account will complement existing field theory texts.

7.1.2 Time-independent perturbation theory: an example

We concentrate here on the simplest of results, so as to indicate the link with earlier material, particularly that of §§3 and 6. Some representative if incomplete references may assist the reader with a special interest here. A fuller analysis of perturbation theory and associated (Goldstone) diagram methods is given in Lindgren and Morrison (1982). The relation between time-independent and time-dependent perturbation is explored by Langhoff, Epstein and Karplus (1984). The effects of degeneracy in perturbation theory (which we shall largely ignore) are discussed in detail not only by Lindgren and Morrison, but also, for example, Soliverez (1967), Silverstone and Holloway (1971), Suzuki and Okamoto (1983).

Brillouin–Wigner perturbation theory

Often, as in some preceding material of §6, we require the calculation of the changes induced by a time-independent interaction $\mathscr{V}$ in the energies E_i^0 and wavefunctions Ψ_i^0 associated with an unperturbed Hamiltonian $\mathscr{H}$. It is not necessary to invoke the full machinery of following sections. The standard approach of time-independent perturbation theory (e.g. Merzbacher 1961) leads to the Brillouin–Wigner expansion (in which the perturbed energy enters the denominators):

$$E_i = E_i^0 + \Sigma_n E_i^{(n)},\ \Psi_i = \Psi_i^0 + \Sigma_n \Psi_i^{(n)}$$

$$E_i^{(n)} = \langle i| \left\{ \mathscr{V} \left(\frac{Q_i}{E_i - \mathscr{H}} \right) \right\}^{n-1} \mathscr{V} |i\rangle,$$

$$\Psi_i^{(n)} = \left\{ \left(\frac{Q_i}{E_i - \mathscr{H}} \right) \mathscr{V} \right\}^{n-1} |i\rangle,$$

where Q_i projects intermediate states off the initial state i. On iteration, we generate the various energy shifts:

$$E_i^{(1)} = \langle \Psi_i^0 | \mathscr{V} | \Psi_i^0 \rangle,$$

$$E_i^{(2)} = \Sigma_{j \neq i} \langle \Psi_i^0 | \mathscr{V} | \Psi_j^0 \rangle \langle \Psi_j^0 | \mathscr{V} | \Psi_i^0 \rangle / [E_i - E_j^0]$$

$$E_i^{(3)} = \Sigma_{j,k \neq i} \langle \Psi_i^0 | \mathscr{V} | \Psi_j^0 \rangle \langle \Psi_j^0 | \mathscr{V} | \Psi_k^0 \rangle \langle \Psi_k^0 | \mathscr{V} | \Psi_i^0 \rangle / [E_i - E_j^0][E_i - E_k^0]$$

and wavefunction components

$$\Psi_i^{(1)} = \Sigma_{j \neq i} |\Psi_j^0\rangle \langle \Psi_j^0 | \mathscr{V} | \Psi_i^0 \rangle / [E_i - E_j^0], \text{ etc.}$$

The matrix elements in these diagrams as well as the sum over intermediate state labels are naturally depicted using the rules of earlier sections. The energy denominator associated with any intermediate state may be denoted by a cross on the appropriate summation line. For a general interaction, we may still choose to include a 2jm symbol explicitly on the diagram for every bra label, so that the interaction node is of definite (negative) parity. The otherwise unspecified operator character of $\mathscr{V}$ is denoted by terminating the operator label line also by a cross. This enables us to write:

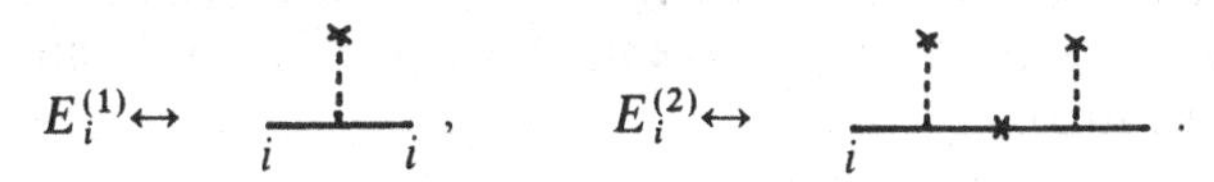

Some comments on the Brillouin–Wigner expansion may be made. First, the perturbed value for the energy appears in the denominator factors. When the perturbative expansion of the shifted energy E_i is systematically inserted in the denominators of that expansion, we obtain an expansion in which only the unperturbed energy appears – the Rayleigh–Schrödinger perturbation expansion. It has a correspondingly complicated structure (Brueckner 1955), but has some compensating advantages. The Green's function method as used later in this chapter leads naturally to the Brillouin–Wigner form.

Second, the link with time-dependent perturbation theory is indicated by rewriting the 2jm symbol as an arrow (Comment (f) on Rule 6, p. 5). (It would make for elegance in this context to define the arrow as the combination of stub and energy denominator.) If now, in conformity with the interaction picture of quantum mechanics, we wish to think of the time-independent energy shift as the result of a dynamic interaction process in which state i is converted temporarily into state j (and state k, etc.) and then back again, the 2jm arrow may be given the extra significance of denoting the sense of the flow of time in this dynamic interaction process. Since every interaction matrix element involves the bra vector for some state, and since each intermediate state $j,k, \ldots$ appears in one bra and one ket and is associated with one energy denominator, just one arrow appears on each line (corresponding to an intermediate state label), and the diagram is essentially a (nonrelativistic counterpart of a) Feynman diagram. This kind of interpretation will be more fully justified in the time-dependent theory given in later sections.

Application of diagram techniques in group theory

Since this notation for perturbative energy shifts is completely compatible with all our rules and definitions, we may manipulate these diagrams, using JLVn etc., to determine the effects of symmetry restrictions.

As a simple example, the invariance of $\mathscr{H} + \mathscr{V}$ under a symmetry group G enables us to label levels by irreps λ of G, and thence to show that the energy shifts of states i $(i \equiv \lambda l)$ within any level are identical (independent of l):

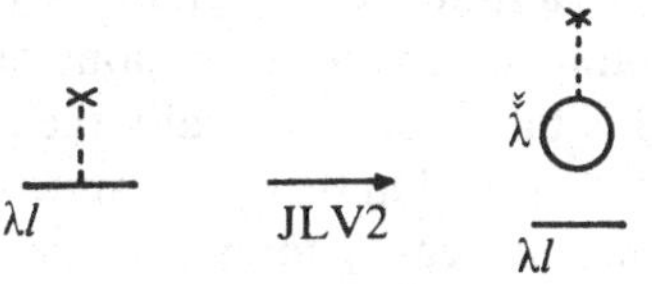

λl $\xrightarrow{\text{JLV2}}$ $\check{\lambda}$ λl .

If, however, $\mathscr{V}$ breaks the symmetry G of $\mathscr{H}$, we may expand it in irreducible tensor operators $\{O^{\mu}_{m}\}$ of G to gain other results.

For example, the first-order energy shift of level i $(i \equiv \lambda l)$ is associated with purely diagonal matrix elements. Their average will cancel unless $\mu = 0$, thanks to JLV1:

$$\lambda \bigcirc^{\mu m} \xrightarrow[\text{JLV1}]{} |\lambda|\delta_{\mu 0}.$$

The term $\mu = 0$ preserves the symmetry, and is independent of component l. Hence the level-splitting terms in $\mathscr{V}$ do not change the centre of gravity of the states in $\{l\}$ in first order.

Application in atomic and solid state physics

We shall later obtain some more interesting diagrams when we note explicitly any coupling that $\mathscr{V}$ may afford between subsystems, e.g. Coulomb coupling between electrons in an atom (§8.2), or electron–lattice coupling in a solid (§7.3). The interaction, if internal to the combined system, will now preserve the full symmetry G. However, it will change the symmetries of the (previously uncoupled) states of the subsystems, and in a manner which is readily visualised using diagram techniques. The heavy theory in the remainder of this chapter is not always required (and when it is required, is needed only for background justification), and the reader may wish to jump to the relevant subsections for applications of diagram techniques in group theory. It will be readily seen that many of the applications of §8 (e.g. §§8.2.2–8.2.4, 8.3.2, 8.4.1, 8.5.2) may be worked through as exercises in time-independent perturbation theory along the lines first illustrated above. Likewise, the material of §6 was presented in anticipation of such a discussion of perturbation theoretic expressions for the effects of physical interactions. In the context of time-independent perturbation theory for atomic or nuclear states, the two-body interactions such as the Coulomb interaction are under discussion. The traditional (Goldstone) diagram technique is very closely linked with the Feynman diagram, differing from it mainly by virtue of a specific notation for core and virtual states, particles and holes, etc. As such, Goldstone diagrams are broadly amenable to the results of §§1–3, including the JLVn pinching theorems. We refer to Lindgren and Morrison (1982), as well as the texts and reviews cited in §8, for a full introduction to Goldstone diagrams. Similarly, the so-called 'quantum electrodynamics' diagrams of quantum optics (Shen 1984) amount to another such nonrelativistic cousin of the Feynman diagram, and is based on essentially the above considerations.

In all such methods, traditional in their own sphere of application, truly systematic derivations are not easily found. While one may readily and quite justifiably (provided one adheres to the rules of §1.2, say) transliterate the Brillouin–Wigner or, for that matter, Rayleigh–Schrödinger perturbation expansions into diagrams, it is strictly impossible to base these diagram methods on quantum field theory, with its dependence on commutation or anticommutation relations, unless the Abriksov projection is included – and this is never done! It is to redress this situation that we now apparently make a detour from group theory into a full development of quantum field theory.

7.2 Nonrelativistic diagrammatic perturbation theory for fermions

7.2.1 Introduction

We review an elementary approach to the Golden Rule of time-dependent perturbation theory in statistical physics using Green's function techniques. As a tutorial exercise, it will illustrate all the machinery of the perturbative approach to quantum field theory at finite temperature in a nonrelativistic context, including the physical significance of the Green's function and the perturbative expansion of the evolution operator.

We follow the Matsubara approach to quantum statistical physics, and use operator products in the definition of the Green's functions which are ordered in imaginary time. For tutorial clarity we shall avoid the very elegant, but very condensed, notation of functional differentiation. We also respectfully ignore an interesting modern rival to the complicated formal manipulations of the Matsubara technique: thermo field dynamics, as developed in cosmology for interpreting relativistic gauge theories such as QED at finite temperature, gives an alternative method. A calculation using thermo field dynamics in a form easily contrasted with the techniques used here is exhibited by, for example, Fujimoto, Morikawa and Sasuki (1986). Since the method of thermo field dynamics has its own obscurities (Ojima 1981), we consider it premature to abandon the Matsubara method.

7.2.2 Definition of the Green's function

The system under consideration throughout this chapter is an atom, ion or molecule – say an ion – and is perhaps a constituent of a host molecule or crystal. We describe its energy levels using a second quantised Hamiltonian of the form

$$\mathscr{H} = \Sigma_i E_i^0 N_i, \tag{7.1}$$

where the number operator $N_i \equiv a_i^\dagger a_i$, $a_i^\dagger$ creates the eigenstate $|i\rangle_0$ from the vacuum $|\rangle$, and a_i (also written a_i^- when convenient) is the corresponding annihilation (hermitian conjugate) operator. We take these operators to obey fermion commutation relations: $\{a_i^\dagger, a_j\} = \delta_{ij}$, $\{a_i, a_j\} = 0$, where the braces denote an anticommutator. We have therefore the formal possibility of obtaining all the states $\{|\alpha\rangle_0\}$ of the fermion Fock space, including such unphysical states as $a_i^\dagger a_j^\dagger|\rangle_0$ (two- or three-particle occupancy), or $|\rangle_0$ (the vacuum, or zero occupancy of an ionic level) as well as the physical states $\{|i\rangle_0\}$. These indeed will all inevitably appear as possible intermediate states in the many-body formalism we develop. The unphysical side effects will have to be eliminated by a suitable projection later in the calculation. This projection will have the effect of changing the statistics from Fermi–Dirac to Maxwell–Boltzmann. For the meantime, it is essential to retain fermion anticommutation relations to develop Wick's theorem and the quantum field theoretic diagram expansion.

We define the density matrix in the general case of an unperturbed Hamiltonian $\mathscr{H}$ with an interaction $\mathscr{V}$ by

$$\rho=\exp[-\beta(\mathscr{H}+\mathscr{V})]/\mathrm{Tr}\{\exp(-\beta(\mathscr{H}+\mathscr{V})]\}, \tag{7.2}$$

so that a thermal average in the presence of perturbations is denoted by $\langle\ldots\rangle\equiv\mathrm{Tr}(\rho\ldots)$. In the absence of interaction we add a suffix 0 to the density matrix and the thermal average chevrons, as well as to eigenstates and eigenvalues of $\mathscr{H}$.

For example, in the one-particle subspace $\{|i\rangle_0\}\subset\{|\alpha\rangle_0\}$, the unperturbed expectation value of the number operator is then simply

$$\begin{aligned}\langle N_j\rangle_0^{(1)}&=[\Sigma_i\langle i|\exp(-\beta\Sigma_i E_i^0 N_i)N_j|i\rangle_0]/[\langle i|\exp(-\beta\Sigma_i E_i^0 N_i)|i\rangle_0]\\&=\exp(-\beta E_j^0)/[\Sigma_i\exp(-\beta E_i^0)],\end{aligned} \tag{7.3}$$

the Maxwell–Boltzmann result.

The Fermi–Dirac distribution may be derived in the same manner, or less laboriously using anticommutation. Since (now in a more general, multiparticle, basis) $\langle N_i\rangle_0=\langle 1-a_i a_i^\dagger\rangle_0$, we may use the cyclic properties of the trace in the second term on the right to rewrite it as

$$\begin{aligned}\mathrm{Tr}[\exp(-\beta\mathscr{H})a_i a_i^\dagger]&=\mathrm{Tr}[\exp(-\beta\mathscr{H})a_i\exp(\beta\mathscr{H})\exp(-\beta\mathscr{H})a_i^\dagger]\\&=\mathrm{Tr}[\exp(-\beta\mathscr{H})a_i^\dagger\exp(-\beta\mathscr{H})a_i\exp(\beta\mathscr{H})]\\&=\exp(-\beta E_i)\mathrm{Tr}[\exp|(-\beta\mathscr{H})N_i].\end{aligned} \tag{7.4}$$

The last line follows since $\exp(-\beta\mathscr{H})a_i\exp(\beta\mathscr{H})=\exp(\beta E_i^0)a_i$; a_i will either annihilate every eigenstate of $\mathscr{H}$, or reduce its eigenvalue by E_i^0. This leads directly to the Fermi–Dirac distribution $\rho_i^0\equiv\langle N_i\rangle_0=1/[\exp(\beta E_i^0)+1]$.

We consider now the effects of an interaction $\mathscr{V}$ perturbing the system. The mixing of the eigenstates of $\mathscr{H}$ may be denoted by the second quantised form:

$$\mathscr{V}=\Sigma_{ij}V_{ij}a_i^\dagger a_j. \tag{7.5}$$

The coefficients are the matrix elements: $V_{ij}=\langle i|\mathscr{V}|j\rangle_0$.

In the Schrödinger picture of quantum mechanics, operators $O\equiv O_S$ are time-independent while wavefunctions have a time dependence:

$$\Psi_S(\tau)=U_S(\tau,\tau')\Psi_S(\tau'), \tag{7.6}$$

where for convenience we use an imaginary time label $\tau \equiv it/\hbar$, and where the evolution operator

$$U_S(\tau,\tau') \equiv \exp[-(\tau-\tau')(\mathscr{H}+\mathscr{V})] \tag{7.7}$$

and describes the time development of the system wavefunction.

In the Heisenberg picture, denoted by tildes, this time dependence is translated to the operators: $\tilde{\Psi}(\tau)=\Psi_S(0)$, while $\tilde{O}(\tau) \equiv U_S(0,\tau)O_S U_S(\tau,0)$.

Let us formally define the thermal Green's function by

$$G_i(\tau) \equiv -\langle T_\tau\{\tilde{a}_i(\tau)\tilde{a}_i^\dagger(0)\}\rangle. \tag{7.8}$$

In this equation, a thermal average over the full Hamiltonian is indicated by the lack of a subscript on the chevron, and the chronological operator is defined by

$$\begin{aligned} T_\tau\{A(\tau)B(\tau')\} &= A(\tau)B(\tau'), & \tau \geq \tau'; \\ &= -B(\tau')A(\tau), & \tau < \tau' \end{aligned} \tag{7.9}$$

(the minus sign only when ordering fermion operators). This definition assumes that τ is real, and so that the time coordinate is imaginary. The Green's function for physical - real - time is defined by analytic continuation in the complex t plane.

With these definitions, $G_i(\tau)$ possesses a formal periodicity. Expanding the Green's function for $0 > \tau > -\beta$:

$$G_i(\tau) = \mathrm{Tr}\{U_S(\beta,0)a_i^\dagger U_S(0,\tau)a_i U_S(\tau,0)\}/\mathrm{Tr}\{U_S(\beta,0)\}, \tag{7.10}$$

and manipulating operators cyclically in the trace, we find that

$$G_i(\tau+\beta) = -G_i(\tau). \tag{7.11}$$

This allows the periodic Fourier representation

$$G_i(\tau) = \beta^{-1} \sum_{p=-\infty}^{\infty} \exp(-\mathrm{i}\omega_p\tau)\mathscr{G}_i(\mathrm{i}\omega_p), \tag{7.12}$$

where the Fourier coefficients

$$\mathscr{G}_i(\mathrm{i}\omega_p) = \tfrac{1}{2}\int_{-\beta}^{\beta} G_i(\tau)\exp(\mathrm{i}\omega_p\tau)\mathrm{d}\tau \tag{7.13}$$

and the discrete frequencies are given by $\omega_p = (2p+1)\pi/\beta$, $p = -\infty, \ldots, \infty$. The normalisation factors in equations (7.12), (7.13) follow from

$$\sum_{p=-\infty}^{\infty} \exp(-2\pi \mathrm{i} p\omega/\Omega) = \Omega\Delta_\Omega(\omega), \tag{7.14}$$

where $\Delta_\Omega(\omega)$ is the Dirac delta function $\delta(\omega')$, with ω' defined as ω reduced, modulo Ω, to the interval $[-\beta/2, \beta/2]$. Note that from equation (7.11) the integrand has the period β in τ, so that the integration need be carried out only over the range 0 to β say, with the factor $\frac{1}{2}$ omitted.

We may interpret the Green's function $G_i(\tau)$ physically. We note that it is the thermal average over $|x\rangle$ of the overlap of two states, $U(\tau,0)a_i^\dagger|x\rangle$ and $a_i^\dagger U(\tau,0)|x\rangle$, which in the Schrödinger representation denote creation of the state $|i\rangle_0$, respectively, before and after the system is allowed to evolve for a time τ (strictly, a time t corresponding to an imaginary value of τ in a process described by the analytic continuation to real time of the original definition of the Green's function; we do not repeat this pedantic point in the following). Hence $G_i(\tau)$ is the probability amplitude for a process in which a one-particle state $|i\rangle$ is created at time 0 and propagates unchanged until time τ. $|G_i(\tau)|^2$ is the probability that the excitation can be created at time 0 and will not have decayed by time τ.

For example, suppose the interaction is absent. Using standard relations such as $f(\mathscr{H})|i\rangle_0 = f(E_i^0)|i\rangle_0$, $\langle a_i^\dagger a_i\rangle = (1-\langle a_i a_i^\dagger\rangle_0) = \rho_i^0$, we find that the Green's function in the absence of perturbations is

$$\begin{aligned} G_i^0(\tau) &= -\exp(-E_i^0\tau)(1-\rho_i^0), && \tau \geq 0, \\ &= \exp(-E_i^0\tau)\rho_i^0, && \tau < 0. \end{aligned} \tag{7.15}$$

We may verify equation (7.11) for example. We note that the causal Green's function is everywhere physically defined. For example, if τ is less than zero, the Green's function $\langle a_i^\dagger a_i(\tau)\rangle_0$ is the probability amplitude of annihilating state $|i\rangle_0$ at time τ and restoring it at time 0. From equation (7.15) we see this to be explicitly represented as consisting of two parts. First we have the probability of potential annihilation – the probability of previous occupancy, i.e. ρ_i^0. Thereafter the probability amplitude varies only in phase, oscillating sinusoidally according to the difference in the time development of the occupied and empty states; in the absence of interaction this is the only consequence of the probing of the system for its propagation. Similarly the unperturbed Green's function vanishes if the fermion operators are both annihilation, or both creation operators, or if the initial and final ionic states were different. Finally, the energy conservation of the system corresponds to time translation invariance; only the difference in the time coordinates of the fermion operators affects the answer. This follows trivially from the cyclic invariance of the trace. Summarising these results:

$$\langle T_\tau\{a_i(\tau)a_j^\dagger(\tau')\}\rangle_0 = \delta_{ij}G_i^0(\tau-\tau'). \tag{7.16}$$

Corresponding to this Green's function, we have the Fourier coefficients

$$\mathscr{G}_i^0(E) = \frac{1}{E-E_i^0}, \tag{7.17}$$

where we write E for $i\omega_p$, and shall define the function by analytic continuation for real E as well.

Clearly $\mathscr{G}_i(E)$ had a pole at the point $E=E_i^0$ on the real axis in the complex E plane. In general the singularities in the analytic continuation of these Fourier coefficients describe the spectrum of the one-particle excitations of the system.

7.2.3 The qualitative effects of an interaction

We may anticipate, for later confirmation, that when $\mathscr{V}$ is small in a certain sense, we have the approximate forms

$$G_i(\tau)=(\rho_i^0-1)\exp[-(E_i^0+\Sigma_i)\tau], \qquad \tau>0,$$

$$\mathscr{G}_i(E)=\frac{1}{E-E_i^0-\Sigma_i}. \tag{7.18}$$

The real part Λ_i of the self-energy Σ_i represents the shift of the one-particle excitation energies $E_i^0 \to E_i^0+\Lambda_i$ due to the interactions. The imaginary part, Γ_i, corresponds to the energy width or inverse lifetime of the level. More precisely, since

$$|G_i(\tau)/G_i(0)|^2=\exp(-2\Gamma_i t/\hbar) \tag{7.19}$$

the probability of the propagation decays with a state lifetime or time constant of $\hbar/2\Gamma_i$, and

$$W_i \equiv 2\Gamma_i/\hbar \tag{7.20}$$

is the rate of transition from state $|i\rangle_0$ due to the interaction $\mathscr{V}$.

7.2.4 Analytic continuation of the Fourier coefficients and the spectral density of excitations

As in the theory of classical wave equations, or linear circuit theory, the Green's function is an impulse response – the history of the life of a particle excitation – which by linearity or superposition, causality and time translation invariance may be used to find the response of the system to any excitation (see Mattuck 1967, Stedman 1968, for elementary discussions of this idea). The relation between the spectral distribution and the Green's function is one example. More generally we note that the expectation values of all one-particle operators in the presence of interaction are obtainable from this one-particle Green's function.

We establish a connection between the value of the full Green's function Fourier coefficients $\mathscr{G}_i(E)$ when analytically continued to the real axis, and the spectral density $\rho_i(E)$ of the one-particle states in the presence of interactions, via the Lehmann spectral representation.

When the creation operator $a_i^\dagger$ is applied to any energy eigenstate $|\alpha\rangle$ of the *perturbed* system with energy E_α say (where we use the Greek letter to denote that this could be the vacuum or multifermion state energy), it will create a state which is no longer an eigenstate of the full Hamiltonian, but will correspond to a linear combination of energy eigenstates, which in the first approximation (§7.2.3) may be described by a broadened level.

Now the Green's function is defined by (taking $\tau>0$ for example):

$$G_i(\tau)=-\langle\exp[\tau(\mathscr{H}+\mathscr{V})]a_i\exp[-\tau(\mathscr{H}+\mathscr{V})]a_i^\dagger\rangle$$
$$=-\frac{\Sigma_\alpha\langle\alpha|\exp[-\beta(\mathscr{H}+\mathscr{V})]\exp[\tau(\mathscr{H}+\mathscr{V})]a_i\exp[-\tau(\mathscr{H}+\mathscr{V})]a_i^\dagger|\alpha\rangle}{\Sigma_\gamma\langle\gamma|\exp[-\beta(\mathscr{H}+\mathscr{V})]|\gamma\rangle}. \quad (7.21)$$

We insert a complete set of perturbed eigenstates $\Sigma_\beta|\beta\rangle\langle\beta|$ after the operator a_i in the numerator of this expression. The various Hamiltonians $\mathscr{H}+\mathscr{V}$ acting on their eigenstates $\{|\alpha\rangle\}$ give rise to the (perturbed) energies $\{E_\alpha\}$, and all exponential operators become c-numbers. Hence

$$G_i(\tau)=\{\Sigma_{\alpha\beta}|\langle\beta|a_i^\dagger|\alpha\rangle|^2\exp(-\beta E_\alpha)\exp[-\tau(E_\beta-E_\alpha)]\}\ \gamma\Sigma_\gamma\exp(-\beta E_\gamma)] \quad (7.22)$$

and hence by integration

$$\mathscr{G}_i(\mathrm{i}\omega_p)=\Sigma_{\alpha\beta}\frac{|\langle\beta|a_i^\dagger|\alpha\rangle^2[\exp(-\beta E_\beta)+\exp(-\beta E_\alpha)]}{[E_\beta-E_\alpha-\mathrm{i}\omega_p][\Sigma_\gamma\exp(-\beta E_\gamma]}. \quad (7.23)$$

The new thermal factor arises from evaluating the limits in the integral over τ at 0 and β, using equation (7.14).

When equation (7.23) is analytically continued ($\mathrm{i}\omega_p\to E$), its imaginary part has a discontinuity at the real axis, thanks to the relation:

$$\lim_{\varepsilon\to 0+}\frac{1}{x-\mathrm{i}\varepsilon}=\mathscr{P}(1/x)+\mathrm{i}\pi\delta(x) \quad (7.24)$$

giving

$$\mathrm{Im}\lim_{\varepsilon\to 0+}[\mathscr{G}_i(E-\mathrm{i}\varepsilon)-\mathscr{G}_i(E+\mathrm{i}\varepsilon)]$$
$$=2\pi[\exp(\beta E)+1]\frac{\Sigma_{\alpha\beta}\exp(-\beta E_\alpha)|\langle\beta|a_i^\dagger|\alpha\rangle|^2}{\Sigma_\gamma\exp(-\beta E_\gamma)} \quad (7.25)$$
$$=2\pi[\exp(\beta E)+1]\mathscr{A}v_\alpha\Sigma_\beta|\langle\beta|a_i^\dagger|\alpha\rangle|^2\delta(E-E_\beta+E_\alpha),$$

where we have used $E=E_\beta-E_\alpha$ (arising from the delta function in the imaginary part of $\mathscr{G}_i(E)$) to simplify the population factor, and $\mathscr{A}v$ denotes a thermal average.

We define $\rho_i(E)$, the *spectral density* for state i, by setting $\rho_i(E)\mathrm{d}E$ to be the thermal average over the chosen initial perturbed eigenstate α of the probability that the state $a_i^\dagger|\alpha\rangle$ has its excitation energy in the range E to $E+\mathrm{d}E$. Hence:

$$\rho_i(E)\equiv\mathscr{A}v_\alpha\Sigma_\beta|\langle\beta|a_i^\dagger|\alpha\rangle^2, \qquad E<E_\beta-E_\alpha<E+\mathrm{d}E; \quad (7.26)$$

and

$$\mathrm{Im}\lim_{\varepsilon\to 0+}[\mathscr{S}_i(E-\mathrm{i}\varepsilon)-\mathscr{S}_i(E+\mathrm{i}\varepsilon)]=2\pi[\exp(\beta E)+1]\rho_i(E). \quad (7.27)$$

7.2.5 Perturbation expansion

In the interaction representation of quantum theory, we write

$$\left.\begin{aligned} \Psi_I(\tau) &= \exp(-\mathscr{H}\tau)\Psi_I(0) = U(\tau,\tau')\Psi_I(\tau') \\ O(\tau) &= \exp(\mathscr{H}\tau)O\exp(-\mathscr{H}\tau) \\ U(\tau,\tau') &= \exp(\mathscr{H}\tau)\exp[(\mathscr{H}+\mathscr{V})(\tau'-\tau)]\exp(-\mathscr{H}\tau') \end{aligned}\right\} \quad (7.28)$$

according to which the observer evolves in the same manner in which the wavefunction would, were the interaction absent. From this viewpoint, the residual evolution of the wavefunction is due entirely to the interaction $\mathscr{V}$. We omit the suffix I for operators in the interaction picture; the time-dependence in their arguments is a sufficient indication of the distinction with Schrödinger-picture operators, and we will no longer have occasion to use the Heisenberg picture.

We note here that using this representation we may write a relation between perturbed and unperturbed density matrices:

$$\rho = \rho_0 \frac{U(\beta,0)}{\langle U(\beta,0)\rangle_0} \quad (7.29)$$

since

$$\begin{aligned} \tilde{a}_i(\tau) &= U(0,\tau)a_i(\tau)U(\tau,0) \\ G_i(\tau) &= -\langle U(\beta,0)T_\tau\{U(0,\tau)a_i(\tau)U(\tau,0)a_i^\dagger\}\rangle_0/[\langle U(\beta,0)\rangle_0. \end{aligned} \quad (7.30)$$

It is readily seen that $\partial U(\tau,\tau')/\partial\tau = -\mathscr{V}(\tau)U(\tau,\tau')$, so that using $U(\tau,\tau)=1$ we may integrate to get

$$U(\tau,\tau') = 1 - \int_{\tau'}^{\tau} \mathscr{V}(\tau_1)U(\tau_1,\tau')\mathrm{d}\tau_1.$$

Iterating this equation, we have

$$U(\tau,\tau') = \Sigma_r(-1)^r \int_{\tau'}^{\tau} \mathrm{d}\tau_1 \int_{\tau'}^{\tau_1} \mathrm{d}\tau_2 \ldots \int_{\tau'}^{\tau_{r-1}} \mathrm{d}\tau_r \mathscr{V}(\tau_1)\mathscr{V}(\tau_2)\ldots\mathscr{V}(\tau_r). \quad (7.31)$$

The integrations in this equation are over only one of $r!$ sectors of the space of the variables $\{\tau_i | i=1,2,\ldots,r\}$, this sector being $\tau \geq \tau_1 \geq \tau_2 \geq \ldots \geq \tau_r$. The integrals over all similar sectors will be the same provided the $\mathscr{V}(\tau_i)$ are permuted to take account of their noncommutativity. This permutation can be implemented using the chronological operator T_τ:

$$\int_{\tau'}^{\tau} \mathrm{d}\tau_1 \int_{\tau'}^{\tau_1} \mathrm{d}\tau_2 \ldots \int_{\tau'}^{\tau_{r-1}} \mathrm{d}\tau_r \mathscr{V}(\tau_1)\mathscr{V}(\tau_2)\ldots\mathscr{V}(\tau_r) \quad (7.32)$$

$$= \frac{1}{r!}\int_{\tau'}^{\tau}\int_{\tau'}^{\tau}\ldots\int_{\tau'}^{\tau} \mathrm{d}\tau_1\mathrm{d}\tau_2\ldots\mathrm{d}\tau_r \mathrm{T}_\tau\{\mathscr{V}(\tau_1)\mathscr{V}(\tau_2)\ldots\mathscr{V}(\tau_r)\}. \quad (7.32)$$

Hence, defining the exponential by its series expansion, we have

$$U(\tau,\tau')=T_\tau\left[\exp\{-\int_{\tau'}^{\tau}\mathrm{d}\tau_1\,\mathscr{V}(\tau_1)\}\right]. \qquad (7.33)$$

The Green's function of equation (7.30) may now be written more simply. With the expansion of equation (7.33) for the various evolution operators, equation (7.30) becomes a set of time-ordered products of operators in the interaction representation. The answer is the same if all operators are commuted in any way, and the chronological operator is applied to the whole expression. In that case the operators $U(\tau,0)U(0,\tau)$ combine to give the identity, and

$$G_i(\tau)=-\frac{\langle T_\tau\{U(\beta,0)a_i(\tau)a_i^\dagger\}\rangle_0}{\langle U(\beta,0)\rangle_0}. \qquad (7.34)$$

The numerator N and the denominator D of this expression each are amenable to a diagram expansion.

Combinatoric problems associated with the topology of the diagram expansion are discussed by, for example, Rosensteel, Ihrig and Trainor (1975), Wise and Trainor (1977) and Prakash (1980).

7.2.6 Wick's theorem and the field theoretic diagram expansion

Wick's theorem relates the thermal average with respect to the unperturbed Hamiltonian of a time-ordered product of an even number of fermion (or boson) creation and annihilation operators in the interaction representation to a sum of products of such products of two operators, paired in all possible ways.

Let the function X_r denote a thermal average over the unperturbed Hamiltonian of a time-ordered product of r fermion operators in the interaction picture:

$$X_r(A,B,\ldots,Z)\equiv\langle T_\tau\{A(\tau_1)B(\tau_2)\ldots Z(\tau_r)\}\rangle_0. \qquad (7.35)$$

Then our essential result may be stated as

$$X_r(A,B,\ldots,Z)=\Sigma_{I\in\{B,\ldots,Z\}}(-1)^p\overline{AI}\,X_{r-2}(B,\ldots H,J,\ldots,Z), \qquad (7.36)$$

where p is the permutation necessary to bring the operator $I(\tau_i)$ to the front of the set $\{B\ldots Z\}$, and the *contraction*

$$\overline{IJ}\equiv X_2(I,J)=\langle T_\tau\{I(\tau_i)J(\tau_j)\}\rangle_0. \qquad (7.37)$$

Iterating this result, for odd r the expression on the left vanishes, while for even r, we have *Wick's theorem*: $X_r(A,B,\ldots,Z)$ is the sum over signed products of all possible contractions in pairs. (For an early reference, see Gaudin 1960.)

Proof

Without loss of generality (i.e. modulo a relabelling of operators) we can assume that the operators in $X_r(A,\ldots,Z)$ are already time-ordered.

If $A(\tau_1)$ is a creation/annihilation operator $a_i^{\pm}(\tau_1)$, the commutation relation $[\mathscr{H},a_i^{\pm}]=\pm E_i^0 a_i^{\pm}$ shows that

$$a_i^{\pm}(\tau_1)=a_i^{\pm}\exp(\pm\tau_1 E_i^0). \tag{7.38}$$

Hence the anticommutation relations give

$$\{a_i(\tau_1),\, a_j^{\dagger}(\tau_2)\}=\exp[-(\tau_1-\tau_2)E_i^0]\delta_{ij}, \tag{7.39}$$

for example, with a similar result (a sign change in the exponential) for the anticommutator of a creation with an annihilation operator, and give zero if both are creation, or both annihilation, operators.

Since the anticommutator of these operators in the interaction representation is a c-number, we can reorder operators within a thermal average. Hence X_r becomes

$$\begin{aligned}&\langle A(\tau_1)B(\tau_2)\ldots Z(\tau_r)\rangle_0\\ &\quad=\{A(\tau_1),B(\tau_2)\}\langle C(\tau_3)\ldots Z(\tau_r)\rangle_0-\langle B(\tau_2)A(\tau_1)\ldots Z(\tau_r)\rangle_0.\end{aligned}$$

Iterating, $A(\tau_1)$ may be commuted through all the other operators to the last position in the trace, generating at each step an appropriate anticommutator:

$$\begin{aligned}&\langle A(\tau_1)B(\tau_2)\ldots Z(\tau_r)\rangle_0\\ &\quad=\Sigma_{I\in\{B\ldots Z\}}(-1)^p\{A(\tau_1),I(\tau_i)\}\langle B(\tau_2)\ldots H(\tau_h)J(\tau_j)\ldots Z(\tau_r)\rangle_0\\ &\qquad+(-1)^{r+1}\langle B(\tau_2)C(\tau_3)\ldots Z(\tau_r)A(\tau_1)\rangle_0.\end{aligned} \tag{7.40}$$

Using cyclic invariance, $A(\tau_1)$ may be brought to the start of the trace $\langle\ldots\rangle_0=\mathrm{Tr}(\hat{\rho}^0\ldots)$. From equation (7.38) we have $A(\tau_1)\exp(-\beta\mathscr{H})=\exp(\pm\beta E_i^0)\exp)(\pm/\beta E_i^0)-\beta\mathscr{H})A(\tau_1)$ (with $\pm$ as $A\equiv a_i{}^{\pm}$), so that the last term on the right side of equation (7.40) becomes proportional to the term on the left side. Collecting these terms, we obtain

$$\begin{aligned}&\langle A(\tau_1)B(\tau_2)\ldots Z(\tau_r)\rangle_0[1+(-1)^r\exp(\pm\beta E_i^0)]\\ &\quad=\Sigma_{I\in\{B\ldots Z\}}(-1)^p\{A(\tau_1),\, I(\tau_i)\}\langle B(\tau_2)\ldots H(\tau_h)J(\tau_j)\ldots Z(\tau_r)\rangle_0.\end{aligned} \tag{7.41}$$

Hence x_r is related to a sum over X_{r-2}-type averages. Iterating, for odd r this must finally vanish, since any X_1 average vanishes (this may be seen from equation (7.38); the interaction operator is proportional to the Schrödinger operator, which annihilates either the bra or ket on which it acts). For even r, we obtain equation (7.36) with the definition

$$\overline{AB}\equiv\frac{\{A(\tau_1),\, B(\tau_2)\}}{1+\exp(\pm\beta E_i^0)} \tag{7.42}$$

(again with $\pm$ as $A=a_i^{\pm}$). From equation (7.39) and the Fermi–Dirac form of the denominator, we readily find that the right side of equation (7.42) is nonvanishing when $B(\tau_2)=a_i^{\mp}$ and is then equal to $\langle a_i^{\pm}a_i^{\mp}\rangle_0$ $\exp\pm E_i^0(\tau_1-\tau_2)$ in agreement with the definition of equation (7.37). This completes the proof.

Hence we may expand the numerator and denominator of equation (7.34) in terms of the sums of products of unperturbed Green's functions.

The pairing of time coordinates between propagators corresponds to the evaluation of both operators in $\mathscr{V}(\tau)$ at time τ, and enables a diagram notation for cataloguing, representing and calculating the various terms. Each unperturbed Green's function $G_i^0(\tau)$ is denoted by a directed line connecting terminals representing the initial (fermion creation) and final (fermion annihilation) time coordinates, which differ by τ in this case, with the arrow directing the line pointing from creation to annihilation:

$G_i^0(\tau)\leftrightarrow$ ——←—— .
τ i 0

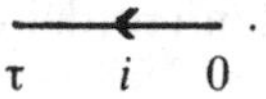

Each intermediate terminal or vertex corresponds to the action of the interaction in altering the state of the ion, and may be taken to literally represent the magnitude of the matrix element V_{ij} associated with the evaluation of these expressions, with i corresponding to the state label of the propagator directed away from the point (i.e. the state created at that time) and j that of the propagator directed to that point. The point as such labels the time coordinate τ of the associated second quantised operators in $\mathscr{V}(\tau)=\Sigma_{ij}V_{ij}[a_i^{\dagger}a_j](\tau)$:

×
$V_{ij}\leftrightarrow$ ←—┴—← .
i j

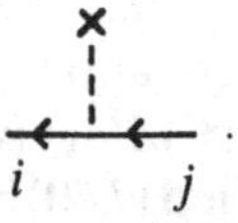

Such intermediate times should be integrated from 0 to β according to equations (7.33), (7.34). The initial (0) and final (τ) labels are not integrated. At any given order r, one has r intermediate times – points in the diagram – to be linked by lines in all possible ways.

For example, at first order the numerator in equation (7.34) contains the term:

$$\mathrm{N}^{(1)}=\Sigma_{pq}V_{pq}\int_0^{\beta}d\tau_1[(-1)/1!]\langle T_{\tau}\{\underbrace{a_i(\tau)a_p^{\dagger}(\tau_1)}\underbrace{a_q(\tau_1)a_i^{\dagger}(0)}\}\rangle. \quad (7.43)$$

The braces underneath denote the two possible pairings arising from an application of Wick's theorem to this expression. These give the two terms, respectively:

$$N^{(1a)} = -V_{ii}\int_0^\beta d\tau_1 G_i^0(\tau-\tau_1)G_i^0(\tau_1) \leftrightarrow \tag{7.44}$$

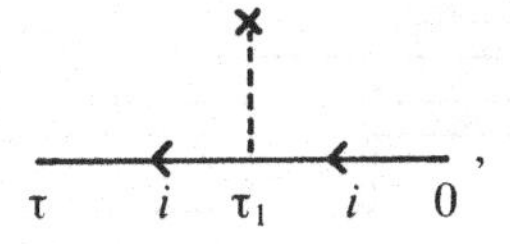

$$N^{(1b)} = [-\Sigma_j V_{jj}\beta G_j^0(0)]G_i^0(\tau)] \leftrightarrow \tag{7.45}$$

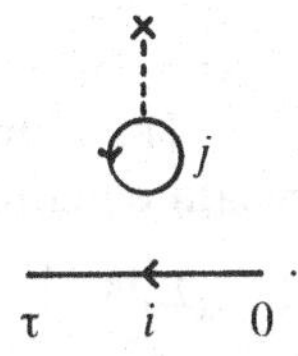

The factor in square brackets in equation (7.45) corresponds to the separate 'bubble' subdiagram; the integral over the intermediate time becomes the factor β, and the electronic loops to the Green's function $G_j^0(0)$ which evaluates to the population factor $(1-\rho_j^0)$.

The corresponding term in the denominator expansion (D of equation 7.34) will be just this bubble diagram:

$$D^{(1)} = \Sigma_{pq} V_{pq}[(-1)/1!]\int_0^\beta d\tau_1 \langle T_\tau\{a_p^\dagger(\tau_1)a_q(\tau_1)\}\rangle_0 \leftrightarrow \tag{7.46}$$

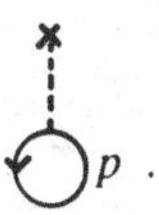

We note that in such terms, whenever a closed loop is formed among the propagators an extra minus sign appears, since an odd permutation of fermion operators is necessary to achieve their canonical order. This sign arises from the sign in the definition of the chronological operator for fermions, without which Wick's theorem would not have its standard form. For example in the above term, we need to interchange $a_p^\dagger$ and a_q to conform with equation (7.8). We can remember this as a rule: *a closed fermion loop contributes a factor* (-1).

At second order we obtain a term

$$N^{(2)} = \Sigma_{pqrs} V_{pq} V_{rs} [(-1)^2/2!] \iint_0^\beta d\tau_1 d\tau_2 \times$$
$$\langle T_\tau\{a_i(\tau) a_p^\dagger(\tau_1) a_q(\tau_1) a_r^\dagger(\tau_2) a_s(\tau_2) a_i^\dagger(0)\}\rangle_0. \tag{7.47}$$

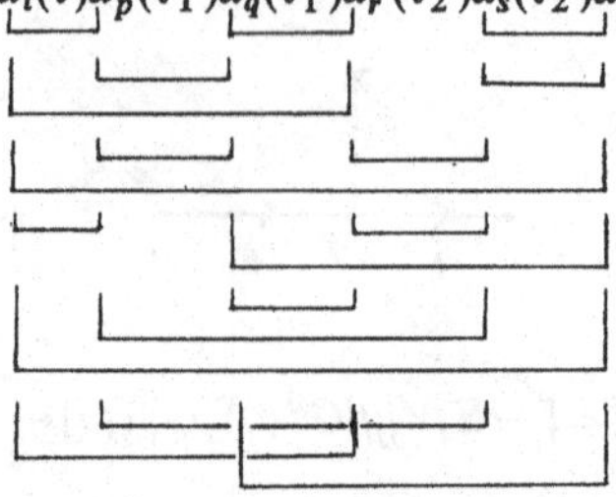

There are now six possible pairings of terms giving rise to potentially nonzero contractions of operators. The term corresponding to the pairing in the first set of braces underneath equation (7.47) is

$$N^{(2a)} = [1/2!] \Sigma_q |V_{iq}|^2 (-1)^2 \times$$
$$\iint_0^\beta d\tau_1 d\tau_2 \langle T_\tau\{a_i(\tau) a_i^\dagger(\tau_1)\}\rangle_0 \langle T_\tau\{a_q(\tau_1) a_q^\dagger(\tau_2)\}\rangle_0 \langle T_\tau\{a_i(\tau_2) a_i^\dagger(0)\}\rangle_0 \tag{7.48}$$
$$= \tfrac{1}{2} \Sigma_q |V_{iq}|^2 \iint_0^\beta d\tau_1 d\tau_2 G_i^0(\tau - \tau_1) G_q^0(\tau_1 - \tau_2) G_i^0(\tau_2) \leftrightarrow$$

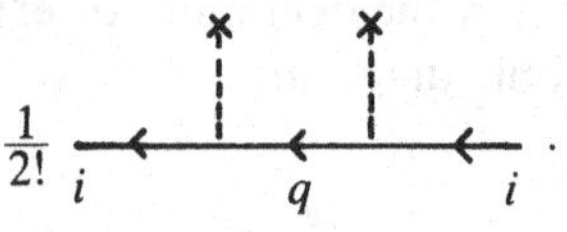

Similarly the other terms in equation (7.47) correspond to:

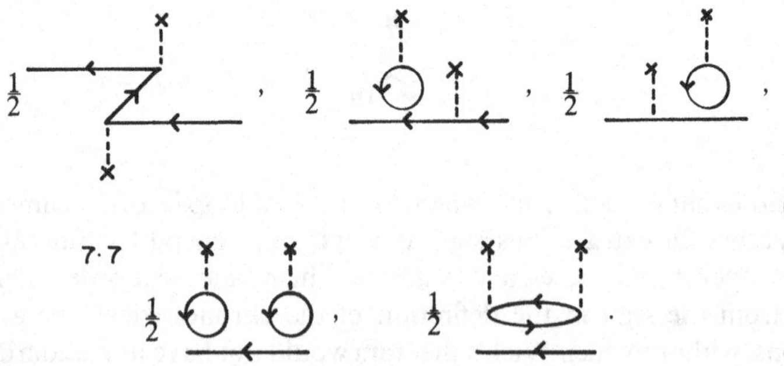

(7.49)

The first diagram is very similar to equation (7.48); however, one line runs backwards. This corresponds to permutations of the r intermediate time labels τ_i. These are in any case dummy labels for integration, and the associated operator permutation is of an even number of fermion operators, and thus involves no sign change; the two terms are equal. This corresponds to their topological equivalence; as in §1, a diagram may be mildly deformed. Since there are $r!$ such permutations, imposing a restriction to topologically distinct diagrams neatly cancels the $[1/r!]$ in the expansion of the evolution operator for such terms (equation 7.32).

The second order contributions to D are:

$$D^{(2)} \leftrightarrow \tfrac{1}{2} \quad , \quad \tfrac{1}{2} \quad . \tag{7.50}$$

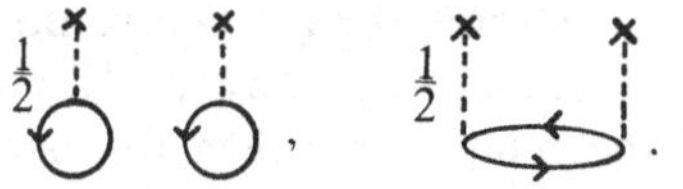

7.2.7 Linked cluster expansion

There are two kinds of diagrams appearing in N: a subset N_c of *connected* diagrams, in which all propagators are joined in one long string starting at 0 and ending at τ, and *disconnected* diagrams, in which in addition to such a connected string there are *vacuum fluctuation* diagrams in which the propagators form one or more closed loops. The latter diagrams are precisely the diagrams in the series for D. The series in N consists of each diagram in N_c coupled with all diagrams in D. Further, the disconnected nature of these diagrams betokens the factorisation of the multiple integrals over intermediate τ variables. Hence the whole series factorises into a linked diagram expansion and into a factor cancelling the denominator expansion: $N = N_c D$, and equation (7.34) reduces to just the *linked cluster expansion* N_c:

$$G_i(\tau) = -\frac{\langle T_\tau\{U(\beta,0)a_i(\tau)a_i^\dagger\}\rangle_0}{\langle U(\beta,0)\rangle_0} \leftrightarrow \tag{7.51}$$

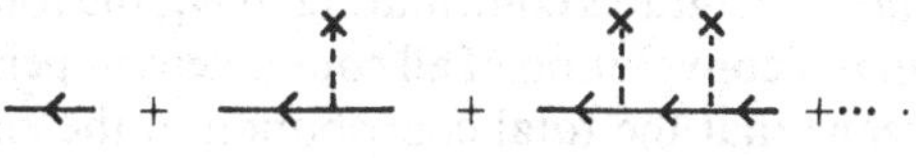

$$+ \quad + \quad + \cdots .$$

We have written only the topologically distinct diagrams, so that we need include only a factor $(-1)^r$ and not $[1/r!]$ when combining the terms from different orders r.

It is at such points that the later restriction to 'physical' states in which only one-fermion states are included in the thermal average introduces a major technical complication. The linked cluster theorem fails under Abrikosov projection. However, the disconnected diagrams appear only in a denominator which does not affect the energy dependence and so the form of the spectral density $\rho_i(E)$. We defer a consideration of this problem until §7.4.

7.2.8. Fourier transformation and Dyson equation

The next simplification is to transform the significance of the diagrams by using energies rather than times as the relevant variables. Integrations over intermediate times (e.g. equation 7.44) have the structure of a convolution:

$$A(\tau)^*B(\tau) \equiv \int_0^\beta d\tau' A(\tau')B(\tau-\tau'). \tag{7.52}$$

On taking Fourier transforms, then, the corresponding Fourier coefficients $\mathscr{G}_i^0(E)$ are multiplied (rather than convolved). The same diagrams may be used to represent the relation between the perturbed and unperturbed Fourier coefficients. Instead of the now redundant and inapplicable τ labels at vertices, an *energy variable* ε (as well as a state label i) may be placed on each propagator, to denote the appropriate argument of the corresponding Green's function $\mathscr{G}_i^0(\varepsilon)$. Each such variable should be summed over (equation 7.12).

A word may be added here about numerical factors, especially the factors β. The expansion of equation (7.36) indicates that no numerical factors are involved in the time domain when stringing the various $G_i^0(\tau)$ together in the manner of equation (7.52). However, in the energy domain, each Green's function brings a factor of $(1/\beta)$ (equation 7.12). In addition, each integration over an intermediate time variable τ takes the form $\int_0^\beta d\tau \exp(\tau\varepsilon)$, where ε is the algebraic sum of energy variables at the vertex labelled τ. Since ε is necessarily of bosonic form (i.e. an even multiple of $i\pi/\beta$) the integral vanishes unless $\varepsilon = 0$; this proves that energy variables must add to zero at each vertex, i.e. they obey *Kirchhoff's law* for current flow in a network. When this condition is satisfied, the integral equals β, cancelling one such denominator factor. Any loop of propagators will involve one free energy variable together with one uncancelled factor of $(1/\beta)$.

For the linear linked diagrams contributing to N_c, the total contribution in the time domain is a convolution of all component unperturbed Green's functions. This means that the total contribution in the energy (Fourier) domain is a simple product of terms for which each energy variable is the same: simply $i\omega_p$ or E.

As an example, we take equation (7.48). We now reinterpret it as a contribution to $\mathscr{G}_i(E)$, and indicate this by an energy variable argument. We omit the $1/\beta$ of equation (7.12) for this energy domain interpretation. This gives

$$N^{(2a)}(i\omega_p) = 1/(2\beta^2)\Sigma_q |V_{iq}|^2 \iint_0^\beta d\tau' d\tau'' \sum_{q,r=-\infty}^{\infty} \mathscr{G}_i^0(i\omega_p)\mathscr{G}_q^0(i\omega_q)\mathscr{G}_i^0(i\omega_r)$$

$$\times \exp[\omega_p(\tau-\tau') + \omega_q(\tau'-\tau'') + \omega_r\tau''].$$

The integrals give $\beta^2 \delta_{pq}\delta_{qr}$, so that

$$N^{(2a)}(E) = \tfrac{1}{2}\Sigma_q |V_{iq}|^2 \mathscr{G}_i^0(E)\mathscr{G}_q^0(E)\mathscr{G}_i^0(E) \leftrightarrow$$

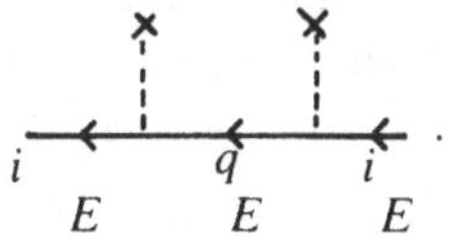

For the vacuum diagrams of D, or the unlinked parts of N, the requirement of integrating over intermediate times in the time domain converts into the requirements that energy variables are the same for all propagators in the same loop, and that each loop variable is summed over all possible values of $i\omega_p$, where $p = -\infty, \ldots, \infty$:

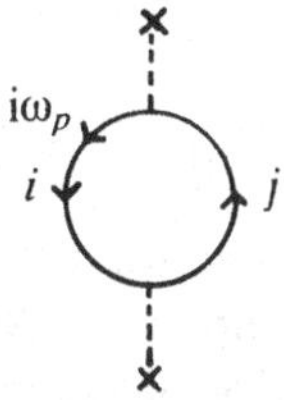

$$\leftrightarrow \sum_{ij} |V_{ij}|^2 \sum_{p=-\infty}^{\infty} \mathscr{G}_j^0(i\omega_p)\mathscr{G}_i^0(i\omega_p).$$

The multiplicative rule for combination of propagators in the Fourier transform case helps us to see a further rearrangement. (Since the convolution operation is associative, the following has a counterpart in the time domain also.) The diagrams in N_c may be further decomposed in the following way:

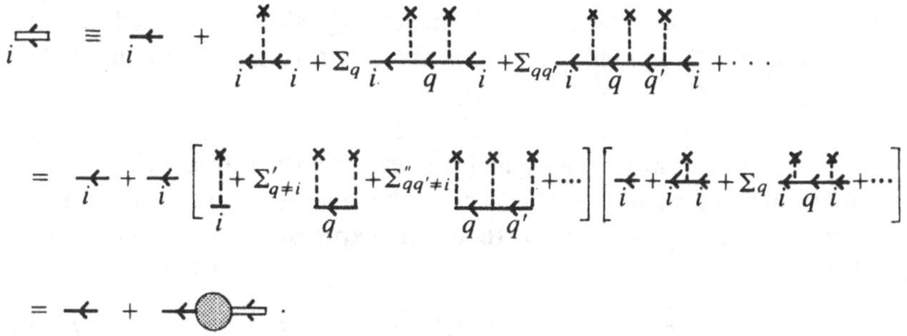

We recognise the series on the right as the original series, and the term on the left as the unperturbed propagator; the term in the middle is called the *self-energy* $\Sigma_i(E)$ (not to be confused with a summation symbol), and consists of the sum over *irreducible* diagrams. In more general applications, irreducibility denotes diagrams which cannot be disconnected by severing one electronic propagator. Here, we make the restriction only that no intermediate propagator have the original state label. This corresponds to the requirement that in perturbation theory we use a projector to prevent any intermediate state coinciding with the initial state. When any N_c contribution contains a propagator $\mathscr{G}_i^0(E)$, we simply terminate the contribution to the self-energy at that point:

$$\Sigma_i(E) \leftrightarrow \quad [\text{diagram}] \quad \equiv \quad [\text{diagram}] + \Sigma'_{q \neq i} [\text{diagram}] + \Sigma''_{qq' \neq i} [\text{diagram}] + \cdots .$$

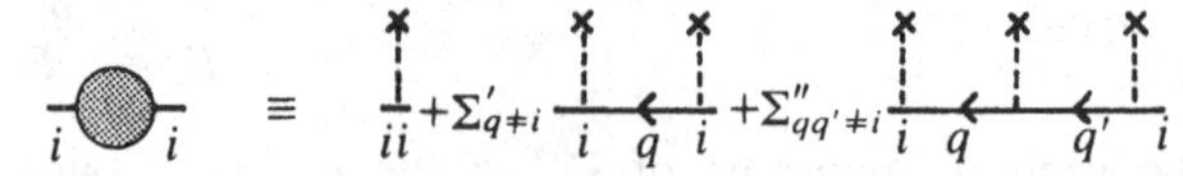

Since the various terms factorise, the above expansion is equivalent to the geometric progression (the *Dyson equation*)

$$\mathscr{G}_i(E) = \mathscr{G}_i^0(E) + \mathscr{G}_i^0(E)\Sigma_i(E)\mathscr{G}_i^0(E)$$
$$= \mathscr{G}_i^0(E) + \mathscr{G}_i^0(E)\Sigma_i(E)\mathscr{G}_i^0(E) + \mathscr{G}_i^0(E)\Sigma_i(E)\mathscr{G}_i^0(E)\Sigma_i(E)\mathscr{G}_i^0(E) + \dots .$$

The sum of this geometric series may be written

$$\mathscr{G}_i(E) = \mathscr{G}_i^0(E)/[1 - \mathscr{G}_i^0(E)\Sigma_i(E)]. \tag{7.53}$$

We have now focussed our interest not only on numerator diagrams and on linked diagrams, but further to their irreducible components. As a result, we have achieved essentially a *denominator expansion* of the Fourier coefficients; the self-energy expansion now appears in the denominator.

This permits a much more direct link with, and justification of, the intuitive ideas of §7.2.3. For, substituting the form of $\mathscr{G}_i^0(E)(=1/[E - E_i^0])$, we finally obtain an expression corresponding to equation (7.18):

$$\mathscr{G}_i(E) = 1/[E - E_i^0 - \Sigma_i(E)]. \tag{7.54}$$

If perturbation theory is valid, in the sense that $V_{pq}/(E_p^0 - E_q^0) \ll 1$, we may identify the real part of $\Sigma_i(E)$ as the energy shift Λ_i caused by the interaction, and the imaginary part as the width Γ_i corresponding to the finite lifetime of the state i.

7.2.9 Perturbation theory and the interpretation of diagrams

Enumerating and calculating the terms in the self-energy gives the standard Brillouin–Wigner perturbation theoretic expressions for energy shift (§7.1.2):

$$\Lambda_i(E) = E - E_i^0 = \sum_{r=0}^{\infty} \langle i| \mathscr{V} \left\{ \frac{Q_i}{E - \mathscr{H}} \mathscr{V} \right\}^r |i\rangle \tag{7.55}$$

where Q_i projects off the state $|i\rangle$, corresponding to our definition of an irreducible diagram. The value of E inserted in this calculation should be interpreted self-consistently as $E_i^0 + \Lambda_i$.

For example, at first order we have the diagram

$$\Sigma_i^{(1)}(E) \leftrightarrow \quad [\text{diagram}], \tag{7.56}$$

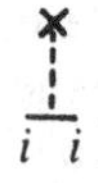

which has only the real part, and first-order energy shift, $\Lambda_i^{(1)} = V_{ii}$.

At second order we have a single diagram contributing to the self-energy,

$$\Sigma_i^{(2)}(E) = \Sigma_{p \neq i} \frac{|V_{ip}|^2}{E - E_p^0} \quad \leftrightarrow \quad \Sigma'_{q \neq i} \quad i \quad q \quad i \quad . \tag{7.57}$$

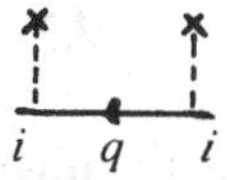

Using equation (7.24) in this relation, and setting $E = E_i^0$ (accurate to zeroth order, and therefore appropriate for estimating an expression already of second order to the same accuracy), we have

$$\left.\begin{aligned} \Lambda_i^{(2)} &= \Sigma_{p \neq i} \frac{|V_{ip}|^2}{E_i^0 - E_p^0}, \\ \Gamma_i^{(2)} &= 2\pi \Sigma_{p \neq i} |V_{ip}|^2 \delta(E_p^0 - E_i^0), \end{aligned}\right\} \tag{7.58}$$

which we recognise as the second-order perturbation energy shift and the Golden Rule expression for the rate of interaction with other levels, and so for the decay of the probability amplitude of the propagation. These may obviously be extended to higher orders.

This demonstration achieves our immediate goal of showing the link between quantum field theory and the Golden Rule expression used in §6 for example. As a proof of the Golden Rule (Stedman 1971b) it avoids the problem of some standard approaches, in which a wrong result for the rate is 'corrected'.

It also illustrates the very close connection between the Golden Rule type of expression for a dissipative process such as a decay rate Γ_i, and the corresponding fluctuation process such as an energy shift Λ_i. Formally, these are imaginary and real parts of the same expression. Physically, these are bound together by virtue of the *fluctuation–dissipation theorem* (Callen 1961, Kubo 1966), or more fundamentally by causality, via the Kramers–Kronig dispersion relations. Even causality is not sacrosanct, but its violations are not important in atomic physics (Bennett 1987, Valentini 1988). The contributions to the shift $\Lambda_i^{(2)}$ may be interpreted physically as the effect of virtual transitions between i and other states p, themselves governed by the Golden Rule transition rate $\Gamma_i^{(2)}$, and associated with a temporary energy deficit corresponding to the denominator. This integrates with the physical interpretation of the Green's function in §§7.2.2–7.2.3 as a probability amplitude for a propagation process limited in magnitude by the tendency of the state to interact.

We have now justified all the essential ingredients of the peculiar physical viewpoint suggested by the interaction representation, and expounded so ably by Feynman (1985) in the QED context for example. Field theoretic diagrams now 'spring to life' with a unique intuitive appeal; they may be associated with dynamic effects in the system, and lend themselves to an anecdotal dissection of their physical meaning. Even diagram-motivated approximation methods have exact counterparts to algebraic approximation methods (Mattuck and Theumann 1971).

7.3 Phonon coupling theory

7.3.1 The interactions

We first consider the interaction $\mathscr{V}$ to be the Hamiltonian for the linear coupling between the electronic states of an ion or molecule and the quantised molecular vibrations or crystal lattice modes (phonons) interacting with the ion. Many of our examples will use this coupling, and many features will be preserved in the other interactions of interest – that between radiation and matter, and also the Coulomb interaction in atomic physics.

In this case we may write:

$$\mathscr{V} = \Sigma_{ijk}(V_{ijk-}b_k + V_{ijk+}b_\mu^\dagger)a_i^\dagger a_j, \tag{7.59}$$

where $V_{ijk\pm}$ are the matrix elements for creation and annihilation, respectively, of a quantum in the harmonic lattice mode k simultaneous with the electronic transition $j \rightarrow i$. The label k denotes both the wavevector and the polarisation of a lattice mode.

Hermiticity demands that $V_{ijk-} = (V_{jik+})^*$, while time-reversal invariance of the interaction demands that $\bar{\mathscr{V}} = \mathscr{V}$ and so that

$$V_{ijk-} = (V_{\bar{i}\bar{j}\bar{k}-})^*, \tag{7.60}$$

where $\bar{i}$, $\bar{k}$ denote the time-reversed electronic and lattice states. A further restriction will be discussed in §7.3.2.

This interaction may be enlarged to consider nonlinear couplings as well, of the form:

$$\mathscr{V}_{\text{nl}} \equiv \Sigma_{ijkln}[V^{(n)}_{ijk+l-\ldots} b_k^\dagger b_l \ldots + V^{(n)}_{ijk-l+\ldots} b_k b_l^\dagger \ldots + \ldots] a_i^\dagger a_j, \tag{7.61}$$

where n phonon operators appear in the nth term. The quadratic coupling ($n=2$) may be interpreted in terms of the nonlinearity of the crystal field in the position of the ligands.

Similarly we may add the anharmonicity of the lattice modes as a separate term, of the form:

$$\mathscr{V}_{\text{anh}} \equiv \Sigma_{k\pm l\pm m\pm\ldots n} W^{(n)}_{k\pm l\pm m\pm\ldots} b_k^\pm b_l^\pm b_m^\pm \ldots . \tag{7.62}$$

When calculating the temperature dependence of any phonon relaxation effect, it is useful to note that all these interactions will involve a phonon frequency dependence given by $(\omega_k)^{\frac{1}{2}}$ for each participating phonon mode k.

7.3.2 Time-even (position) coupling

We specialise the discussion to a time-even phonon interaction, in which both the electronic and the lattice operators in the interaction satisfy the property $\bar{\mathscr{O}}^\dagger = \mathscr{O}$. This means that the phonon creation and annihilation operators appear in the combination $[b_k + b^\dagger{}_k]$ appropriate for a lattice position operator, rather than the combination $[b_k - b_k^\dagger]$ appropriate for the conjugate (and time-odd) lattice momentum operator.

In physical terms, we assume that the interaction occurs in a quasistatic manner, the instantaneous position of the ligands determining the electrostatic perturbation, as modified by quantum effects such as overlap and exchange, on the electronic states. In making this assumption we ignore the dynamic effects of ligand motion, such as the Zeeman interaction associated with the velocity and hence the electric current of the ligand electrons. What data there are tend to support this assumption; ion–lattice coupling constants seem to be of the same order of magnitude for ultrasonic and for phonon (true lattice mode) frequencies. On the theoretical side, purely classical calculations support the assumption strongly. Such model calculations are however not a realistic indication, and indeed the assumption of time-even phonon coupling seems to be a very ill-tested one, for all its universal acceptance. The electrostatic energy differences between different states of say a lanthanide ion within a J manifold are an order of magnitude too small to explain the crystal field. It is necessary to invoke the strong effects on the apparent directional (really, lanthanide orbital) dependence of the crystal field due to overlap and covalency to explain the experimental data. These quantum corrections are even more important for the electrostatic and time-even dynamic coupling (Newman 1971). No analysis of the effects of static and dynamic quantum corrections and possible amplifications of the Zeeman interaction for example is available. There is even some evidence of the physical importance of a time-odd coupling in the Barnett effect (Fletcher and Pooler 1982).

Disregarding all such doubts, we choose plus signs for all time-reversal phases for all subsystem operators in the arguments of §3.6.4 and so constrain the interactions to have the symmetry:

$$V^{(n)}_{ijk\eta l\zeta\ldots} = V^{\underline{(n)}}_{jik-\bar{\eta l}-\zeta}. \tag{7.63}$$

As a result, the interaction has a time-even character, and new selection rules based on Comment (b) of Problem 3.12 apply (§8.4.1).

Incidentally, before one may write $-\mathbf{k}$ for k, as is normally done in the literature on ion–phonon coupling, one should decide on a relative phase convention for the counterpropagating lattice modes. Our formalism, while more abstract, avoids and is independent of these choices.

7.3.3 Calculations of phonon effects on the self-energy

The phonons complicate the analysis of §7.2 in the following ways. We must add the unperturbed lattice vibration Hamiltonian

$$\mathcal{H}_p = \Sigma_k \mathcal{E}_k (b_k^\dagger b_k + \tfrac{1}{2}) \tag{7.64}$$

to our previous $\mathscr{H}$, now relabelled $\mathscr{H}_e$, to form the new unperturbed Hamiltonian $\mathscr{H}$. $\mathscr{E}_k = \hbar\omega_k$ and is the phonon mode energy.

Since the field operators $\{b_k\}$ satisfy boson statistics, we have to define a boson Green's function $D_k(\tau) \equiv -\langle T_\tau\{b_k(\tau)b_k^\dagger(0)\}\rangle$, in which now the definition of the chronological operator does not include a sign change on reordering. This Green's function is now *periodic* (instead of antiperiodic) in τ with period β, so that under Fourier transformation we find the characteristic Fourier frequencies to be given by

$$\omega_q = 2q\pi/\beta,\ q = -\infty, \ldots, \infty \tag{7.65}$$

in place of equation (7.14). We show in §7.5 how Bose–Einstein statistics may be derived from this choice of frequency variable.

The unperturbed Fourier coefficient is then readily proved to have the same form as for fermions:

$$\mathscr{D}_k(E) = \frac{1}{E - \mathscr{E}_k} \leftrightarrow \quad \underset{k}{\sim\!\sim\!\sim} \,. \tag{7.66}$$

For the phonon system alone this leads to the same form for Wick's theorem, with various sign changes appropriate to Bose–Einstein statistics at intermediate points in the analysis; for example, closed loops no longer generate a minus sign.

In the unperturbed system, the additivity of the Hamiltonian $\mathscr{H}$ makes all density matrices factorise, and the result of any calculation is simply the product of the results for each subsystem, electronic and vibrational. When, however, the interaction $\mathscr{V}$ is included as in the formalism of §7.2.5 for the time development of the system, both electronic and vibration operators in the interaction representation enter each expectation value. Wick's theorem still amounts to an expansion of each such term as a sum over all possible contractions. The only survivors amongst these will be those in which for each contraction an annihilation and creation operator are paired within either the electronic or lattice subsystem.

For example, we take again the second-order term in the expansion of the numerator, first in the time domain:

$$N^{(2)} = \Sigma_{pqrsk\pm l\pm} V_{pqk\pm} V_{rsl\mp} \iint_0^\beta d\tau_1 d\tau_2 \times$$

$$[(-1)^2/2!]\langle T_\tau\{a_i(\tau)a_p^\dagger(\tau_1)a_q(\tau_1)a_r^\dagger(\tau_2)a_s(\tau_2)a_i^\dagger(0)b_k^\pm(\tau_1)b_l^\mp(\tau_2)\}\rangle_0. \tag{7.67}$$

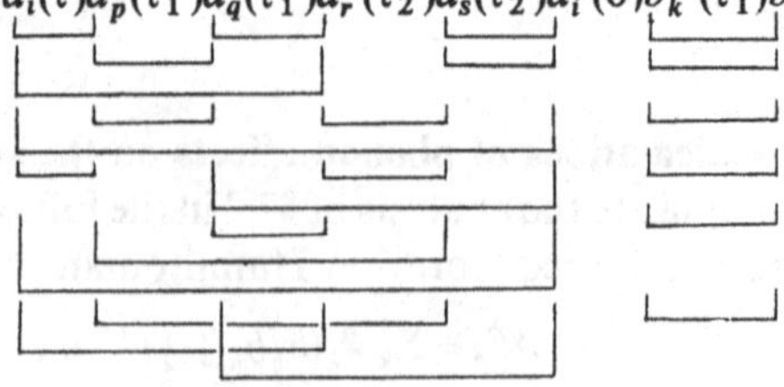

The six possible pairings indicated above lead to the six diagrams:

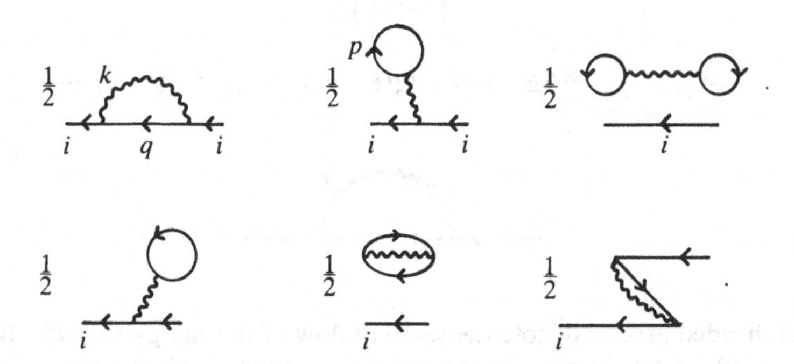

Arrows on the phonon propagators are omitted; both senses are to be used, and the results summed.

The first term gives the final contribution:

$$N^{(2a)}=\tfrac{1}{2}\Sigma_{qk\pm}\iint_0^\beta d\tau_1 d\tau_2 G_i^0(\tau-\tau_1)G_q^0(\tau_1-\tau_2)G_i^0(\tau_2)[|V_{iqk\pm}|^2D_k^0(\pm[\tau_1-\tau_2])]. \qquad (7.68)$$

We note the following. The two terms in the summation over sign correspond to phonon creation and destruction, respectively. The appearance of the phonon propagator gives the nested convolutions a more complicated form than merely the associative combination that we had for linear diagrams; this reflects the nontrivial topology of the linked diagrams.

In the frequency domain, the same diagram has a similar interpretation, in which the lines depict the corresponding $\mathscr{G}_i^0(\varepsilon_i)$ and $\mathscr{D}_k^0(\varepsilon_k)$. These energy variables are of fermionic or bosonic type (i.e. even or odd multiples of $i\pi/\beta$) according to the propagator to which they relate.

As proved in §7.2.8, the structure of the nested convolution integrals in such expressions, e.g. equation (7.68), shows that these energy variables satisfy Kirchhoff's laws for current, i.e. sum algebraically to zero, at all vertices. As a result, each loop of propagators (of either sort) in a diagram (whether connected or disconnected), implies the need for one free energy variable for summation. Within connected diagrams (which as we shall show in §7.4 is the only part necessary for calculations) this free variable is conveniently taken to be a phonon energy variable appropriate to the loop.

Again we take the first second-order contribution to the numerator as an example. In the Fourier or energy domain, combining topologically equivalent diagrams (so omitting a factor $\tfrac{1}{2}$), we have a contribution:

$$N^{(2a)}(i\omega_p)=$$

$$(1/\beta^3)\Sigma_{pk\pm}\iint_0^\beta d\tau' d\tau'' \sum_{q,r,s=-\infty}^{\infty} \mathscr{G}_i^0(i\omega_p)\,\mathscr{G}_i^0(i\omega_r)\,\mathscr{G}_p^0(i\omega_s)|V_{ipk\pm}|^2$$

$$\times\, \mathscr{D}_k^0(\pm\varepsilon_q)\exp[(\tau-\tau')\omega_p+(\tau'-\tau'')(\omega_q+\omega_s)+\tau''\omega_r].$$

The integrals give $\omega_p - \omega_q - \omega_s = 0$, $\omega_r = \omega_p$, with two factors of β, so that

$$N^{(2a)}(E) = \frac{1}{\beta}\Sigma_{pk\pm} \sum_{q=-\infty}^{\infty} \mathscr{G}_i^0(E)\,\mathscr{G}_i^0(E)\,\mathscr{G}_p^0(E-\varepsilon_q)[|V_{ipk\pm}|^2\,\mathscr{D}_k^0(\pm\varepsilon_q)] \leftrightarrow \tag{7.69}$$

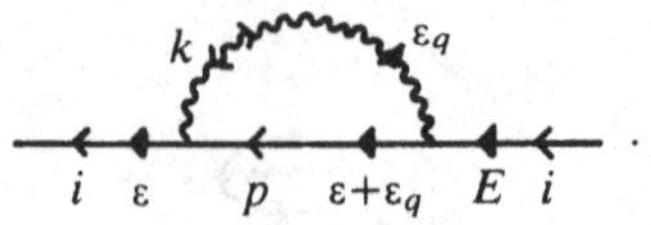

k ε_q
i ε p $\varepsilon + \varepsilon_q$ E i

Full-headed arrows denote the sense of flow of the energy variable. If the two kinds of arrows have opposite direction on say a phonon propagator, the counterpart will be $\mathscr{D}_k^0(-\varepsilon)$. Some comments on the efficient calculation of these sums will be deferred to §7.5.

The phonon propagator itself may be expanded perturbatively in the same interactions (the electronic couplings and the anharmonicity of the lattice):

Once again a Dyson equation may be constructed:

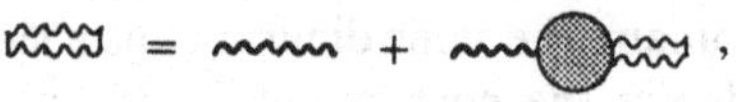

with the self-energy now termed the *polarisation* part:

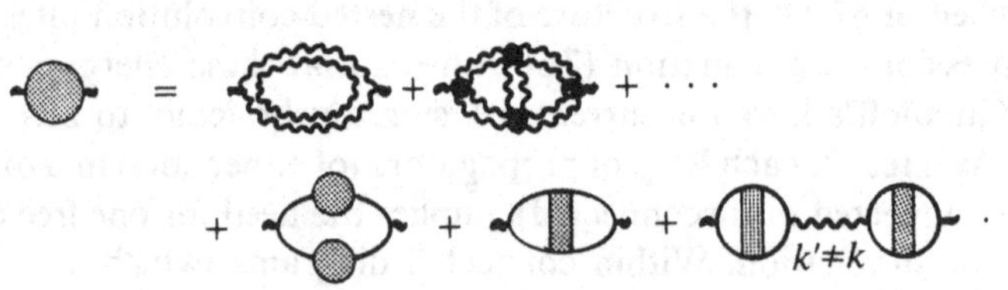

$k' \neq k$

Once again, the real and imaginary parts give the perturbation-induced shift and width of the lattice modes.

7.4 The Abrikosov projection

7.4.1 Introduction

The problem we tackle in this section is the following. If the traces over complete sets of states in §7.2 are restricted to the physical one-particle states $a_i^\dagger|\rangle_0$, the various theorems, in particular the linked cluster theorem, fail. The introduction of an anticommutator between two fermion operators implies for its accuracy sums over intermediate states which include all states in the Fock space $\{|\rangle_0, a_i^\dagger|\rangle_0, a_i^\dagger a_j^\dagger|\rangle_0, \ldots, |i, j, \ldots\}$ generated by the fermion operators. Equally, the logical result of the use of anticommutators is the derivation of Fermi–Dirac statistics. However, as mentioned in §7.2, we expect Maxwell–Boltzmann statistics.

Of many possible solutions to this problem, we discuss only the Abrikosov method. This leads to a minimal change in the diagram formalism, in which the physical interpretation of the parts of a diagram, beloved of all devotees of Feynman and discussed at the end of §7.2, is preserved. In addition, the Abrikosov technique turns out to be at least as convenient as the alternative methods – drones, quasispin, etc. (McKenzie and Stedman 1976, Psaltikas and Cottam 1980, 1981) – for calculation in the general case.

Finally, and most importantly for us, the resulting diagrams correspond almost directly to the group theoretic diagrams introduced earlier in this book. This enables us to link together the two apparently different techniques. This is the central reason for including the material of this chapter in this book. It shows that it is possible to discuss the group theory of time-dependent and time-independent perturbation theory at finite temperatures and with the full and standard quantum field theoretic formalism as starting point in one coherent manner, without the need for justifying separate techniques for different parts of the problem, such as the introduction of Goldstone diagrams as is customarily done in atomic physics.

If our interest is limited to time-independent perturbation theoretic expressions for energy shift, there is no strong reason to develop the full formalism. One is free to invent diagrams to describe the terms in the Rayleigh–Schrödinger or Brillouin–Wigner series, preferably avoiding some of the oddities with which the literature on diagram techniques is littered, but rather being guided by the general form of the field theoretic formalism. Examples have been mentioned in §7.1, and it is indicated there how it is then possible to assign group theoretic significance to the lines in the same manner we do later in the Abrikosov technique. Such methods are limited in their applicability; the quantum field theoretic development is universal in its applicability. It seems worthwhile in the long term to make the necessary capital investment (§7.2 and this section).

The essential idea of the Abrikosov method is to retain the fermion formalism, but to add a large constant to the energy of the one-fermion state:

$$E_i^0 \to E_i^0 + \lambda. \tag{7.70}$$

Since λ is independent of i, this will not affect the relative energies or the relative populations of the one-fermion physical states $\{|i\rangle_0 = a_i^\dagger|\rangle_0\}$;

Maxwell–Boltzmann statistics will be preserved in this subspace. However, as $\lambda \to \infty$, there is a corresponding reduction, of the order of $\exp[-(n-1)\beta\lambda]$, in the relative contribution of n-particle states $\{a_i^\dagger a_j^\dagger a_k^\dagger \ldots |\rangle_0\}$ for $n > 1$. In effect, their unphysical contributions are 'frozen out'.

The remaining problem is that, while this argument indeed destroys the effect of the multiparticle states, by the same token it enhances the importance of the zero-particle (vacuum) state, which is equally unphysical in the applications we discuss. We wish our ion to be definitely present, and in some particular level! It is agreed that it is actually made of fermions – quarks, protons and neutrons, electrons – and that the vacuum is at root a possible state. However, the energy penalty in contemplating ion pair creation and annihilation is vastly in excess of the energies of interest in the optical spectroscopy of condensed matter systems! In addition, our formalism will not have the relevant strong and electroweak physics sufficiently fully embodied to cope with such quantum field theoretic effects. In practice, therefore, we wish to limit consideration to those intermediate states which preserve the ion in some level. These have the real practical interest. The problem is that any prescription which uses energetic considerations alone to kill the multiparticle states enhances the effects of the vacuum state.

It turns out that, while this complication is sufficiently fundamental to invalidate the linked cluster theorem, the numerator expansion at least will not need special handling of the vacuum state, since in every contributing term the form $a_i^\dagger a_j$ for the second quantised operators in each interaction term ensure that an annihilation operator always lies to the right in the time ordering, and so always acts first to kill the vacuum contribution. This is not so for the denominator terms; the unphysical effects of the vacuum will appear not in determining the spectral shape (the energy dependence, defined by the numerator), but merely an overall normalisation (the denominator), and so for virtually all practical purposes may be neglected. The problem then becomes much less fearsome.

Indeed the Abrikosov projection even becomes a simplification of the original fermionic quantum field theoretic expansion. We show later that various diagrams that need to be considered in the fermion case vanish under Abrikosov projection.

To prove all this, we give a more systematic account of the Abrikosov projection technique.

7.4.2 Physical and Fock space expectation values

We wish to evaluate the expectation value with respect to the physical one-particle subspace of some electronic operator f relevant to the coupled electron–phonon system:

$$\mathscr{F} = \frac{\mathrm{Tr_{phys}}\{\exp[-\beta(\mathscr{H}+\mathscr{V})]f\}}{\mathrm{Tr_{phys}}\exp[-\beta(\mathscr{H}+\mathscr{V})]}. \tag{7.71}$$

This may be written, using equation (7.29) and the separability of the unperturbed Hamiltonian, as $\mathscr{F} = \mathscr{N}_1/\mathscr{D}_1$, where

$$\begin{aligned} \mathscr{N}_1 &\equiv \mathscr{L}_1^{-1}\mathrm{Tr}_l[\exp(-\beta\mathscr{H}_l)\mathrm{Tr}_1\{\exp(-\beta\mathscr{H}_e)U(\beta,0)f\}] \\ \mathscr{D}_1 &\equiv \mathscr{L}_1^{-1}\mathrm{Tr}_l[\exp(-\beta\mathscr{H}_l)\mathrm{Tr}_1\{\exp(-\beta\mathscr{H}_e)U(\beta,0)\}] \\ \mathscr{L}_1 &\equiv \mathrm{Tr}_l\exp(-\beta\mathscr{H}_l)\mathrm{Tr}_1\exp(-\beta\mathscr{H}_e) \end{aligned} \tag{7.72}$$

and Tr_n represents the trace (i.e. diagonal sum) over *unperturbed* n-fermion states $\mathrm{Tr}_n X \equiv \Sigma_{\{r_i\}}\langle|a_{r_n}\ldots a_{r_1}Xa_{r_1}^\dagger\ldots a_{r_n}^\dagger|\rangle_0$.

Let us now replace $\mathscr{H}_e$ by $\mathscr{H}_e^\lambda \equiv \mathscr{H}_e + \lambda\mathscr{N}$, where $\mathscr{N} = \Sigma_i a_i^\dagger a_i$ and is the number operator. This transformation will not affect operators such as the evolution operator $U(\beta,0)$ or the interaction $\mathscr{V}(\tau)$ in the interaction representation, since the fermion operators in these operators are number-conserving and commute with the number operator (there are as many annihilation as creation operators, viz. one of each). $\mathscr{V}(\tau)$ is independent of λ. Hence the only effect of the additive term is to change the thermal exponential term to $\exp[-\beta(\mathscr{H}_e+\lambda\mathscr{N})]$, and thus induce within any trace over n-fermion states (Tr_n) a factor $\exp(-\beta n\lambda)$.

The value $\langle f\rangle$ obtained from the full diagram expansion is given by

$$\overline{\mathscr{F}} = \frac{\Sigma_n\mathrm{Tr}_n\{\exp[-\beta(\mathscr{H}+\mathscr{N}\lambda)]U(\beta,0)f\}}{\Sigma_n\mathrm{Tr}_n\{\exp[-\beta(\mathscr{H}+\mathscr{N}\lambda)]U(\beta,0)\}} = \mathrm{N/D}, \tag{7.73}$$

with

$$\begin{aligned} \mathrm{N} &\equiv \mathrm{Z}^{-1}\sum_{n=1}^{\infty}\Lambda^n\mathscr{N}_n, \quad \mathscr{N}_n \equiv \mathrm{Tr}_l[\exp(-\beta\mathscr{H}_l)\mathrm{Tr}_n\{\exp(-\beta\mathscr{H}_e)U(\beta,0)f\}] \\ \mathrm{D} &\equiv \mathrm{Z}^{-1}\sum_{n=0}^{\infty}\Lambda^n\mathscr{D}_n, \quad \mathscr{D}_n \equiv \mathrm{Tr}_l[\exp(-\beta\mathscr{H}_l)\mathrm{Tr}_n\{\exp(-\beta\mathscr{H}_e)U(\beta,0)\}] \\ \mathrm{Z} &\equiv \sum_{n=0}^{\infty}\Lambda^n\mathscr{Z}_n, \quad \mathscr{Z}_n \equiv \mathrm{Tr}_l\exp(-\beta\mathscr{H}_l)\mathrm{Tr}_n\exp(-\beta\mathscr{H}_e), \end{aligned} \tag{7.74}$$

where $\Lambda \equiv \exp(-\beta\lambda)$ and all $\mathscr{N}_n$, $\mathscr{D}_n$, $\mathscr{L}$ are independent of λ. The term $n=0$ in N vanishes since any electronic operator f annihilates the unperturbed electronic vacuum $|\rangle_0$. For the same reason, $\mathscr{D}_0 = \mathscr{L}_0$ since in this term only the zero-interaction part (equal to unity, from equation 7.31) of the evolution operator $U(\beta,0)$ can contribute.

However, what is required as the physical answer is $\mathscr{F} = \mathscr{N}_1/\mathscr{D}_1$.

We expand N and D in powers of Λ: $\mathrm{N} = \Sigma_{n=1}^{\infty}\mathrm{N}^{(n)}\Lambda^n$, $\mathrm{D} = \Sigma_{n=0}^{\infty}\mathrm{D}^{(n)}\Lambda^n$, we obtain $\mathrm{N}^{[1]} = \mathscr{N}_1/\mathscr{Z}_0$, $\mathrm{D}^{(1)} = \mathscr{D}_0/\mathscr{Z}_0 = 1$, $\mathscr{D}^{(1)} = (\mathscr{D}_1 - \mathscr{Z}_1)/\mathscr{Z}_0$. We solve these for $\mathcal{N}_1$ and $\mathcal{D}_1$ in terms of N and D, and note that from the linked cluster theorem (equation 7.51) $\mathrm{N}=\mathrm{N_c D}$:

$$\begin{aligned} \mathcal{N}_1 &= \mathrm{N_c D}\ \mathcal{Z}_0/\Lambda + \mathrm{O}(\Lambda) \\ \mathcal{D}_1 &= (\mathrm{D}-1)\ \mathcal{Z}_0/\Lambda + \mathcal{Z}_1 + \mathrm{O}(\Lambda) \end{aligned} \tag{7.75}$$

$$\frac{\mathcal{N}_1}{\mathcal{D}_1} = \frac{\dfrac{N_c\mathrm{D}}{\Lambda \mathcal{Z}_1} + \mathrm{O}(\Lambda)}{1 + \dfrac{(\mathrm{D}-1)}{\Lambda \mathcal{Z}_1} + \mathrm{O}(\Lambda)}$$

The terms $\mathrm{N_c}$ and $(\mathrm{D}-1)$ in numerator and denominator of this expression are both of O (Λ) (since $\mathcal{N}_0=0$). The product $\mathrm{N_c D}$ deviates from $\mathrm{N_c}$ only at O (Λ_2) since the bubble diagrams in D are of O (Λ); hence we can replace D in the numerator by unity, the difference being absorbed in the additive $\mathrm{O}(\Lambda)$ term. These additive terms of $\mathrm{O}(\Lambda)$ in numerator and denominator will each go to zero in the limit $\beta\lambda \to \infty, \Lambda \to 0$. In this limit we therefore may write equation (7.75) in the form

$$\mathcal{F} = \frac{\mathcal{P}(\mathrm{N_c})}{1+\mathcal{P}(\mathrm{D}-1)}, \tag{7.76}$$

where we define the Abrikosov projection operator as:

$$\mathcal{P}(\mathcal{O}) \equiv \mathcal{Z}_1^{-1} \lim_{\beta\lambda\to\infty} (\Lambda^{-1}\ \mathcal{O}). \tag{7.77}$$

This shows that the numerator need only include linked diagrams, and then only the $\mathrm{O}(\Lambda)$ part of these. Unlinked vacuum excitation diagrams are present in the denominator at the same order as the contribution of the vacuum state; the potential dominance of the vacuum term here is neutralised by the denominator Λ factor in equation (7.77).

We generalise this argument to the case when the operator f contains a term which is not of the form of a single-particle electronic operator $a^{\dagger}a$, and so has a nonvanishing expectation with respect to the unperturbed electronic vacuum. Such a term emerges unscathed from the thermal average of equation (7.72), and appears additively on the right side of equation (7.76). The remaining terms (including at least one $a^{\dagger}a$) in f are subject to the full cosmetic treatment of equations (7.76), (7.77). On recombining their results, it often occurs that both types of terms survive Abrikosov projection.

We take two examples. First, the perturbed phonon propagator of §7.3.3 will contain terms due to lattice anharmonicity etc. which do not depend on the electronic system; these are not affected by the discussion of this section. It will also contain terms with one or more electronic bubble on the phonon propagator; these need projection as above. When this is performed, only the lowest order terms will survive, namely those corresponding to a single electronic bubble.

Second, an evaluation of the perturbed electronic Green's function $G_i(\tau)$ in the time domain involves the calculation of an average $\mathscr{F}$ of a kernel operator given by $f = \mathrm{T}_\tau\{U(\beta,0)a_i(\tau)a_i^\dagger\}$. This has the vacuum expectation value $\exp(-\tau E_i^0)$ associated with the anticommutator of $a_i(\tau)$ and $a_i^\dagger$. This term should in principle contribute at the end of the calculation. However, the consistent application of the Abrikosov limit $\lambda \to \infty$ demands that this term vanish through its E_i^0 dependence. The remaining terms in f survive the full-blown Abrikosov projection. It is curious that this corresponds in equation (7.15) to retaining the (projected) ρ_i^0 factor rather than the unit factor in the coefficient $(1-\rho_i^0)$; one would naively have expected the opposite! Since the Fourier coefficient of equations (7.17), (7.18) has no explicit population dependence, Abrikosov projection is relatively straightforward in the energy domain.

7.5 Calculational methods

We summarise here a variety of hints with which a newcomer may quickly develop facility in calculating perturbative contributions to one-particle self-energies, widths and shifts.

In practice the analysis of §7.4 means that we may ignore the unlinked terms in the numerator expansion. For ionic states, this means that we may restrict attention to terms such as:

The central diagram in this figure has an extra electronic integration through the electronic loop. This makes it of one higher order in Λ than the other figures; it too may be ignored in the Abrikosov limit.

We shall see the physical effects of this restriction by evaluating this series and comparing with the original form. To do this, we employ the standard trick of converting the sum over any free energy variable to a contour integration, using Cauchy's theorem:

$$\sum_{p=-\infty}^{\infty} f[(2p+1)\mathrm{i}\pi/\beta] = -2\pi\mathrm{i}\beta \oint_{\mathrm{C}} \frac{f(z)\mathrm{d}z}{\exp(\beta z)+1},$$

$$\sum_{p=-\infty}^{\infty} f[2p\mathrm{i}\pi/\beta] = 2\pi\mathrm{i}\beta \oint_{\mathrm{C}} \frac{f(z)\mathrm{d}z}{\exp(\beta z)-1}, \tag{7.78}$$

where f is any function which is analytic in the contour C, which in turn encloses in an anticlockwise sense all the poles of the denominator in the integrand along the imaginary axis of the complex z plane. These poles are precisely the points specified by $i\omega_p$, $p = -\infty, \ldots, \infty$:

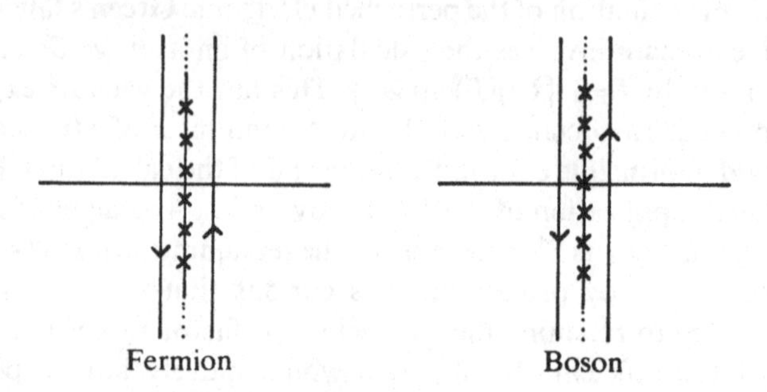

In verifying equation (7.78) it is only necessary to note that the residues of the denominators are given by $(\eta \equiv z - i\omega_p)$:

$$\lim_{\eta \to 0} \frac{\eta}{\exp(i\omega_p\beta + \beta\eta) \pm 1} = \mp \lim_{\eta \to 0} \frac{\eta}{\exp(\beta\eta) - 1}$$

(the upper sign for fermions, and the lower for bosons) and so are, respectively, $-1/\beta$, $1/\beta$. Notice that we substitute into this expression the value for z on the imaginary axis, namely $i\omega_p$ with ω_p being an odd or even multiple of π/β, i.e. the value of z prior to analytic continuation. In all following work it must be remembered that the evaluation of residues when energy variables are involved is to be carried out with these variables on the imaginary axis.

In practice $f(z)$ will be the product of electronic and phonon propagators. In each integration z will be some free energy variable. The poles of such functions are on the real axis, where the energy variables may equal the real excitation energies. If we now analytically continue to the real axis, the contour integration may be performed by finding the residues at the poles of $f(z)$:

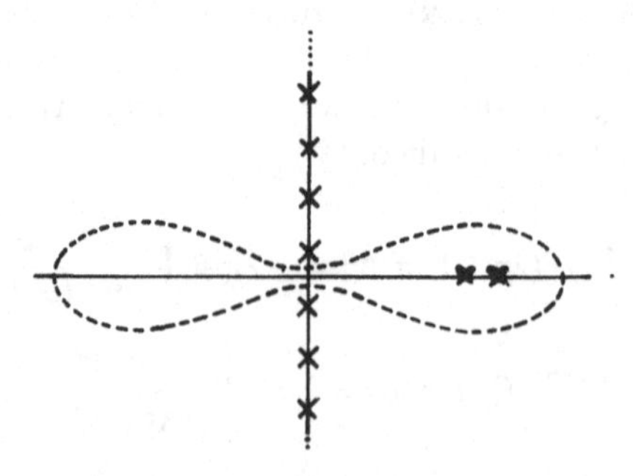

(The zero frequency pole in the bosonic case may need special treatment; see Stevens and Toombs 1965, also Kwok and Schultz 1969.) The factor $2\pi i$

cancels on re-expressing the contour integration in terms of residues. The factor β in equation (7.78) cancels the residual factor $1/\beta$ derived from the frequency space conversion, as in equation (7.69) for example.

Suppose that we are evaluating a perturbed fermionic (and ultimately ionic) propagator. There will be at least one electronic energy variable present – the 'current flow' through the string of electronic propagators threading the diagram. (There is only one such string in the Abrikosov limit.) It is not an integration variable, but will ultimately be analytically continued to the real axis (equation 7.27). We arrange by convention that the only free energy variables associated with loops in the diagram and therefore requiring integration are boson energy variables of the excitations in those loops. (If intra-bosonic interactions such as lattice anharmonicity are included in the interaction, not all boson energy variables are independent, but we can still arrange that all independent energy variables are bosonic.) In this case we will prove that within the Abrikosov limit we need never integrate over an electronic energy variable within loops. However, this restriction is not possible before we take the Abrikosov limit.

In the fermion case, the change in the sense of the contour from anticlockwise to clockwise upon analytic continuation compensates the minus sign in equation (7.78). The residue will require evaluation at the poles of all propagators, fermion and boson. Suppose we consider the contribution from the pole of an electronic Green's function $\mathcal{G}_i^0(z)$ whose energy variable is chosen to be the free variable z. This entails the substitution $z=E_i^0$ into the other propagators and into the expression $[\exp(\beta z)+1]^{-1}$. From the latter we obtain a factor $[\exp(\beta E_i^0)+1]^{-1}$, precisely the Fermi–Dirac population factor ρ_i^0.

When the Abrikosov projection (equation 7.77) is applied to the formalism, this goes over to the Maxwell–Boltzmann expression $\exp(-\beta E_i^0)/\Sigma_j\exp(-\beta E_j^0)$, since the additive enhancement of energies by λ kills the term $+1$ in the Fermi–Dirac factor, and the factor $1/\Lambda$ in equation (7.74) has the effect of removing the λ-dependence in the final answer.

For boson integrations, a similar result obtains. In the applications we discuss, the free energy variables may all be chosen to be boson variables. When the factor of equation (7.78) is evaluated at the pole of some phonon propagator $\mathcal{D}_k^0(z)$, i.e. for $z=\mathcal{E}_k$, we obtain the phonon Bose–Einstein population factor, together with a minus sign for the clockwise–anticlockwise inversion of the contour. Abrikosov projection has no effect on such terms.

This factor must also be evaluated at the poles of the electronic propagators, although the free variables are bosonic. This gives terms in which the population factors are more complicated, and involve a mixture of electronic and phonon unperturbed energies and energy variables (the latter always to be evaluated initially on the imaginary axis). Often these need very careful simplification, and the end result is an apparently miraculous emergence of the right combination of electronic and phonon populations to give this term the relevant significance.

These more complicated terms from the poles of the electronic propagators may in fact be eliminated using the following convention. We require first that the only free energy variables are bosonic, and, second, that the sense of flow of these free and bosonic energy variables in the Feynman diagram be given (by an available and unambiguous convention) a sense *opposite* to that of the energy variables in the string of electronic propagators threading through the corresponding diagram. This is illustrated in equation (7.69). In this way we ensure that bosonic energy variables enter the electronic propagators with a plus sign. This in turn means that when we evaluate the contribution of this pole during a bosonic variable integration, the electronic energy E_i^0 appears in the relevant substitution prefaced by a plus, rather than a minus, sign. As a result, under Abrikosov projection its contribution relative to other terms in the series will be of O (Λ) and so vanish in the limit.

We therefore distinguish, by a heavy arrow, the sense of the energy variable as fixed by the above convention, and the sense of the bosonic propagation from creation to annihilation (equation 7.69). In any perturbative calculation, processes involving phonon creation and annihilation are to be summed and will appear in the combination

$$V_{ijk+}V_{pqk-}\mathscr{D}_k^0(\varepsilon_k)+V_{ijk-}V_{pqk+}\mathscr{D}_k^0(-\varepsilon_k)\leftrightarrow$$

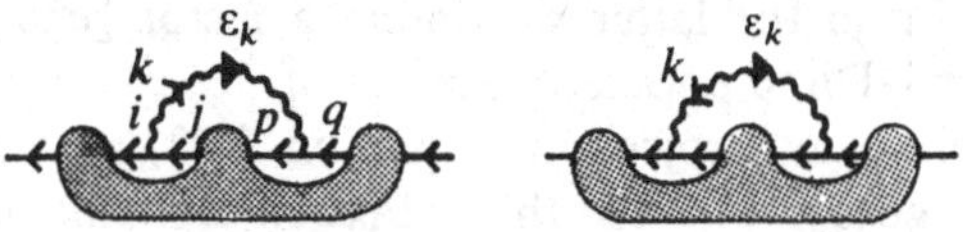

within the algebraic counterpart to the diagram, along with electronic propagators such as $1/[z+\varepsilon_k-E_i^0]$ (z being the electronic energy variable) and the embryonic population factor $[\exp(\beta\varepsilon_k-1]^{-1}$. To evaluate the residues at the poles of the terms in equation (7.76), we substitute $\varepsilon_k=\pm\mathscr{E}_k$, respectively, in to these other factors, and note the presence of an extra overall negative sign in the Green's function in the second case, since it contains $-1/[\varepsilon_k+\mathscr{E}_k]$. From the embryonic population factor we derive the factors $n_k^-\equiv n_k=[\exp(\beta\mathscr{E}_k)-1]^{-1}$ and $(-n_k^+)$, where $n_k^+=n_k^-+1$; the negative sign from the population factor in the second case cancels with that from the Green's function. These terms clearly describe the probability of phonon annihilation and (spontaneous plus stimulated) creation according to the physics to which the diagram corresponds. From the above electronic propagator we get a factor $1/(z\pm\mathscr{E}_k-E_i^0]$, respectively, which by the time z is continued to the real (and perturbed) electronic energy E (equation 7.27) gives the energy mismatch appropriate to this intermediate state.

7.6 One-particle self-energy contributions

As an example of this formalism, we calculate the leading terms in the self-energy for the case of linear and nonlinear ion–phonon coupling together with anharmonicity amongst the phonons.

7.6.1 First-order quadratic coupling

In first order the linear term vanishes. It necessarily involves a change of lattice mode occupancy and has no diagonal elements. The quadratic coupling however has a first-order contribution to the self-energy.

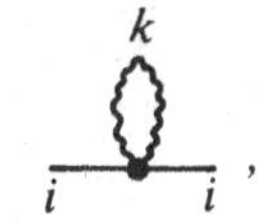

$$\Sigma_{ik\pm} V^{(2)}_{iik\pm k\pm} \oint \mathrm{d}\varepsilon\, {}^{\dagger}\mathscr{D}^0_k(\pm\varepsilon)/[\exp(\beta\varepsilon)-1] = \Sigma_{ik\pm} V^{(2)}_{iik\pm k\mp} n^{\pm}_k$$
$$= \Sigma_{ik} V^{(2)}_{iik+k-}(2n_k+1). \tag{7.79}$$

This contribution to the shift can be interpreted as the extent to which the average value of the electronic energy lies above the nominal values, as a result of the zero-point and temperature-dependent excursion of the lattice motion on a concave energy curve.

7.6.2 Linear coupling at second order

The single contribution from the linear interaction at second order evaluates as the expression (from equation 7.66, but ignoring the $\mathscr{G}^0_i(E)$ propagators which do not belong to the self-energy)

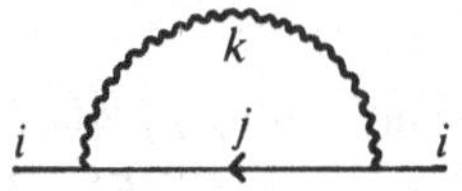

$$\leftrightarrow \Sigma_{jk}\left\{|V_{ijk+}|^2\left[\frac{n_k}{E-E^0_j-\mathscr{E}_k}\right]+|V_{ijk-}|^2\left[\frac{n_k+1}{E-E^0_j+\mathscr{E}_k}\right]\right\}. \tag{7.80}$$

Comments

(a) It is a useful contrast to calculate the corresponding value for true fermion systems, in which the Abrikosov projection is not employed.

The first addition is that we should consider the extra terms in the computation of the above diagram from the poles of the electronic propagator $(E+\mathscr{E}_k-E^0_j)^{-1}$. These have the form

$$\left\{|V_{ijk+}|^2\left[\frac{1}{\varepsilon_k-\mathscr{E}_k}\right]+|V_{ijk-}|^2\left[\frac{1}{\varepsilon_k+\mathscr{E}_k}\right]\right\}\left\{\frac{1}{e^{\beta\varepsilon_k}-1}\right\}\Bigg|_{\varepsilon_k=E_j^0-E}$$

$$=(-\rho_j^0)\left\{|V_{ijk+}|^2\left[\frac{1}{E-E_j^0-\mathscr{E}_k}\right]+|V_{ijk-}|^2\left[\frac{1}{E-E_j^0+\mathscr{E}_k}\right]\right\},$$

the electronic population factor with its negative sign arising from the substitution $E=(2p+1)\mathrm{i}\pi/\beta$ (i.e. the value on the imaginary axis) into the apparently bosonic factor of the above expression.

We have also to consider the second-order linear bubble diagram:

$$\leftrightarrow-(1/\beta)\Sigma_{jk\pm}V_{iik\pm}V_{jjk\mp}\Sigma_{\varepsilon_p}\mathscr{D}_k^0(0)\mathscr{G}_j^0(\varepsilon_p)$$

$$=\Sigma_{jk\pm}V_{iik\pm}V_{jjk\mp}\frac{1}{\mathscr{E}_k\beta}\oint\frac{\mathrm{d}\varepsilon}{(\varepsilon-E_j^0)(e^{\beta\varepsilon}+1)}=\Sigma_{jk\pm}V_{iik\pm}V_{jjk\mp}\rho_j^0/\mathscr{E}_k.$$

One negative sign arises since there is a closed loop of fermion propagators; a cancelling sign arises from $\mathscr{D}_k^0(0)$.

Combining these we get (Schrieffer 1964):

$$\Sigma_i(E)=\Sigma_{jk}\left\{|V_{ijk+}|^2\left[\frac{n_k+\rho_j^0}{E-E_j^0+\mathscr{E}_k}\right]+|V_{ijk-}|^2\left[\frac{n_k+1-\rho_j^0}{E-E_j^0-\mathscr{E}_k}\right]\right\}$$
$$-\Sigma_{\pm}\rho_n V_{iik\pm}V_{jjk\mp}/\mathscr{E}_k. \tag{7.81}$$

Now consider how the electronic population factors might affect the line shift. Taking the imaginary part,

$$\Gamma_i=2\pi\Sigma_{jk}\{|V_{ijk+}|^2(n_k-\rho_j^0)\delta(E_i^0-E_j^0+\mathscr{E}_k)$$
$$+|V_{ijk-}|^2(n_k+1-\rho_j^0)\delta(E_i^0-E_j^0-\mathscr{E}_k)\}. \tag{7.82}$$

Equation (7.82) has a direct physical interpretation, most easily seen by taking zero temperature ($n_k\rightarrow 0$), when only phonon creation processes are possible. The second term reflects the possibility of increasing the transition rate from electronic state i by populating an empty electronic state j (the latter is empty with the Fermi–Dirac probability factor $1-\rho_j^0$ which appears in the expression) and creating a phonon, while the first term gives rather a decrease in the transition rate and so in the linewidth caused by a potential repopulation in the state i from the filled (with probability ρ_j^0) state j, also with the creation of a phonon, since now the sign of $\mathscr{E}_k$ in the argument of the delta function has changed. This interpretation is inapplicable to the ionic interpretation of electronic states, where the filling of state i requires the emptying of state j. The terms proportional to ρ_j^0 cancel under Abrikosov projection. This may be verified, in view of the nontrivial effect of Abrikosov projection on population factors in the time domain (see the remarks at the end of §7.4), by comparison with the imaginary part of equation (7.80).

(b) The contribution this makes to the shift may be evaluated as for the quadratic term, by substituting the unperturbed energy E_i^0 for the energy variable E (strictly, by an analytic continuation of E to E_i^0) and by taking the real part, in this case a principal value integral about the denominator pole:

$$\Lambda_i = \Sigma_{jk}\left\{|V_{ijk+}|^2\left[\frac{n_k}{E_i^0 - E_j^0 - \mathscr{E}_k}\right] + |V_{ijk-}|^2\left[\frac{n_k+1}{E_i^0 - E_j^0 + \mathscr{E}_k}\right]\right\}. \quad (7.83)$$

There is now an imaginary part, from equation (7.24):

$$\Gamma_i = 2\pi\Sigma_{jk}\{|V_{ijk+}|^2 n_k \delta(E_i^0 - E_j^0 - \mathscr{E}_k) + |V_{ijk-}|^2 (n_k+1)\delta(E_i^0 - E_j^0 + \mathscr{E}_k)\}. \quad (7.84)$$

This corresponds to putting the intermediate, previously virtual, excitation (electronic state $|j\rangle$, together with the addition or subtraction of one quantum from lattice mode k) on the mass shell, that is by taking the limit in which the creation of this intermediate state from electronic state $|i\rangle$ conserves energy. We recognise this as a special case of the Golden Rule for transition rates whose general form was discussed in §7.2.9.

(c) In general, every topologically distinct way of cutting through a self-energy diagram isolates a possible intermediate state that may resonate with the original state, and thus a contribution to the transition rate out of any state.

(d) The value of the energy variable substituted into this equation in order to evaluate the diagram in a consistent perturbative manner is equal to the unperturbed energy strictly only to zeroth order accuracy. For greater accuracy, the energy substituted for E must be progressively corrected by the real part of this and the subsequent contributions to the full self-energy.

7.6.3 Linear, quadratic and anharmonic coupling to fourth order

Computing the diagrams in this category:

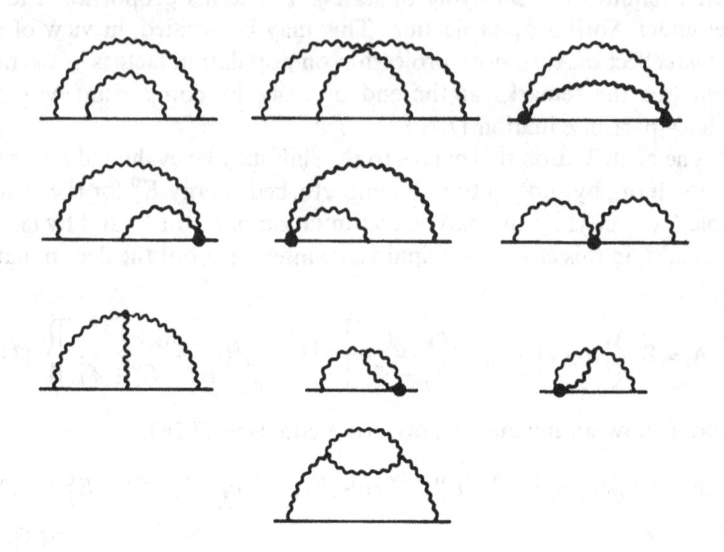

and retaining only the Raman terms (one phonon absorbed, the other created) for simplicity, we obtain

$$\Sigma_i(E)=\Sigma_{jkl}\frac{n_k(n_l+1)}{E-E_j^0+\mathscr{E}_k-\mathscr{E}_l}|T_{ijkl}|^2, \tag{7.85}$$

where the transition amplitude T_{ijkl} may be written as a sum $\Sigma_{r=1}^3 T_{ijkl}^{(r)}$ of the various amplitudes corresponding to all types of interaction linking the initial and the intermediate state, at least in the limit that $E\to E_i^0$. The square modulus structure of this expression is a consequence of the reflection symmetry of the diagram series.

The diagram amplitudes

contribute, respectively:

$$T_{ijkl}^{(1)}=\Sigma_p\left\{\frac{V_{pik}V_{jpl}}{E_i^0-E_j^0+\mathscr{E}_k}+\frac{V_{pil}V_{jpk}}{E_i^0-E_j^0-\mathscr{E}_j}\right\},$$

$$T_{ijkl}^{(2)}=2V_{ijkl}^{(2)}, \tag{7.86}$$

$$T_{ijkl}^{(3)}=\Sigma_m W_{klm}^{(1)}V_{jim}\left\{\frac{1}{\mathscr{E}_k-\mathscr{E}_l+\mathscr{E}_m}-\frac{1}{\mathscr{E}_k-\mathscr{E}_l-\mathscr{E}_m}\right\}$$

Each of these T-matrix terms on its own corresponds to a reflection symmetric diagram. Reflection–asymmetric diagrams correspond to the cross (or interference) terms in the expansion of the square modulus of equation (7.85).

Comments

(a) It would be convenient if we could have anticipated this form of the final result, and write the self-energy contribution as the sum over a square modulus of a T-matrix or amplitude summation. In general this is rendered difficult by virtue of the necessity of applying the Abrikosov projection to the final form of each self-energy diagram, and by the need for specifying the full diagram to obtain the correct population factors. Nevertheless, all our final results will be of this form.

(b) There is a new feature in the computation of diagrams involving lattice anharmonicity. Not all bosonic energy variables are free variables of summation, and one has to keep track of a wider variety of terms in nested summations, the evaluation of one phonon propagator at the pole of another being in turn relevant for a later stage of the calculation. A useful relation in reducing population factors after two such nested summations/integrations is:

$$n(\dot{\mathscr{E}}_k - \dot{\mathscr{E}}_l)[n(\dot{\mathscr{E}}_k) - n(\dot{\mathscr{E}}_l)] = -n(\dot{\mathscr{E}}_k)n(-\dot{\mathscr{E}}_l),$$

where $n(\varepsilon) \equiv [\exp(\beta\varepsilon) - 1]^{-1}$.

8

Applications

I prepared some lectures and I went to New Zealand to try them out – because New Zealand is far enough away that if they weren't successful it would be all right. . . . You will have to make an awful lot of little arrows on a piece of paper. It takes seven years . . . to train our physics students to do that in a tricky, efficient way. . . .

Feynman (1985)

8.1 Introduction

We offer a pot-pourri of specimen applications of the techniques developed in earlier chapters. Many of these are from condensed matter theory and use our development of quantum field theory in §7. The allied group theoretic methods of §3 will be illustrated in conjunction with the field theoretic diagram below in a wide variety of applications, though certainly not as wide a range of topics as may be obtained from the literature. We shall not include a review of the application of the approach of §§2–3 in exchange theory in metals (Kuramoto 1983) and in lattice gauge theory (Creutz 1983, Hamer, Irving and Preece 1986), for example. Applications in fields such as atomic physics and relativistic quantum field theory are illustrated.

Particularly in the less familiar areas of solid state case we explain the physical context of the illustrations at some length. Sometimes this is vital for understanding later applications of diagrams. For example, the discussion of the separation of no-phonon line from vibronic sideband in the field theoretic formalism is of fundamental importance in discussing Raman scattering and Ham reduction factors in Jahn–Teller systems.

In much of this chapter we shall not trouble to include all but essential signs, numerical factors, reduced matrix elements, nonstandard normalisations etc. Our aim is to show the structure of a wide variety of applications, usually in the simplest or a few representative cases. If the discussion and problems of the earlier chapters have been mastered, including these details will not be a problem.

8.2 Atomic and ligand field physics

8.2.1 Atomic physics

Since this is by far the most well reviewed of all the techniques discussed in this book, we concentrate on listing references and quoting a few examples.

In addition to the books and major reviews referenced in §2.1, many authors have discussed graphical representations (using the Jucys approach) of the effects of spin–orbit interaction (e.g. Huang and Starace 1978, Huang 1979) and configuration interaction (e.g. Briggs 1971, Adams, Paldus and Čižek 1977, Paldus, Adams and Čižek 1977). Judd (1967) gives a brief but innovative introduction. Sandars (1969a,b) illustrates the application to hyperfine structure. Other related works include ElBaz, Massot and Lafoucriere (1966), ElBaz and Nahabetian (1977), ElBaz (1981, 1985, 1986), Billy and Lhuillier (1982, 1983), Mason and Starace (1982). Newman and Wallis (1976) discuss the computer implementation of the Jucys method. The effects of symmetry principles in atomic spectroscopy are reviewed and developed by Wybourne (1970), for example.

There is a superabundance of alternative diagram techniques, of which we merely list a few: Bishton and Newman (1968), Yang and Wang (1974), Skinner and Weil (1978), Grewe and Keiter (1981), Mukhopadhnay and Pickup (1984), Pickup and Mukhopadhyay (1984), Burgos and Bonadeo (1987). (This list could be expanded almost indefinitely, and we apologise to authors whose work is omitted.) Some of these are more highly developed than others. It is not our aim (since it would be a separate and lengthy task) to review all of these in any systematic manner. However, one may roughly characterise most of these methods in one of the following categories. Each may have affinities with some standard perturbation theoretic method such as the Feynman/field theoretic diagram, and/or the Goldstone diagram (Crichton and Yu 1973, Kuo *et al.* 1981), or the Hugenholtz (1965) diagram. If not, it may be relevant to alternatives to the Abrikosov method developed in §7; it will then probably be of a relatively unphysical appearance, as compared to the Feynman diagram. If all of these categories prove irrelevant, it is likely that the method represents a decidedly individual style, while still abiding by many of the rules of §1. Finally, we mention that Judd (1983) discusses one very elegant alternative to the Jucys method, based, as is the Dirac formulation of quantum mechanics, on projective geometry.

One-particle matrix elements: spin–orbit coupling

Problem 8.1

Prove the diagonal sum rule for Landé factors in Russell–Saunders coupling:

$$M_L M_S \lambda(LS) = \Sigma_i m_{l_i} m_{s_i} \zeta(nl). \qquad (8.1)$$

Solution

The constants $\zeta(nl)$ are the coefficients of the one-particle operators $\mathbf{l}_i\cdot\mathbf{s}_i$. For simplicity consider two-electron terms, with antisymmetrised wavefunctions of the form $|\, nlm_l m_s\, nlm'm_s'|$ (in each case $s=\frac{1}{2}$) coupled finally through appropriate Clebsch–Gordan coefficients $\langle lm_l l'm_l'|LM_L\rangle$ $\langle sm_s sm_s'|SM_S\rangle$ to a definite L, M_L, S, M_S. The matrix elements of the operator $\Sigma_i\zeta(nl)\mathbf{l}_i\cdot\mathbf{s}_i$ within such terms have the form:

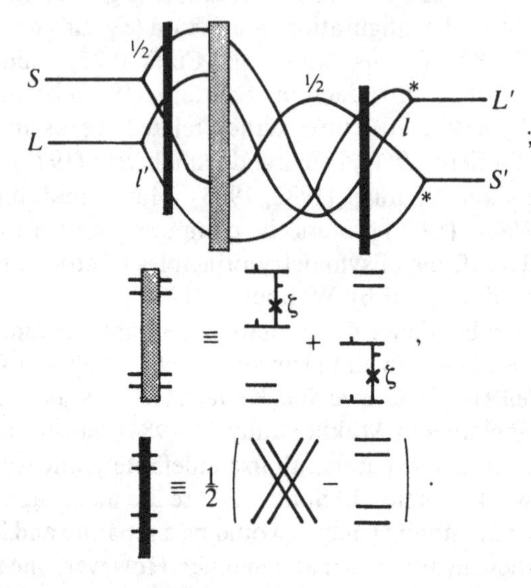

The antisymmetriser associated with the determinant induces two kinds of terms in the answer. In the *direct* terms, either the crossed or uncrossed term in each symmetriser (the last of the above diagrams) as selected. The *exchange* terms are the cross terms in the expansions of the two symmetrisers, and differ essentially by the products of 3j phases. Using JLV3 we may reduce this to

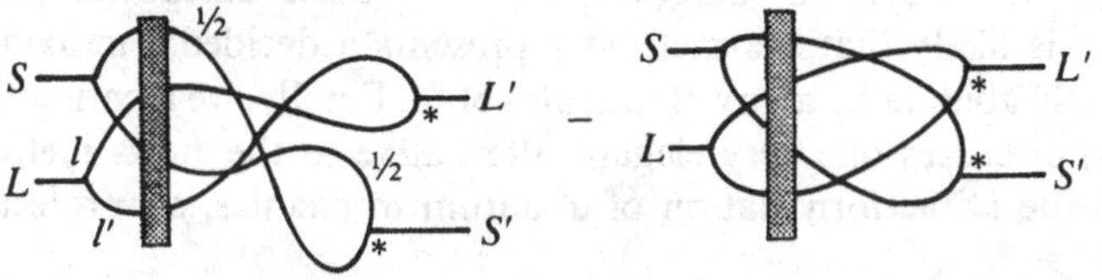

$$=[1-\{S'\,\tfrac{1}{2}\,\tfrac{1}{2}\}\,\{L'\,l\,l'\}]$$

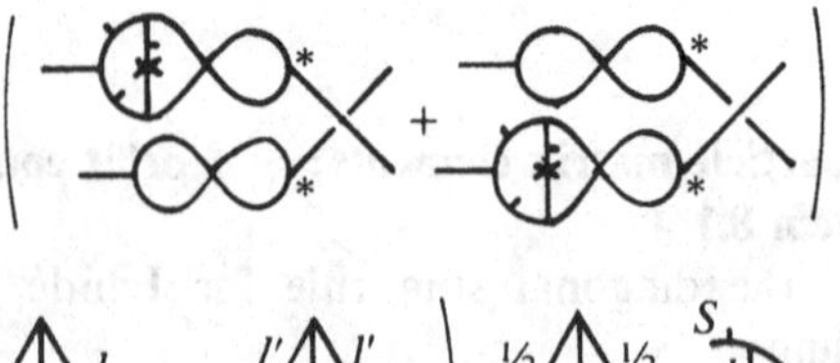

$$=[1-\{S'\,\tfrac{1}{2}\,\tfrac{1}{2}\}\,\{L'\,l\,l'\}]$$

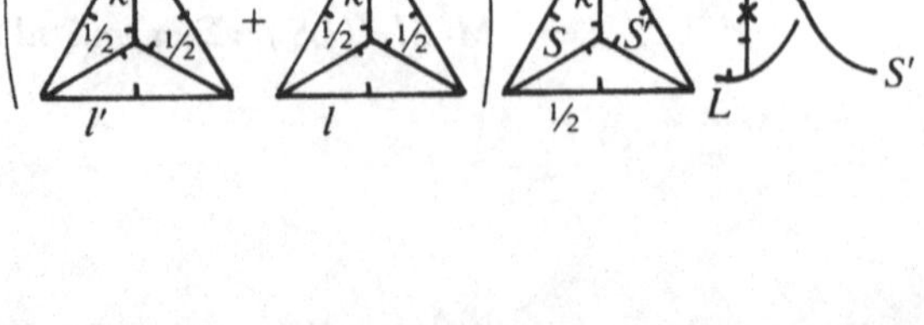

This shows that the matrix element of $\Sigma_i \zeta(nl)\mathbf{l}_i \cdot \mathbf{S}_i$ is equivalent (again from JLV3) to the matrix element of the double vector operator $\mathbf{L}\cdot\mathbf{S}$ up to a factor λ which includes the sum of bilinear products of 6j symbols from the recoupling, and the ratio of reduced matrix elements of the various angular momentum operators. Considering the diagonal matrix element ($M_L = M_L'$, etc.) gives the stated result.

Two-particle matrix elements: Coulomb interaction

We may expand the two-particle Coulomb interaction operator in a sum over products of one-particle operators, using the identity:

$$\frac{1}{r_{12}} = \sum_{k=0}^{\infty} \frac{r_<^l}{r_>^l + 1} P_l(\cos\theta_{12})$$

$$= \sum_{k=0}^{\infty} \frac{r_<^l}{r_>^{l+1}} \frac{4\pi}{2l+1} \sum_{q=-k}^{k} Y_q^k(\theta_1,\phi_1) Y_q^k(\theta_2,\phi_2)^*. \tag{8.2}$$

Using the generalised conjugation lemma, the complex conjugated spherical harmonic may be written as the harmonic transformed by a 2jm symbol, and the Coulomb interaction takes the rotationally invariant form (we do not specify the normalisation of the irreducible tensor operators C_q^k but lump any conversion factors in the constants a_k):

$$\frac{1}{r_{12}} = \sum_{k=0}^{\infty} \sum_{q=-k}^{k} a_k \begin{pmatrix} k\,k \\ q\,q' \end{pmatrix} C_q^k(\theta_1,\phi_1) C_{q'}^k(\theta_2,\phi_2) \leftrightarrow \tag{8.3}$$

Problem 8.2

Reduce to their essentials the Coulomb orbital matrix elements of the direct (D) and the exchange (E) type between (uncoupled) electron states in an open shell, nlm_lm_s, $nl\overline{m}_l\overline{m}_s$ when interacting with a closed shell $n'l'$ of electrons.

Solution

The matrix elements of interest are of the form

$$\langle | nlm_lm_s \, \Pi_{m_l'm_s'} n'l'm_l'm_s' | (1/r_{12}) | nl\overline{m}_l\overline{m}_s \, \Pi_{m_l'm_s'} n'l'm_l'm_s' | \rangle. \tag{8.4}$$

Orthonormality of the closed shell states guarantees that the two participating closed shell states must be the same in bra and ket, and on summation over all possibilities we obtain

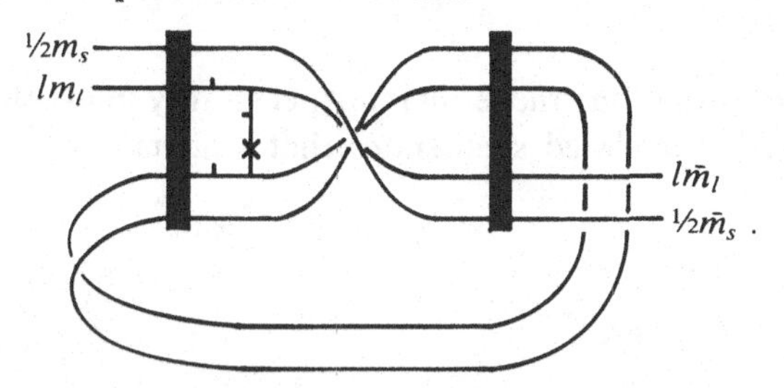

On expansion of the antisymmetrisers this becomes the sum of a direct and of an exchange term, the latter from terms of mixed symmetry in the expansion. For the direct term:

$$D \equiv \Sigma_{m_l m_s'} \langle nlm_l, n'l'm_l' | \frac{1}{r_{12}} | nlm_l, n'l'm_l' \rangle \langle m_s | \overline{m_s} \rangle \langle m_s' | m_s' \rangle \leftrightarrow \tag{8.5}$$

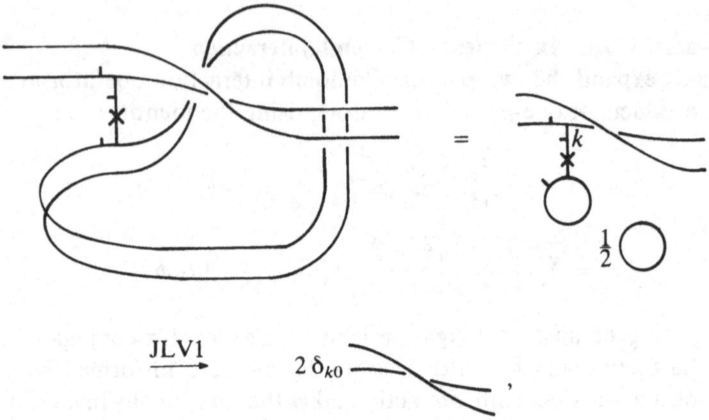

$= \quad k \quad \frac{1}{2}$

$\xrightarrow{\text{JLV1}} \quad 2\,\delta_{k0}$,

which reduces from JLV1 to a diagonal form, contributing only for $k = 0$. The exchange term is

$$E \equiv \Sigma_{m_l m_s'} \langle nlm_l, n'l'm_l' | \frac{1}{r_{12}} | n'l'm_l', nl\overline{m_l} \rangle \langle m_s | m_s' \rangle \langle m_s' | \overline{m_s} \rangle \leftrightarrow \tag{8.6}$$

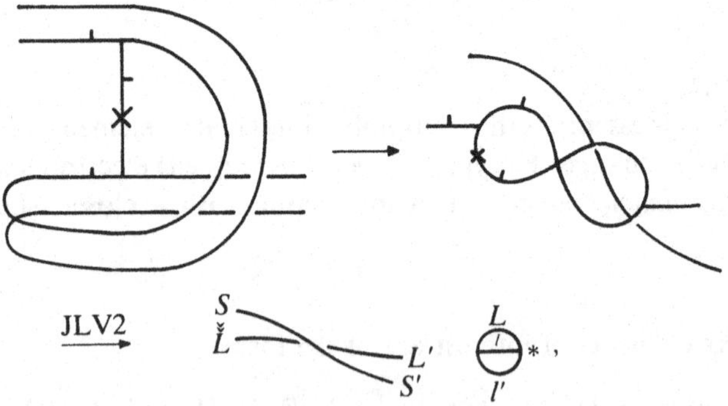

$\xrightarrow{\text{JLV2}} \quad S\ L \quad L'\ S' \quad L\ l\ l' \ *$,

which again gives (from JLV2) a diagonal coupling between the open-shell states. However, all k now contribute.

Comments

(a) In other situations, the result is not necessarily spin-independent. It is only the sum over closed shell states that renders the above exchange matrix elements diagonal in spin ($\delta_{m_s \bar{m}_s}$). The essential features of the spin dependence arising from antisymmetry will be discussed in connection with exchange theory in §8.5.

(b) The proof, now including spin dependence, may be adapted to *LS* or *jj* coupling straightforwardly, by an appropriate coupling of the terminals. For example, the direct matrix elements arising from the interaction of two open shell electrons *nl*,*nl'* in Russell–Saunders coupling takes the form:

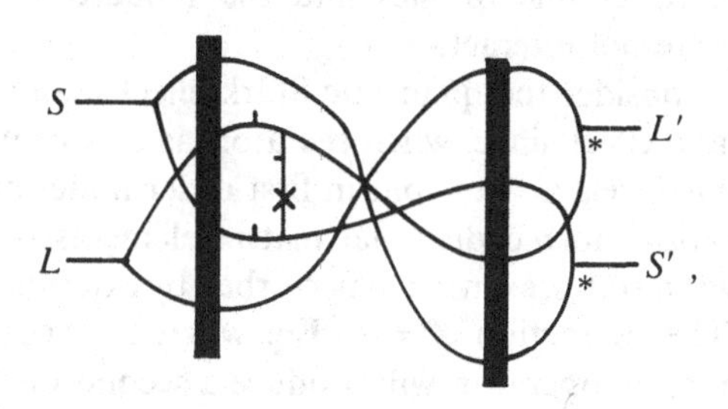

giving a direct term:

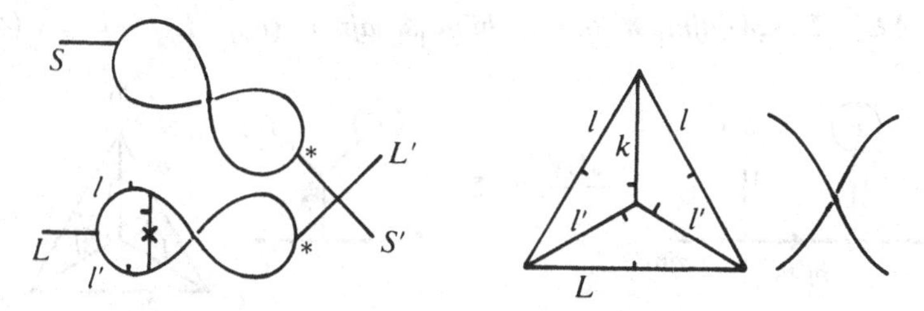

and an exchange term

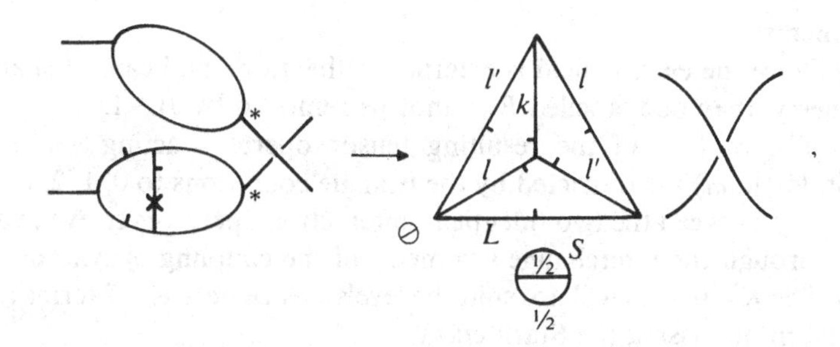

so that while now all k contribute, the interaction is still diagonal in *LS* labels. However, the appearance of a 3j phase renders the interaction spin-dependent; the exchange terms for triplet (where $\{\tfrac{1}{2}\tfrac{1}{2}1\}=1$) and singlet $\{\tfrac{1}{2}\tfrac{1}{2}0\}=-1\}$ states have opposite signs. This is conspicuous in the expansion of term energies in Slater parameters (Condon and Shortley 1935, Shore and Menzel 1968, Sandars 1977).

(c) Many other examples may be found in the books and articles cited above. Sandars (1969a,b, 1977) discusses second-order and three-particle contributions, for example.

8.2.2 External field effects on open-shell electrons

External field effects illustrate the interplay between the rotational symmetry basic to atomic physics and the reduced symmetry in the presence of an external interaction.

For example, consider the quadratic Stark effect (Sandars 1969a). We take the quadratic effect since, when the atom is at a centre of inversion symmetry, atomic levels do not split in first order under an electric field; parity considerations forbid diagonal matrix elements of an odd-parity interaction between states, even when perturbed by external interactions, of definite parity. The interaction $\mathscr{H} = -c\mathbf{E}\cdot\mathbf{p}$, where c is a constant and $\mathbf{p}$ the electronic momentum operator, will produce a second-order perturbative energy shift of the form

$$\Delta E_a = \Sigma_{bj'm'}[\langle ajm|\mathscr{H}|bj'm'\rangle\langle bj'm'|\mathscr{H}|ajm\rangle]/(E_{aj} - E_{bj'}) \leftrightarrow \quad (8.7)$$

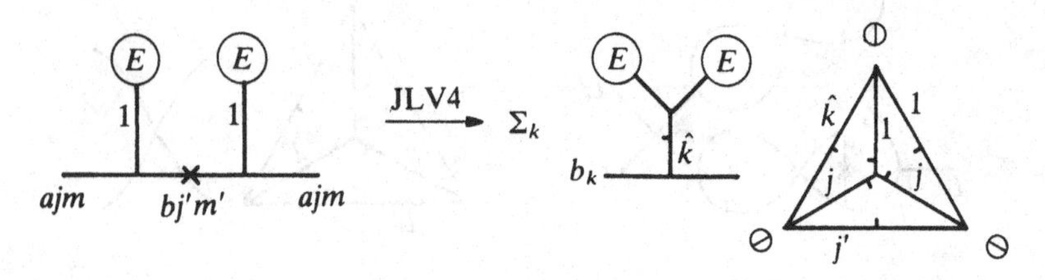

Comments

(a) Since the electric field is external to the atom, and cannot share its symmetry, the node labelled E cannot be removed by JLV1.

(b) The rank κ of the resulting tensor operator acting within the manifold $\{|ajm\rangle\}$ is restricted by the triangle conditions to 0, 1, 2, and by symmetry between the two (identical) interaction operators to even values only, through the interchange symmetry of the coupling 3j symbol.

(c) The $K=0$ term will not split the levels; hence only $K=2$ terms are of interest in discussing the Stark effect.

8.2.3 Ligand field theory

If the electric field arises from the ligands surrounding the atom or ion in a crystal lattice, it has the lattice symmetry, and we may decompose the full contribution into symmetry-allowed terms in which $K=2$ reduces to the identity irrep of the point group:

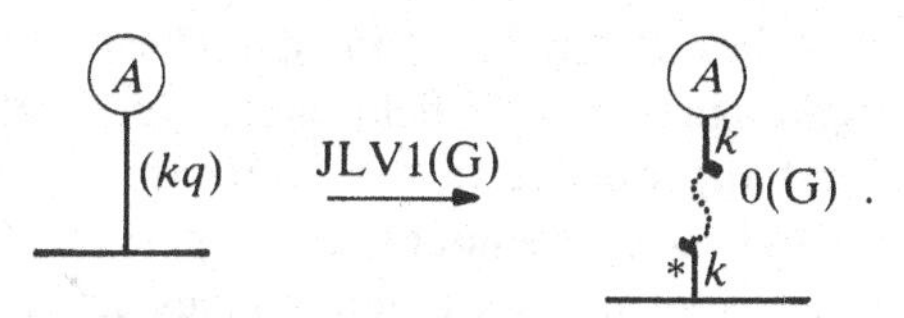

This ligand field or crystal field expansion is used to parametrise the energy levels of paramagnetic (open-shell) ions in substitutional and interstitial sites in crystals. The SO(3)–SO(2) labels kq on the line distinguish in principle the associated *crystal field parameters* $\{A_q^k\}$ (Wybourne 1965, Newman 1971), although site symmetry makes them not fully independent. The relative sizes of these parameters denote the relative size of the components of a spherical harmonic expansion of the equivalent electrostatic field, so far as the energies of the electronic levels are concerned. Since the interactions pertain to one-particle orbital matrix elements, the q values are limited by the triangle condition on the relevant 3jm symbol to $2l$, where l is the orbital angular momentum quantum number for the open-shell electron.

Crystal field parameters of the same k (rank) may be related amongst themselves by the actual site symmetry. The coefficients in these relations are themselves related to the coefficients in the basis transformation coupling the kq labels to point group invariants in the above diagram.

To a very good approximation (which essentially entails ignoring three-centre integrals and the like in molecular orbital theory) the effects of the various ligands may be superposed. An assumed (full rotational) symmetry of the ligand about the ion–ligand axis will ensure that the interaction energy of each ligand with any 4f orbital is the geometrically weighted sum of the interaction energies with the axially symmetrical 4f orbitals. In this *superposition model* (Newman 1971) it then becomes necessary to analyse only the axial case for a single ligand, when $q \equiv 0$ (only $|k0\rangle$ will reduce to an invariant under the group $C_{\infty h}$); the effective ligand field in a general geometry follows from the angular systematics. In diagrams:

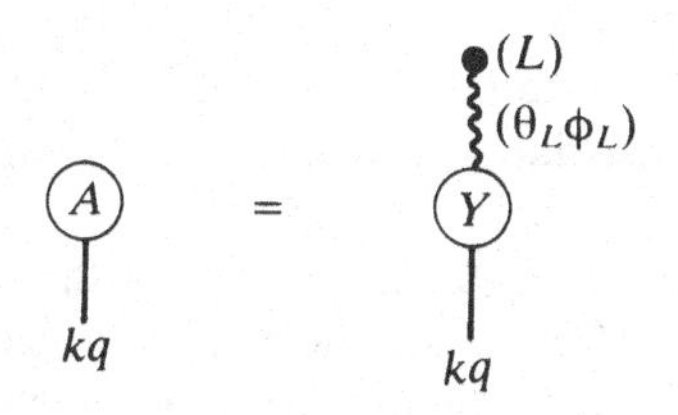

where we represent the spherical harmonic $Y_q^k(\theta,\phi)$ by the node labelled Y, and the sum over ligands L by a black dot. Since the sum over ligands has the site symmetry, the black dot is an invariant node and we may use JLV1 to express the independent crystal field parameters as the disconnected diagram on the right. The cross includes all normalisation factors, but more importantly the single-ligand crystal field parameters (called by Newman 'intrinsic' parameters) appropriate to the axial symmetry for one ligand, and therefore restricted to k-dependent, q-independent factors $\bar{A}_0^k$. This reduction in the number of free parameters gives considerable practical value to the superposition model.

The higher q values in particular are much more important for lanthanide ions than an electrostatic calculation suggests (Newman 1971). This is because the purely classical electrostatic energy differences between the energies of the various 4f orbital states are amplified by the order-of-magnitude differences in the overlap of the 4f states with the ligand electronic shell states. This amounts to saying that the equivalent electrostatic ligand-induced potential has an enhanced dependence on angle over that predicted by an electrostatic model. It is precisely the differences in the energies of the various 4f orbitals which probe the angular dependence of the equivalent potential. Hence the $q=6$ parameters, reflecting as they do the finer details of this angular dependence, dominate the ligand field theory and dynamic lattice interaction for lanthanides. It is partly this strong dependence on orbital which is characteristic of the superposition model, and which helps to ensure its approximate validity.

The application of the superposition model has been extended in various directions. It has permitted more accurate estimates of the dynamic as well as the static field; it has been utilised in S-state ion splittings; it has been extended to a discussion of the parametrisation of two-particle (or correlation) ligand fields, and to the spin-correlated crystal field; references may be traced from, for example, Ng and Newman (1986), Yeung and Newman (1986).

8.2.4 Intensity parametrisation for lanthanides

We shall as an example consider another such application of the superposition model, to the optical absorption intensity parameters in lanthanide systems (Newman and Balasubramanian 1975, Reid and Richardson 1983, 1984). Judd (1962) and Ofelt (1962) (see also Wybourne 1965) developed a tensorial approach to the parametrisation of the intensity of lanthanide spectral lines in the case when the E1 transition became parity-allowed through interference with the static or dynamic odd-parity components of the ligand field. The optical interaction coupled with the odd-parity (odd-k) ligand field parameters will give matrix elements within a $4f^n$ configuration.

First we consider the tensorial description of each amplitude. We may use JLV4 to rewrite the amplitude or matrix element in terms of a coupled operator:

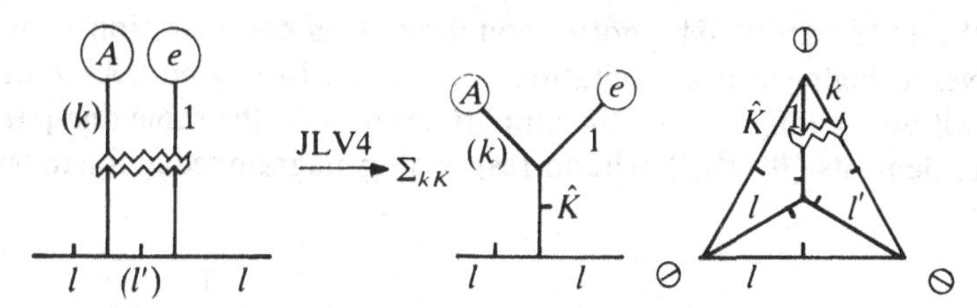

The 6j symbol so factored off retains a dependence on k. We are left with a joint operator acting between lanthanide electronic states, whose rank K is even from the form of the above symmetrisation; the 3j phase $\{Kk1\}$ must be unity.

The application of JLV4 here for the full rotation group has assumed the SO(3) invariance of the network, and so that the energies associated with the intermediate states are the free-ion energies. More precisely this assumes that the effect of crystal field splitting in the excited configurations on the denominators corresponding to the intermediate state propagators can be ignored.

In similar vein, we may perform a more extensive reduction. Consider now, as in §6.2.1 a full-scale contribution to the combined intensity of all transitions from one J manifold to another J':

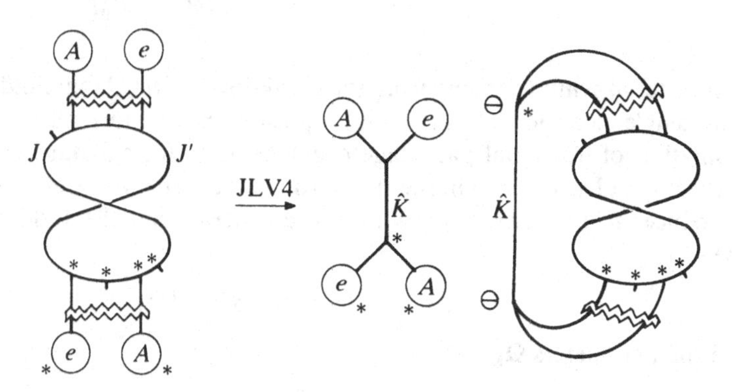

In the expression on the right, again we have ignored the effects of crystal field splitting on the energy denominators, and regarded the central part of the diagram as an invariant node under full rotations. Hence, JLV4 for SO(3) allows us to split off the crystal field and polarisation vector nodes. Each J term in a multielectronic lanthanide ion may be written through LSJ coupling in terms of single electron states. We assume that the only configuration admixture occurring is that considered explicitly in each amplitude, through the single-particle crystal field and optical interaction operators, each of which is first order in each of the (odd-parity) crystal field and the (odd-parity) E1 photon coupling. This configuration coupling involves a single-electron excitation $l \to l'$, where $l + l'$ is odd. As J and J' vary within a configuration, the amplitudes involve the same one-particle matrix elements $\langle l || U^k || l \rangle$. The corresponding diagram reduction takes the form:

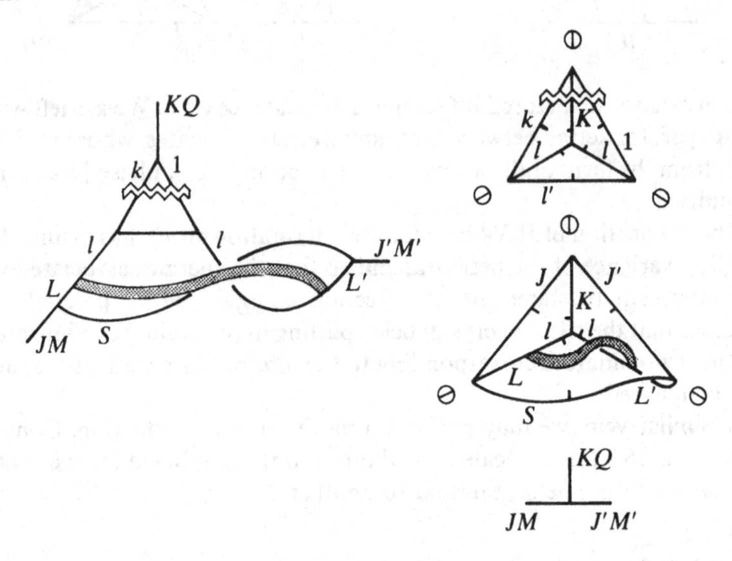

The monster nj symbol, arising from the breakdown of the J manifold into parent single-electron orbital states, appears in the formalism as a combination of fractional parentage coefficients, as for standard crystal field theory, and is included in the standard manner. The whole expression then reduces to the form (somewhat more general than the Judd–Ofelt form)

$$D_{JJ'} = e^2 \Sigma_K \Omega_K |\langle \psi(SL)J | U^K | \psi(S'L')J' \rangle|^2 \tag{8.8}$$

where the parameters Ω_K

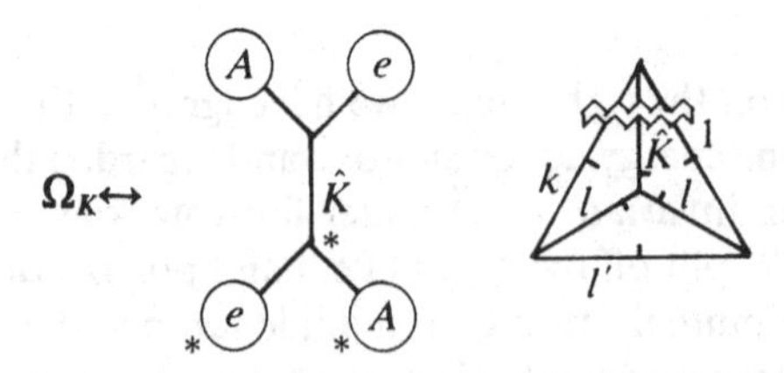

enshrine the J-dependence of the transition intensities ($K = 2, 4, 6$, from the symmetrisers and the necessity of the associated 3j phases $\{k1K\} = 1$). The triangle rules on the 3jm symbols require $K = k \pm 1, k$.

Within the superposition model, however, $K=k$ terms do not appear for the following reason. The crystal field parameters are then linearly dependent on the single-ligand parameters appropriate for the $C_{\infty v}$ symmetry of the single-ligand situation, together with the necessary rotation to laboratory axes from the one-ligand axes. The rotation merely mixes the components q, and does not affect the rank or parity argument which follows. Since the identity $0(C_{\infty v})$ is found in the reduction only of the irreps $0^+, 1^-, 2^+, 3^-, 4^+, 5^-, 6^+, \ldots$ of O(3) (see Appendix), we cannot have $k=K$ terms under the simultaneous restrictions that K is even and that k^- $(O(3))\downarrow 0(C_{\infty v})$. This may be used as an indirect test of the validity of the superposition model for the static field, or of the presence of dynamic effects such as ligand polarisability or phonon interaction which would destroy the assumed axial symmetry of a single ligand.

A similar tensorial analysis leading to a useful parametrisation scheme may be developed for two-photon spectroscopy within lanthanide manifolds (Axe 1964). Once again JLV4 for SO(3) may be used, within the closure approximation (i.e. of ignoring crystal field splitting effects in intermediate denominators), to separate geometrical factor and dynamical constant. The basic structure of the dynamical constant may be written in terms of fractional parentage coefficients and the single-electron matrix elements of a unit tensor operator of even rank K, times rank-dependent constants Ω_K which are now limited to $K=0$, 2 by the triangle constraints. The diagram formulation parallels the analysis of ligand-field-assisted absorption intensities.

More recently the topic has attracted attention since it has been found necessary to develop the formalism more, including third-order effects in the static fields as well as dynamic effects and ligand polarisability, in order to explain TPA intensities in S-state ions. Some important initial papers are those of Judd and Pooler (1982), Downer and Bivas (1984). References may be traced from Sztucki and Strek (1986).

8.3 Optical absorption: lineshift and linewidth

8.3.1 Introduction

In §6 we considered the effect of the choice of polarisation and geometry on the optical absorption spectrum of an electronic system coupled to phonons, etc. Here we fill out this calculation in various ways.

First we examine the moments of the spectrum, transforming the relevant operator traces into the unperturbed product basis, and then developing special diagrams for the representation of the various terms. These constitute a good tutorial exercise in diagram techniques; they have strong links with both the field theoretic and the group theoretic diagrams. Applying the JLVn theorems to these diagrams gives some old and some new results.

Second, we justify the Golden Rule approach assumed in §6 by outlining its derivation from the full field theoretic development. In this we discuss two related approaches based, respectively, on the one-particle photon Green's function and on the two-particle electronic Green's function. The elaboration of this for the discussion of Raman scattering involves some difficult points (see Kawabata 1971, McKenzie and Stedman 1978, Saunders and Young 1980, for example).

Third, we perform a detour into physics and field theory, rather than group theory, to distinguish between the no-phonon line (or quasipure electronic transition) and its phonon-induced shift and broadening on the one hand and the phonon sidebands, in which the photon energy is distributed between electronic and lattice excitations, on the other. This distinction may be made by virtue of identifying the appropriate resonances in the field theoretic expansion. It is an important step in clarifying the physical content of the field theoretic diagrams and the later applications to the Jahn–Teller effect, Raman scattering, vibronic transitions etc., which form further applications discussed here.

8.3.2 Moment analysis

The moments of an optical absorption spectrum involve traces over optical interaction operators and population (Hamiltonian) factors. For example, the second moment of the absorption spectrum may be written as

$$\langle\omega^2\rangle=\int_0^\infty d\omega\,\omega^2\frac{2\pi}{\hbar}\Sigma_{ab}|\mathrm{M}_{ba}|^2(\rho_b-\rho_a)\delta(E_b-E_a-\hbar\omega), \tag{8.9}$$

where a, b represent upper and lower electron–phonon-coupled states in the optical transition, M is the optical interaction operator, and the two population terms ρ_a, ρ_b, where ρ_i is proportional to $\langle i|\exp[-\beta(\mathscr{H}+\mathscr{V})]|i\rangle$, reflect the interplay between absorption and stimulated emission. In this expression ω^2 may be replaced by E_b-E_a through the energy conservation delta function, and therefore by matrix elements of the Hamiltonian. The sum over states (a, b) can then be cast in the form of a trace (e.g. Payne and Stedman 1977), giving:

$$\langle\omega^2\rangle=\frac{2\pi}{\hbar}\mathrm{Tr}\hat{\rho}\mathrm{M}[\mathscr{H},[\mathscr{H},\mathrm{M}]], \tag{8.10}$$

which through the invariance of a matrix trace under a unitary basis transformation may now be evaluated in the unperturbed electron–phonon product basis. Higher-order moments are defined by similar expressions with more nested commutators, and replacing $\mathscr{H}$ by $\mathscr{V}$ gives information on central moments. Strictly speaking, such formulae are exact only for odd moments, or for even moments in an optical approximation, when only the ground electronic level is populated (Payne and Stedman 1977). The simplest example is the zeroth moment, or area under the spectrum. If the ω^2 factor is omitted in the above argument, we obtain similarly

$$\langle\omega^0\rangle = \frac{2\pi}{\hbar}\mathrm{Tr}\hat{\rho}\mathrm{M}^2, \tag{8.11}$$

which leads in the above approximation to the sum rule that the spectral area is temperature-independent.

Let us in any case approximate the population factors by those appropriate to the pure electronic system in the absence of interaction; we thus assume that both photons and phonons have little effect on the electronic level populations. This is reasonable for optical-frequency transitions in the absence of deliberate pumping. We may then replace the density matrix ρ by its unperturbed value ρ_0 which, since it involves purely diagonal terms in the product basis, may be ignored as far as the group theory is concerned. (Without this assumption we should be using equation (7.29) and thus expanding ρ to all orders of the interactions.) The optical interaction M is purely electronic and the only phonon dependence is now in the explicit dependence of $\langle\omega^2\rangle$ on the interaction.

One simple consequence is the sum rule that $\langle\omega^0\rangle$, being now independent of phonon interactions and operators, is constant with changing temperature. Another is that the second moment of the spectrum is of second order in the phonon interaction. We shall therefore obtain a contribution to the trace only if the phonon operators in these two interaction terms give a diagonal matrix element. They must have the mixed form $b_k^\dagger b_k$ or $b_k b_k^\dagger$ for some mode. In general all second quantised operators including the electronic operators must form nonzero contractions after the manner of §7.4.2, except that we may set the time τ to zero. The second moment may therefore be described by the sum of the Feynman-like diagrams:

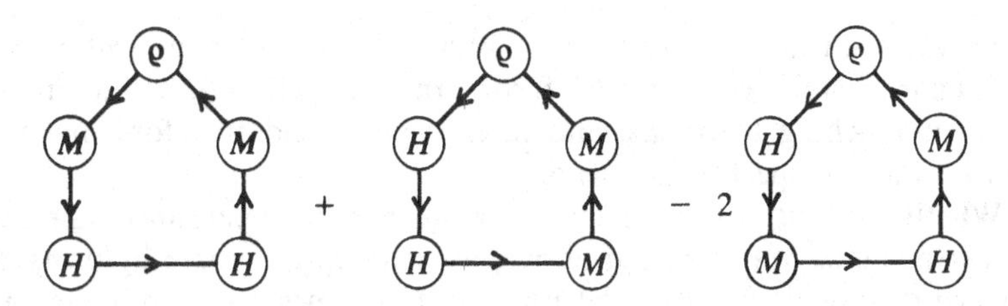

Comments

(a) The lines here signify merely the matching of mode or state labels, while their directionality distinguishes the end appropriate to the creation operator. As such these diagrams accurately reproduce the group theoretic structure of the calculation in the same manner as Feynman diagrams, and for the same reason: each vertex corresponds to a matrix element of a Hamiltonian term, and the linkages satisfy the rules of §1.

(b) In general the nth moment involves analogous terms with n commutator brackets, and the corresponding diagram would have n interaction nodes in a ring.

(c) In contrast with the diagram rules of §§1–3 for crossing lines when linking nodes, these diagrams are planar, with no line crossing. This is a characteristic of traces.

Some links may be discerned between this notation and the diagram notation for moment calculations of Grange and Richert (1976).

We may now ask some physics questions, and may seek their resolution using diagrammatic group theory. For example, while the spectrum depends only on the square modulus of the matrix elements of the optical interaction *in the coupled basis*, in this still true in the unperturbed basis? Or do the various pure electronic transition amplitudes interfere? On the face of it, they certainly do, since the ring structure of the above diagram in the uncoupled basis clearly is not equivalent to the square modulus structure which in the same notation would be written

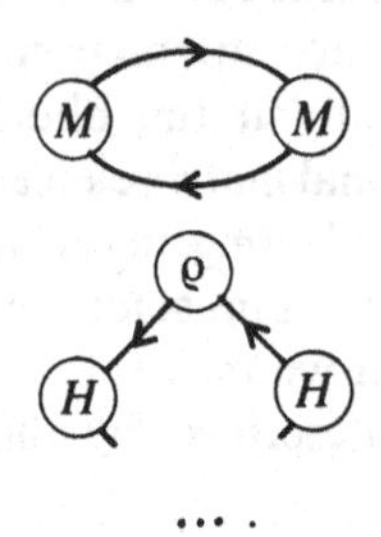

However, the ring diagram *for some particular low-order spectral moment* could conceivably reduce to this form if the symmetry allowed it, via any relevant pinching theorems, and interference could therefore make no contribution to spectral moments.

Whether interference contributes or not may be determined using the JLVn theorems. As discussed above, we may approximate by using the unperturbed density matrix, whereupon it becomes not merely diagonal but also fully invariant with respect to the symmetry group of the physical system. (This would no longer be true if the interaction were included, not that the interaction is itself asymmetrical – in fact $\mathscr{V}$ is also invariant with respect to the symmetry group – but that its invariance properties depend on all constitutuent operators; hence if some phonon operators are contracted against others from the other nodes in the diagram, to that extent the matrix elements of ρ do not satisfy Schur's lemma.) The nodes ρ

may therefore be pinched out by JLV2. The four diagrams now reduce to the quadrilateral form in which for two cases the phonon interaction vertices are adjacent, and for two others they are diagonally opposed:

Since the nodes $\mathscr{V}$ are invariant, we may use JLVn theorems to reduce the former of these terms to a function of the square modulus of the matrix elements of the optical interaction:

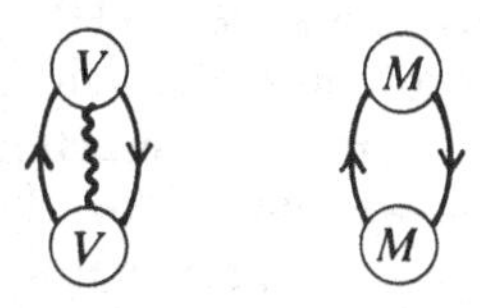

However, the same reduction fails to go through in the second topology. We conclude that interference between pure electronic transitions affects even the second moment of the spectrum.

8.3.3 The link between optical absorption and photon or electronic Green's functions

Optical absorption obviously arises from the coupling of electronic states $|i\rangle$ (themselves perturbed by phonons, etc.) to photons, firstly via the $\mathbf{A}\cdot\mathbf{p}$ interaction of §6.1. This, with its gauge-transformed version $\mathbf{E}\cdot\mathbf{r}$, is similar in form to the linear phonon coupling of §7.3 and involves photon creation and annihilation operators $c_k^{\pm}$ (§6.1), where the mode label k includes a polarisation label: $\mathbf{A}\cdot\mathbf{p} \supset \Sigma_k \mathbf{e}_k\cdot\mathbf{p}\,\exp(\mathrm{i}\mathbf{k}\cdot\mathbf{r})c_k + \text{h.c.}$ When the one-particle electronic operators $\mathbf{p}$, $\mathbf{r}$ are written in second quantised form, this interaction is bilinear in electronic creation and annihilation operators, having the form $\mathrm{M} \equiv \Sigma_{pqk} M_{pqk} a_p^\dagger a_q c_k + \text{c.c.}$ The matrix element M_{pqk} is the counterpart to the linear coupling matrix element V_{ijk} of the linear phonon coupling, and corresponds to a sum over terms in the multipole expansion. The quadratic coupling term proportional to $\mathbf{A}^2$, required in some gauge choices (Lamb *et al.* 1987, Reid 1988), corresponds in form to the nonlinear phonon coupling, and may be included in the following with minimal structural change.

There are two ways of writing the optical absorption spectrum in terms of the discontinuity of a Green's function.

One-particle photon propagator

In determining the number of photons of different energies we are estimating a spectral density of the type evaluated in §7.2.4. This density may therefore be evaluated from the appropriate one-particle Green's function, in this case a photon Green's function:

$$\mathscr{G}_{kk'}(E)=\int_0^\beta \langle T_\tau\{c_k(\tau)c_{k'}^\dagger\}\rangle \exp(E\tau)\mathrm{d}\tau \leftrightarrow \tag{8.12}$$

The cumulative effects of interaction with electronic states and also the interaction between the electronic states and the phonons may be evaluated by a perturbative expansion of this Green's function in exactly the manner detailed in §7, with the two types of interaction acting simultaneously.

The diagram expansion we obtain for this will inevitably involve electronic bubbles on the photon propagator. Each such diagram loop corresponds to an integration over an energy variable, and to an extra fermionic population factor in the final answer. The only terms to survive the Abrikosov limit will be the unperturbed photon propagator, and the term involving just one electronic bubble with phonon dressing:

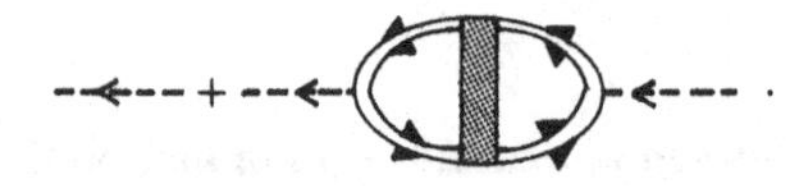

This is justified for the phonon case at the end of §7.4.2.

For weak electron–phonon coupling, the most important perturbation term is the diagram:

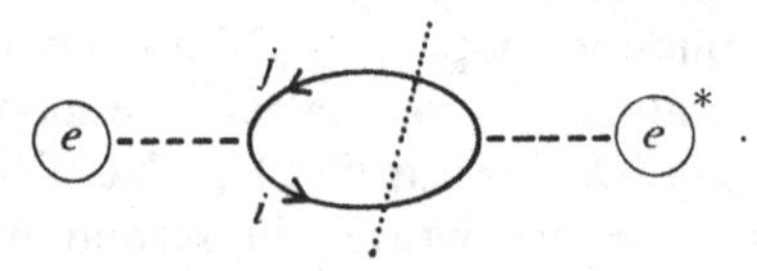

The resonance in this diagram corresponds to the effect of real or virtual photon absorption at an electronic transition. From §7.2 it leads to a Golden-Rule lifetime and a corresponding energy shift for the photon under electronic interaction:

$$I(\mathbf{e})=(2\pi/\hbar)|M_{ijk}|^2\delta(E_i-E_j-\hbar\omega_k). \tag{8.13}$$

This justifies the use of the Golden Rule in §6.1 from the viewpoint of §7 when setting up the formalism for the polarisation dependence of optical absorption.

Two-particle electronic Green's function

Alternatively, we may concentrate on the electronic and phonon operators. Since the electron–photon coupling vertices have such a proscribed form – thanks to the Abrikosov projection and to the second-order perturbation evaluation of the effects of photon interaction – we could from the beginning have factored out the matrix elements of the photon operators in a product basis, and concentrated on the evaluation of the matrix elements of the electronic operators in a coupled electron–phonon basis.

According to the Golden Rule, the probability of the absorption of a photon in mode k is determined by a thermal average of the square modulus of these electronic matrix elements, proportional to

$$\overline{|\langle M\rangle|^2} = \Sigma_{ijpq} M_{ijk} M^*_{pqk} \mathscr{A}v_a \langle a|a_i^\dagger a_j a_q^\dagger a_p|a\rangle, \tag{8.14}$$

where $\mathscr{A}v$ represents a thermal average as in §7.2. In §7.2.4 we linked a Golden Rule expression, involving the averaged square modulus of the matrix element of a *single* electronic operator $a_i^\dagger$, and the discontinuity in the corresponding *one*-particle Green's function. When this argument is adapted for the matrix elements of a *bi*linear operator expression, we find that the Green's function whose discontinuity across the real axis is of interest for finding the spectrum is a *two*-particle Green's function of the form (Barrie and Rystephanick 1966, Stedman 1971a):

$$\mathscr{G}_{ijpq}(E = \int_0^\beta \langle T_\tau\{(a_q^\dagger a_p)(\tau) a_i^\dagger a_j\rangle \exp(E\tau) d\tau \leftrightarrow \tag{8.15}$$

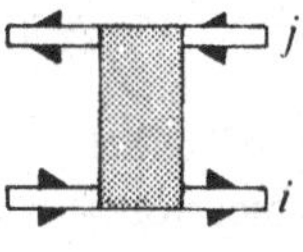

The appropriate spectral density is now to be evaluated at an energy E appropriate to differences in the energies of the two electronic states, respectively, associated with the creation and the annihilation operator. To be precise, for a sharp electronic feature in the optical spectrum at a photon energy E, we may restrict attention to states with $E_i^0 - E_j^0 = E_p^0 - E_q^0 = E$. The limitation to such a resonance may be denoted by the dashed line across the diagram.

8.3.4 Perturbative expansion of the two-particle Green's function

The perturbative analysis of §7 may now be developed for this two-particle Green's function. In this subsection we shall not often be considering the effect of symmetry or state degeneracy, and may be thought to wander rather far from our topic of group theory. However, it is essential to determine precisely the physical significance of the various classes of field theoretic diagrams and their various resonances. While symmetry consequences for no-phonon line shift and broadening are almost trivial, we shall be plunged into a much less trivial problem on looking at the phonon sidebands or at the effects of a further electronic operator on the spectrum in a system with degenerate electronic levels participating in the optical transition (the so-called Jahn–Teller situation).

Unperturbed two-particle propagator

We find an unperturbed diagram of the form

$$\mathscr{G}^0_{ijpq}(E) \leftrightarrow \quad \overset{j}{\longleftarrow} \atop \underset{i}{\longrightarrow},$$

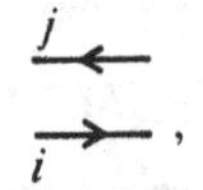

whose resonance, determined by a near-vertical line cutting both single-particle propagators, is at the photon energy. This may be seen by noting that E is the energy variable flow from right to left through the diagram. In algebraic notation the unperturbed two-particle Green's function is a convolution in energy space of two unperturbed one-particle Green's functions:

$$\mathscr{G}^0_{ijpq}(E) = \sum_{s=-\infty}^{\infty} \delta_{ip}\delta_{jq}\mathscr{G}^0_i(E-\varepsilon_s)\mathscr{G}^0_p(\varepsilon_s) \tag{8.16}$$

and evaluates to be

$$\mathscr{G}^0_{ijij}(E) = \frac{\rho^0_i - \rho^0_j}{E - (E^0_i - E^0_j)}. \tag{8.17}$$

The apparent Λ-dependence of these electronic population factors is cancelled under Abrikosov projection through the $1/\Lambda$ factor in equation (7.74).

Perturbative effects

Of course multitudinous phonon corrections are possible, and may be characterised either as one-particle dressing effects:

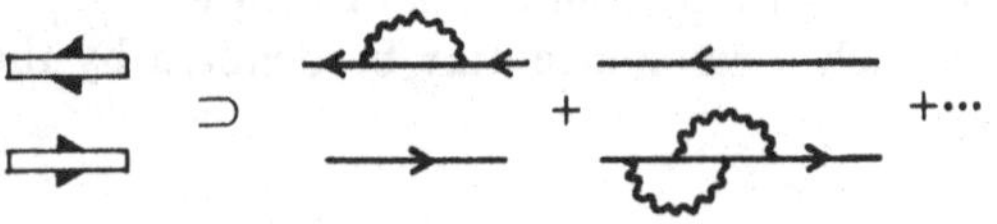

$$\supset \quad + \quad +\cdots.$$

or as *two-particle vertex scattering* corrections coupling the two propagators:

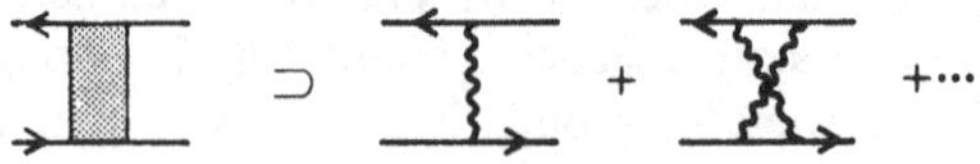

$$\supset \quad + \quad +\cdots.$$

These will produce resonances not only when the photon energy E equals an electronic energy difference, but also when it equals this plus the energy of some lattice excitation or de-excitation.

No-phonon line

We may identify the contributions to the 'no-phonon' line as all terms in the expansion containing one or more resonances at the electronic energy difference:

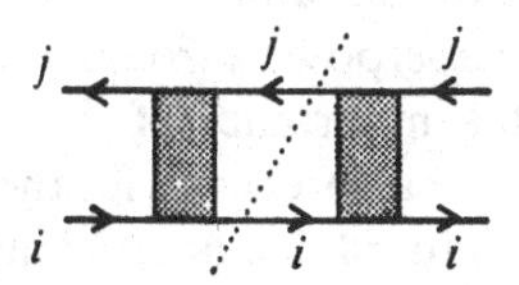

In perturbation theory we may sum such terms as for Dyson's equation (§7.2.8) or for the Bethe–Saltpeter equation to derive the modified form:

$$\mathscr{G}_{ijij}(E)=\frac{\rho_i^0-\rho_j^0}{E-(E_i^0-E_j^0+\Sigma_{ij}(E))} \tag{8.18}$$

for the Green's function. The one-particle dressing diagrams alone will supply subtractive contributions to the two-particle self-energy $\Sigma_{ij}(E)$:

$$\Sigma_{ij}(E)=\Sigma_i(E)-\Sigma_j(E), \tag{8.19}$$

where the one-particle self-energies are those calculated in §7.4. However, the two-particle vertex scattering corrections will modify this through some nonseparable terms.

When the resonance structure of these diagrams is examined near the no-phonon transition line contribution (Stedman 1971a, 1972, 1974), we find the following results.

First, the nonseparable terms affect only the phonon-induced widths of the transitions. The one-particle shifts combine subtractively (the *Ritz combination rule*) $\Lambda_{ij}=\Lambda_i-\Lambda_j$, in determining the shift of the optical transition frequency. Without the vertex corrections, linewidths combine according to the *Weisskopf–Wigner relation* $\Gamma_{ij}=\Gamma_i+\Gamma_j$, the plus sign arising from the effects of the change of sign in the energy variable argument E for the second one-particle Green's function (equation 8.19) when evaluating the imaginary part. However, vertex corrections affect this last relation.

Second, the nonseparable width contributions correspond to relaxation processes in which the electronic state lifetimes are limited by elastic scattering. In a Raman scattering process, for example, this would limit interest to cases in which the final electronic state was degenerate with the initial state, and in which the two phonons had equal energies. Only such processes preserve sufficient resonances in the diagram to survive the Abrikosov limit; a more physical reason is given below.

Third, when combined with the one-particle width contributions $\{\Gamma_i\}$, these elastic scattering vertex contributions act as interference terms to induce a final contribution of the form of $|T_i - T_j|^2$, where T_i is the amplitude for the elastic scattering process as derived from the one-particle Green's function. The essential reason for this is that the elastic scattering amplitudes for the two electronic states involved in the transition each lead from the same initial state to the same (or an indistinguishably different) final state, and so must be superposed through the quantum superposition principle. The experimental importance of elastic or phase relaxation in imparting coherence to the decay processes for the two electronic states has been observed in phonon-induced line broadening in solids (Stedman and Cade 1973), and is well known in magnetic resonance and quantum optics.

Fourth, the interference effects alluded to in §8.3.2 may be expected when the various possible electronic transitions are degenerate, so as to produce overlapping resonances of the field theoretic contribution.

Degeneracy effects

The effect of level degeneracy, and so point group symmetry, on these results is an elementary but still a worthwhile exercise in diagram group theory. Since the shift is essentially determined by the one-particle self-energies, and since the latter constitute an invariant subdiagram (all phonon propagators are internal), they may be factored out at least for group theoretic purposes by JLV2:

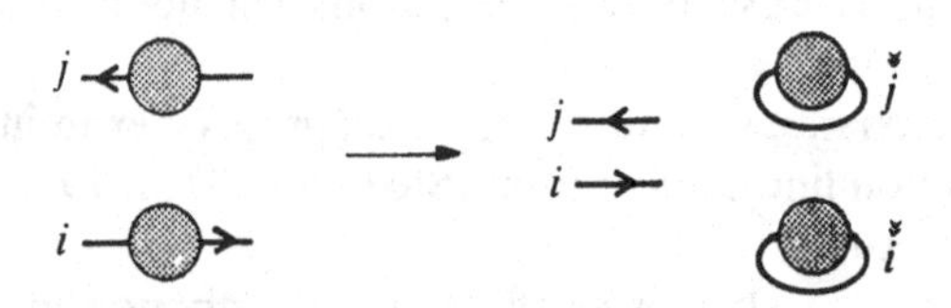

with the consequence that the shifts are identical for all transitions between two levels even if they contain more than one (degenerate) state. It is not possible, therefore, for the Jahn–Teller effect to cause level splittings of ionic systems and optical transitions, if by Jahn–Teller effect we mean the presence of phonon interaction in a system with orbital degeneracy, and if by splitting we mean no-phonon line splitting. We are not thinking about the triggering role of strains or of a cooperative transition here. This simple theorem (Stedman 1972), while discrediting some efforts at interpreting line splitting, places no limit on the possibility of phonon-induced and degeneracy-associated structure in the vibronic band. Indeed much study has gone into such structure, particularly in the case of strong phonon coupling (O'Brien 1969, Ham 1972, for example).

The width calculation gives some similar results. All states in a level have the same width when the degeneracy is associated with a symmetry of the broadening interaction. The nonseparable terms somewhat complicate the simple analysis for shifts.

8.3.5 Phonon sidebands

Further resonances are possible when the photon energy equals the energy of an electronic transition in conjunction with some lattice excitation or de-excitation. These are identified by appropriate cuts through the diagram:

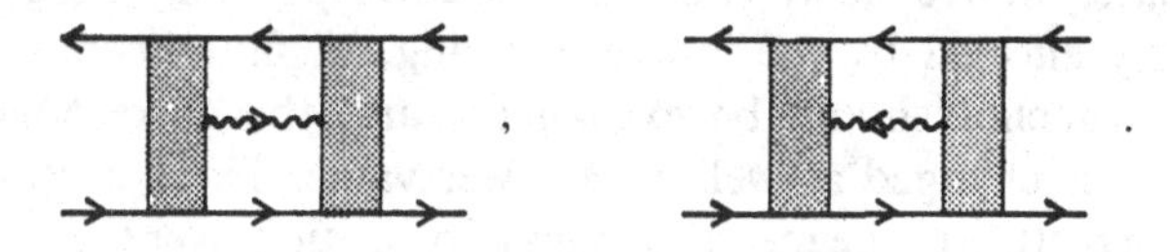

These are the n-phonon sidebands of the electronic line.

Several approaches exist for understanding their origin. One may think semiclassically of the harmonic lattice motion as introducing frequency modulation of the signal radiated by the electronic transition; the no-phonon line is the carrier, and the phonon-assisted transitions are the sidebands. This picture can be extended to discuss the broadening of the no-phonon line by anharmonicity in similar terms. This method gives essentially the same results as the following one, except for the Stokes shift in band energies.

A more obviously quantum mechanical approach is to use the Franck–Condon view of the electronic transition as being on a much smaller time scale than the lattice motion. The probabilities of no-phonon versus multiphonon transitions are determined essentially by the overlap of the relevant ground state and excited state wave functions in lattice mode coordinate space.

We consider one ionic state $|i\rangle$ in interaction with one lattice mode k. A linear coupling V_{iik} means classically a linear energy shift in the ionic state with the amplitude Q_k of the lattice mode. The system will accordingly relax to some nonzero value of this amplitude, which is eventually stabilised by the quadratic dependence of the simple harmonic mode energy on mode displacement. Classically we may depict this energy–amplitude variation on a configuration coordinate diagram:

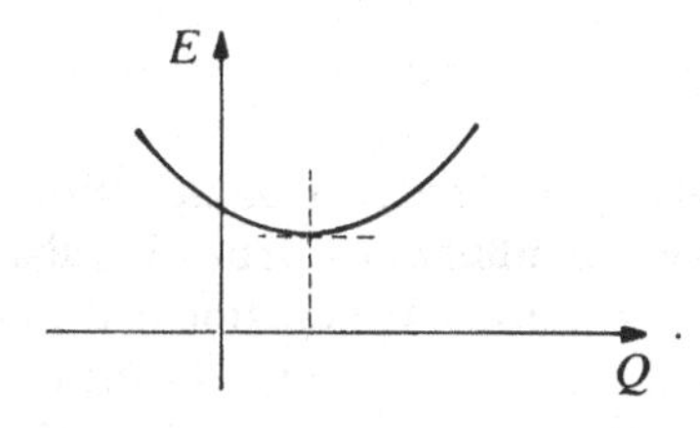

Quantum mechanically, we must allow for a distribution of possible Q_k values at any time in accordance with the Schrödinger wavefunction of the oscillator, whether the ground or some excited state, as appropriate.

The effect of lattice interaction on an optical transition may be approximated by considering transitions between two such curves, corresponding to initial and final ionic states. Classically, the initial state should be regarded as relaxed to its local minimum. In the case of absorption, the initial state is the lower of the two, and the 'instantaneous' optical transition should carry the system to that point on the upper branch immediately above the well bottom. The corresponding energy difference is obviously affected by the lattice coupling. In fluorescence, 'upper' and 'lower' branches should be exchanged, and the corresponding energy difference is changed as well, to a lower value. These correspond in the quantum picture to the mean energies of transition over the whole vibronic band, which differs in absorption and fluorescence through the exchange of Stokes and anti-Stokes sidebands (i.e. high-energy and low-energy bands are, respectively, greater or less in intensity according to which includes stimulated emission).

The total band may be split into its components as follows. In the Franck–Condon picture each sideband has an intensity determined by the superposition of all possible vertical transitions, i.e. the overlap of the two relevant vibrational wavefunctions for the upper and the lower electronic state. The no-phonon line corresponds to the well-bottom-to-well-bottom transition, and its intensity is given by the overlap of the ground state wavefunctions relative to each stabilised state:

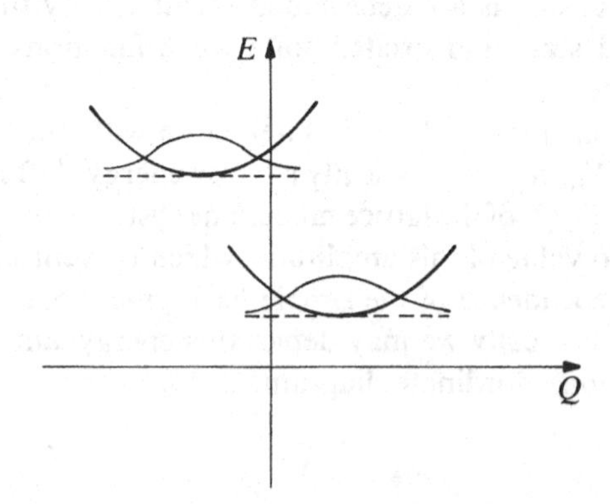

together with the overlap of the two one-phonon wavefunctions, etc. Similarly the one-phonon annihilation sideband in absorption corresponds to the overlap of the one-phonon wavefunction in the lower branch with the zero-phonon or ground state wavefunction in the excited branch:

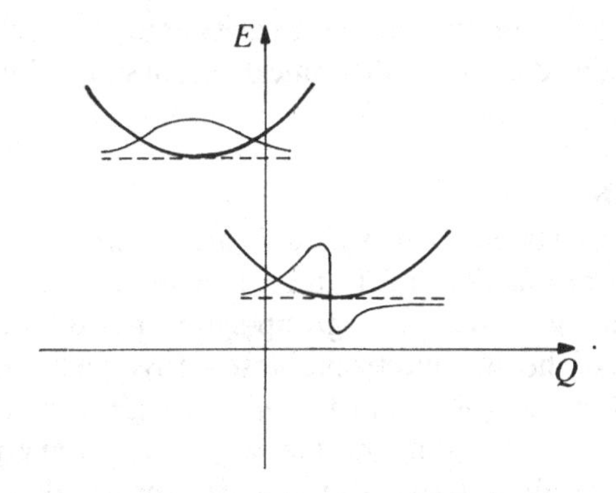

These considerations, together with the analytic values for the overlaps of displaced harmonic oscillator wavefunctions, show that the relative oscillator strength of the n-phonon (creation and annihilation, treated together) sideband is of the form of a Poisson distribution $(S^n/n!)\exp(-S)$. S is related to the linear interaction parameters, the phonon frequency and the (temperature-dependent) quantum amplitude of vibration. S is also the analogue of the modulation index of the semiclassical approach. The $\exp(-S)$ factor is related to the overlap between the two ground states. Since the area under the spectrum is temperature-independent in the optical approximation (§8.2.2), this formula also gives the absolute magnitude of the n-phonon sideband. In particular, in the case $n=0$, the no-phonon line has an oscillator strength given by the Debye–Waller factor $\exp(-S)$ familiar in X-ray crystallography. Since S is very small for Mössbauer transitions, the sidebands can be ignored for lattice phonons, but can be rendered visible in the Mössbauer transition in the case of ultrasound. As the coupling is increased, the no-phonon line intensity reduces exponentially, and in strong coupling systems is swamped by the sidebands.

Suppose that one particular lattice frequency – perhaps a local mode, or a plateau in the frequency–wavevector surface giving rise to a peak in the phonon density of states – is particularly strongly coupled to the electronic transition, so as to give a prominent feature in the sidebands. Such features may be sharp enough to be limited in shift and width by further phonon interaction effects. The arguments used above for the no-phonon line may be adapted to the vibronic lines (Stedman 1978), first by identifying the appropriate resonance in the field theoretic diagram as described above, and then by identifying the higher-order corrections which contribute to its broadening.

When the optical transition occurs within a configuration (or between configurations of the same parity) on an ion at a centrosymmetric site, the E1 multipole is parity forbidden for the no-phonon line. The vibronic peak may however be allowed in E1 if the lattice mode has odd parity. Such features are expected to dominate. These parity arguments may be traced in the corresponding field or group theoretic diagram simply by adding parity labels to each line.

Degeneracy effects

When the electronic state and lattice mode are degenerate we may expect vibronic splitting of this spectral feature. It assumes the form of several peaks, corresponding to the point-group-symmetrised states obtainable from the product of the final electronic state (of symmetry λ_e say) and the strongly coupled mode (of symmetry λ_l) through Kronecker product decomposition: $\lambda_e \times \lambda_l \supset \mu$. Put another way, the unitary matrix which diagonalises the product basis also reduces the product basis to its irreducible parts. In diagram group theory, this amounts to an application of JLV4 to a diagram with two electronic and two lattice mode external legs (Stedman 1976b):

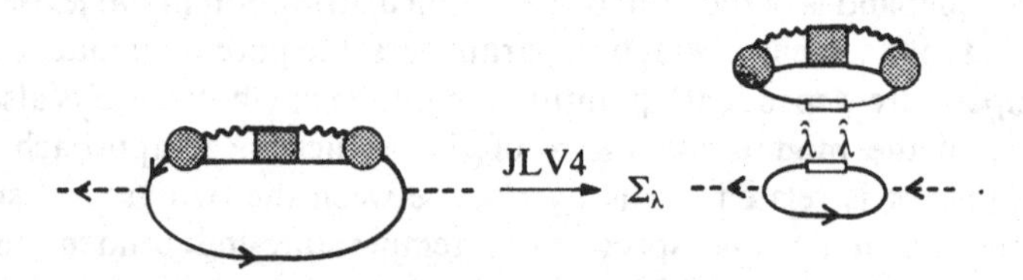

8.4 Jahn–Teller interaction

8.4.1 The Jahn–Teller theorem

First we give a simple diagram argument for the lattice instability in the presence of orbital degeneracy. Fuller reviews are given by Sturge (1967), Englman (1972) and Ham (1972); for a proof of the theorem see Raghavacharyulu (1973). We discuss the interaction of a set of orbitally degenerate ionic states with the modes of vibration of the molecule or local cluster of which the ion is a part. The transformation character of the set of ionic states under the molecular symmetry group is specified by some irrep λ say, whose dimension is greater than unity.

Consider the interaction of each of the states $|\lambda l\rangle$ in the ionic level λ with a set of molecular vibrations $\{Q_{\kappa k} | k = 1, 2, \ldots, |\kappa|\}$ whose symmetry labels are (the irrep or vibrational level) κ and (the partner or degenerate member) k.

The energy of any ionic state $|\lambda l\rangle$ will be affected in first order if the diagonal matrix element V_{llk} is nonvanishing. The sum of all these energy shifts over the set of ionic states is

$$\Sigma_l V_{llk} \leftrightarrow$$

κk
V
λ ,

which by JLV1 vanishes unless the molecular vibration has the molecular symmetry ($\kappa=0$).

In all but linear molecules, there is at least one molecular vibration which is asymmetric and which couples to any degenerate electronic state in first order. All that is required for this is confirmation that the nonidentity irreps in the Kronecker decomposition of $[\lambda \times \lambda]_\tau$ (τ being the time-reversal phase of the state corresponding to λ) also correspond to possible symmetry coordinates of a molecule of the stated symmetry, for all choices of λ. The τ-symmetrisation is included to allow for the time-even nature of the phonon coupling according to the argument of §7.3.2, and has the important effect of eliminating rotations from consideration. The work in the proof of the Jahn–Teller theorem amounts to the detailed checking, by exhaustion, of the existence of such asymmetric modes κ with linear coupling.

If such a mode couples linearly, we may choose an electronic basis in which their matrix elements have a diagonal part. The application of JLV1 above then shows that the signs of these diagonal parts must be mixed, for them to sum algebraically to zero. This means that as the lattice distorts according to some mode κk ($Q_{\kappa k}>0$), the various electronic states in the degenerate level will split in energy and will have mean energy zero. Some electronic state will then be lowered in energy. Each electronic state may be lowered in energy by an appropriate choice of lattice mode and sense of distortion. Hence the lattice may be expected to distort spontaneously about any electronic state, lowering the energy to first order in the displacement, until the distortion becomes large enough for the second-order energy shifts (which normally define the equilibrium positions of the ions) to become significant.

The Jahn–Teller effect initially proved as elusive as Sherlock Holmes's 'dog in the night' (Sturge 1967). In point of fact, there is no overall loss of symmetry, unless some trigger such as lattice strain or a cooperative phase transition of many Jahn–Teller active ions induces it. The fundamental reason for this is that the interaction with the lattice itself has the full symmetry of the undistorted lattice. We have used this result previously, in assuming the invariance of the node associated with the matrix elements V_{ijk} of the phonon interaction. This symmetry may be visualised by applying a lattice symmetry operation to the distortion field associated with any vibration and simultaneously to the electronic state:

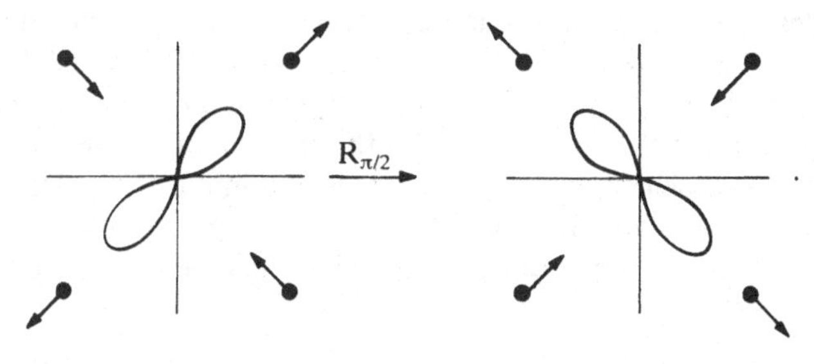

If the distortion field is thought of as acting on the matter that it finds at any angle, we see that the effect of the distortion is indistinguishable under a transformation which is a symmetry operation of the undistorted lattice, since the same relative ion–ligand motions are generated in that case. Hence the change in electrostatic energy with a distortion (including all quantum i.e. overlap- and exchange-induced corrections) is the same in either situation.

We have already seen that the energies remain unsplit, through the application of JLV2 to a self-energy diagram. Electronic degeneracy does not affect no-phonon line shift as such (§8.3.4). All electronic states in any level suffer the same phonon-induced shift of energy, since each may induce a distortion in a corresponding lattice mode. In the (grossly oversimplified) configuration coordinate diagram approach of §8.3.5, we may picture the situation for coupling of a doubly degenerate level to one lattice mode as follows:

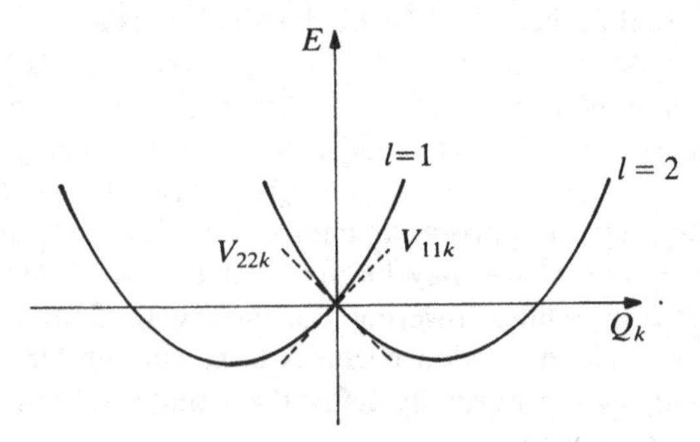

Hence, the well bottoms, defining the no-phonon line frequencies, are seen to be identical just because the mean of the gradients of the curves in the configuration coordinate diagram at the undistorted position is zero.

8.4.2 Ham effect – an example

We discuss the physical origin of the Ham reduction factors. This will motivate the field theoretic calculation of the Ham factors, and their symmetry analysis using diagram techniques in §8.4.3.

The configuration coordinate diagram arguments in §§8.3–8.4 indicate that in supposedly pure, no-phonon electronic transitions, we may nevertheless expect the effect of lattice interaction to affect the intensity in proportion to the overlap of the zero-point wavefunctions of the lattice modes. These functions are defined with respect to the origins specified by the new position of classical stability for the corresponding electronic state, i.e. the Q value at which the curve reaches its minimum. The overlap is not unity since the two states involved in the transition may stabilise at different Q values. If one level in the transition is Jahn–Teller active, this possibility becomes a certainty.

Just as the matrix elements of the dipole moment operator are reduced by lattice coupling, so are the matrix elements of any purely electronic operator, and by the same type of lattice wavefunction overlap factor, within the Jahn–Teller active level. The numerical value of these *reduction factors* depends principally on the symmetry of the Jahn–Teller active level and that of the electronic operator. Indeed they are essentially products of: 6j symbols, various energy factors corresponding to the excitation energies of intermediate states in the diagram, and the reduced matrix elements of the ion–lattice interaction, summed over lattice modes.

Indeed, Jahn–Teller theory has been a happy hunting ground for group theorists for this among other reasons. An enormous amount of ingenuity and effort has been expended by many groups in analysing the effects of the actual lattice symmetry, or of some higher and more artificial symmetry, for the explanation of various exact or approximate relations or equivalences between reduction factors, or for systematic effects and trends in configuration-coordinate type diagrams for the dependence of the energies of the vibronic states on ion–lattice coupling strength, etc.

Let us take a simple example (Stedman 1983b). It may be viewed as a paradigm of the appearance of higher symmetries in what is essentially a classical mechanics problem, and its quantum mechanical extension will be shown to correspond to the diagrammatic approach of the following subsection.

Consider a planar molecule of five ions, the Jahn–Teller active ion at the centre of a square, and the four other and identical ligands at its corners. A p_x and p_y orbital on the central ion will be degenerate in the absence of lattice (square) distortion; the electric field generated by the ligands will induce identical Stark shifts on the p orbitals. The symmetry group includes D_4, the four-fold rotations in the plane, the two-fold rotations out of the plane about the horizontal, vertical and diagonal axes (see (*a*) below).

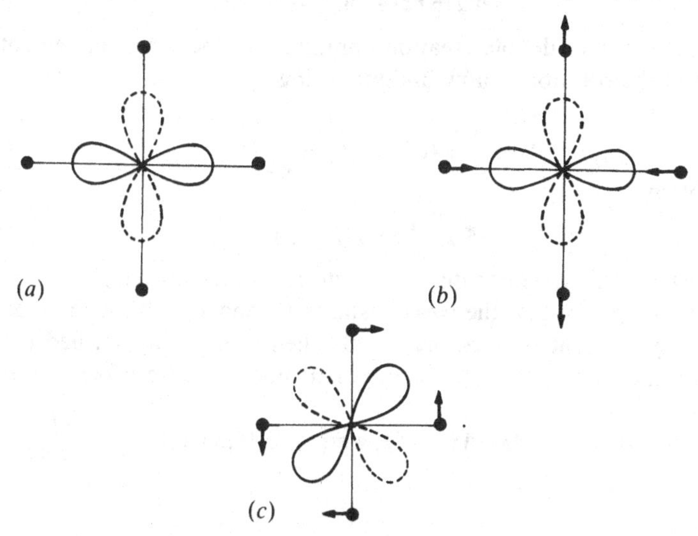

Such a spatial degeneracy of two wavefunctions is vulnerable, by the Jahn–Teller theorem, to the possible distortions of the molecule. There are in fact two Jahn–Teller active modes of this system, (*b*) and (*c*) above. These have the symmetries B_1, B_2 (or in Butler notation 2, $\tilde{2}$) under D_4. The two electronic states transform as the E(1) irrep of D_4. Clearly the B_1 mode with the sense of amplitude given in the diagram will raise the energy of the p_x orbital and lower that of the p_y orbital. This instance of the Jahn–Teller theorem in action corresponds to the occurrence of B_1 in the reduction of $[E \times E]_+$. The opposite signs of the energy shifts for the two orbitals illustrate the theorem of §8.4.1, that the mean gradient of the linear couplings to an asymmetric mode is zero. A symmetric (breathing) mode, in which all ligand move outwards, say, would lower both p orbital energies. Should the occupied electronic state be p_x, the square will spontaneously distort to give a negative amplitude of the B_1 mode ($Q_1 < 0$), gaining energy from the stabilisation. Equally, occupancy of the p_y orbital would imply a stabilisation by the same energy shift, when the square distorts to a positive amplitude Q_1 of the B_1 mode. The states remain unmixed in the distorted configuration, and the relative signs of the energy shifts show that the associated Hamiltonian may be written in the form:

$$\mathscr{H}_1 = V_1(a_x^{\dagger}a_x - a_y^{\dagger}a_y)Q_1, \tag{8.20}$$

where V_1 represents a coupling strength. Similarly B_2 is also present in the reduction of the above Kronecker product, and the diagram (*c*) also corresponds to a Jahn–Teller active mode. This is more easily visualised if we use rotated functions p_x', p_y' as in (*c*); the pinching action of the ligand motion on one orbital indicates a linear dependence of its energy on Q_2, and the opposite shift for the other (rotated) orbital, again without admixture. The corresponding Hamiltonian may then be written as:

$$\mathscr{H}_2 = V_2(a_x'^{\dagger}a_x' - a_y'^{\dagger}a_y')Q_2, \tag{8.21}$$

where the primes denote creation/annihilation operators for the rotated states. If the rotation is now undone, using

$$a_x' = \frac{1}{\sqrt{2}}[a_x + a_y],\ a_y' = \frac{1}{\sqrt{2}}[a_x - a_y] \tag{8.22}$$

we obtain

$$\mathscr{H}_2 = V_2(a_x^{\dagger}a_y + a_y^{\dagger}a_x)Q_2. \tag{8.23}$$

We now use this system to discuss the effect of assuming a higher symmetry than D_4. Suppose that the two constants V_1 and V_2, although associated with very different physics, are equal. Then it is readily verified that the Hamiltonian $\mathscr{H} \equiv \mathscr{H}_1 + \mathscr{H}_2$ is invariant under the transformation:

$$\mathbf{a}^+ \to \mathbf{M}(\theta)\mathbf{a}^+,\ \mathbf{Q} \to \mathbf{M}(2\theta)\mathbf{Q};\ \mathbf{a} \equiv \begin{pmatrix} a_x \\ a_y \end{pmatrix},\ \mathbf{M}(\theta) \equiv \begin{pmatrix} \cos\theta & \sin\theta \\ -\sin\theta & \cos\theta \end{pmatrix}. \tag{8.24}$$

In short, $\mathscr{H}$ is invariant under the higher group SO(2) of rotations in a plane. (Another application of this invariance in classical mechanics is given by Wulfman 1986.)

Normally such rotations are quite abstract, but in this example it is interesting to note that Q_1 and Q_2 differ essentially by a rotation in real space:

$R_{\pi/2}$ $R_{-\pi/4}$ $R_{\pi/2}$ $R_{\pi/2}$ $R_{\pi/2}$

The higher symmetry may be visualised quite explicitly as follows. Let the distortion directions on each ligand be rotated in the same sense by an angle θ, and the orientation of the central p orbital by $\theta/2$ in the opposite sense. Then the electrostatic energy of interaction with the central ion is unchanged. One can devise a pulley-and-string model to mimic this joint rotation:

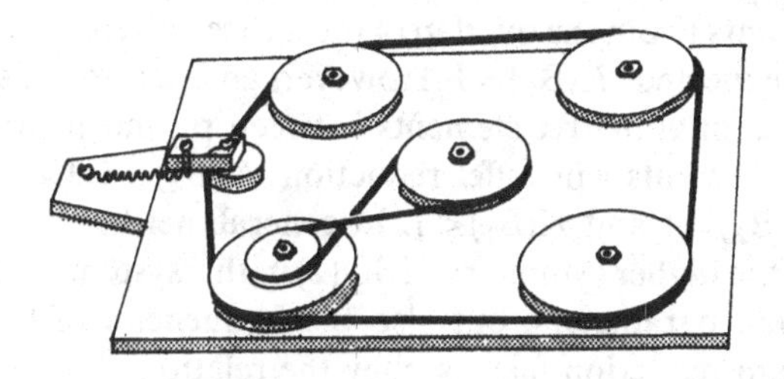

In perspex, its action is conspicuous on an overhead projector.

While airing this explicit example, we may illustrate further points from our earlier general analysis. Rotations do not couple in first order, since the appropriate irrep (A_2, or $\tilde{0}$) appears in $[E \times E]_-$, violating the rule of §7.3.2 for a time-even coupling.

However, our generalised selection rule equally shows that a time-odd interaction of this symmetry would couple the p orbitals. Such an interaction would correspond to the Zeeman, rather than the Stark, effects of the moving ligands – an effect which is almost universally ignored (§7.3.2). We may see this physically as follows. In this case the moving ligands may in a rotational mode be thought to generate an approximately circular current loop, and consequently a magnetic field which to a first approximation may be thought of as a constant field **B** perpendicular to the plane:

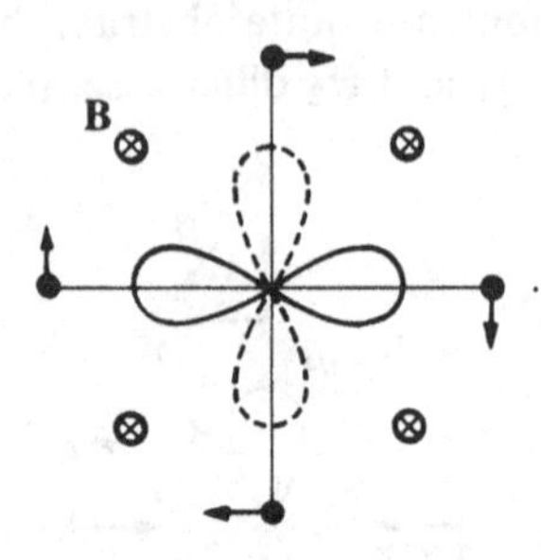

The obvious spatial transformations of this field and the wavefunctions show that the off-diagonal matrix elements such as $\langle p_x|\mathbf{B}\cdot\mathbf{L}|p_y\rangle$ are antisymmetric (i.e. the negative of $\langle p_y|\mathbf{B}\cdot\mathbf{L}|p_x\rangle$) and therefore imaginary, from hermiticity. It is just the time-odd operator **L**, as opposed to the time-even electric field in the standard mechanism, which is imaginary in the required way.

When an external electronic interaction O^{μ} of symmetry μ acts within the p orbital doublet, the reduction factor appropriate to its matrix elements will be written $K(\lambda)$. If the interaction V_1 predominates, the electronic states will stabilise in the original form (x,y) in which an operator of symmetry B_1 is diagonal anyway. Therefore no restabilisation of the lattice is involved when comparing the states linked by an operator O^{B_1}, and no reduction in its matrix elements (from the overlap of the lattice wavefunction before and after) need be expected: $K(B_1)=1$. However, an operator of symmetry B_2 has only off-diagonal matrix elements between p_x and p_y wavefunctions, and its matrix elements will suffer reduction: $K(B_2)<1$. Conversely, if V_2 dominates, $K(B_2)=1$ and $K(B_1)<1$. In general, neither are unity.

However, if the higher symmetry of SO(2) in this system equates not only the Hamiltonian parameters, but also the frequencies of the B_1 and B_2 modes, the Ham reduction factors obey the relation

$$K(B_1)=K(B_2). \tag{8.25}$$

This result is hopefully already plausible from the physical slant of the above discussion. We shall prove this result formally as an example of the diagram formalism in the next subsection.

8.4.3 Ham reduction factors – field theory approach

Of the various ways of estimating these factors, we are here concerned principally with the field theoretic approach pioneered by Gauthier and Walker (1976) and its adaptation and extension to include the symmetry or group theoretic aspects by Payne and Stedman (1977, 1983a,b,c). All merely technical details will be avoided; these references provide full details.

We consider an interaction $\mathscr{H}_s$ containing operators O^μ of symmetry μ in a Jahn–Teller active (spatially degenerate) electronic manifold of symmetry λ, interacting with phonons, including Jahn–Teller active phonons, of various symmetries $\{\kappa\}$. The matrix elements of O^μ are related by JLV3 to the appropriate Clebsch–Gordan coefficients:

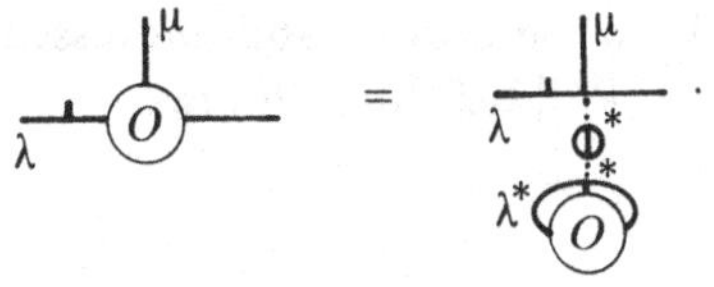

If we add phonon interaction to this picture, we have in principle any number of contributions to the vertex. However, each of these terms in the self-energy also reduces via JLV3 to a multiple of the Clebsch–Gordan coefficient, since all the associated matrix elements correspond to irreducible tensor operators (the ion–lattice interaction has the symmetry of the undistorted site) and the network is an invariant node:

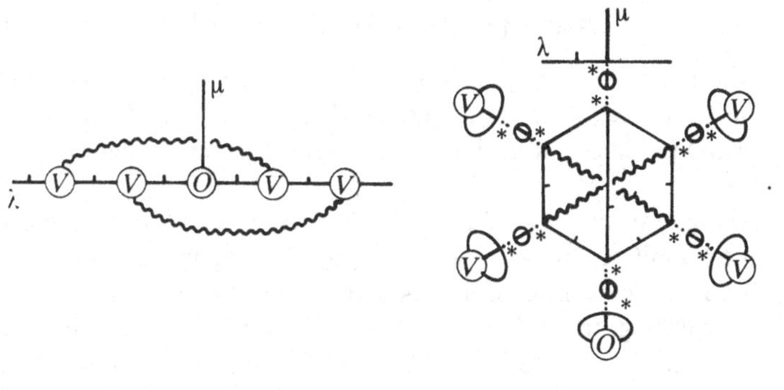

Scaling factors $d(\mu)$

The Ham reduction factor may be derived from the various factors distinguishing the last two sets of diagrams. We will make the relationship with Ham factors more precise shortly. For the moment, we write the ratio of any phonon-dressed self-energy diagram D to its undressed counterpart as $d(\mu)$.

Comparing the above dressed and undressed self-energy diagrams as an example, we note three main sources of these factors $d(\mu)$. First, the dressed diagram will include more interactions, and so more reduced matrix elements $\langle\lambda||V^\kappa||\lambda\rangle$ of these phonon-related interactions. κ ranges over the symmetry of the dressing phonons.

Second, the field theoretic expression for a self-energy includes energy denominators appropriate to the energy mismatch between the initial and each intermediate state. If (again we take the simplest case) the intermediate electronic state is in the λ level, this means a product of denominators $F(\{\omega_k\})$ depending only on lattice mode frequencies in combinations dictated by the topology of each diagram in the series. Finally the group theory of the diagrams contributes the nj symbol $J(\lambda\mu)$ obtained on applying JLV3 to the dressed diagram. Its value is dependent only on the symmetry μ, not the reduced matrix element, of the external interaction; the latter is common to both the dressed and undressed situation. Schematically, then, $d(\mu) = \Pi\langle\lambda||V^{\kappa}||\lambda\rangle F(\{\omega_k\})J(\lambda\mu)$.

Scaling versus reduction factors: linking $K(\mu)$ and the $d(\mu)$

These various $d(\mu)$ for different diagrams D are not quite the reduction factors $K(\mu)$ themselves. We have to convert from one to the other, first, by adding all diagrams; second, by adding unity, essentially to scale the self-energy perturbation terms against the unperturbed term, and third, by scaling the ratio of the result to that for the case of a symmetric interaction $\mu = 0$ (for which the reduction factors are always unity, since the operators are always diagonal):

$$K(\mu) = [1 + \Sigma_{\mathrm{D}} d(\mu)]/[1 + \Sigma_{\mathrm{D}} d(0)]. \tag{8.26}$$

In case these steps seem implausible to the reader, we give a fuller (but still heavily abbreviated; Payne and Stedman 1983a give more detail again) and more formal justification in the next paragraphs. The reader is invited to skip them.

Following Gauthier and Walker (1976), we pure $\mathscr{H}_{\mathrm{s}}$ in the zero-order electronic Hamiltonian $\mathscr{H}_{\mathrm{e}} \to \hat{\mathscr{H}}_{\mathrm{e}} \equiv \mathscr{H}_{\mathrm{e}} + \mathscr{H}_{\mathrm{s}}$, diagonalise this first, then apply the ion–lattice interaction as a perturbation. The new zero-order electronic propagator $\mathscr{G}_i^0(E) \equiv 1/(E - \hat{\mathscr{H}}_{\mathrm{e}})$ is nondiagonal in the old electronic basis $\{|i\rangle\}$. It satisfies a Dyson equation (the first figure below) which is diagonal in a transformed bases $\{|\hat{i}\rangle\}$ corresponding to the solutions of

$$(E - \hat{\mathscr{H}}_{\mathrm{e}})\hat{i}\rangle = 0. \tag{8.27}$$

When phonon interaction is added, the new states are no longer diagonal and the dressed electronic Green's function obeys a new Dyson equation containing a self-energy operator $\hat{\Sigma}(E) \equiv \hat{\Lambda}(E) - \mathrm{i}\hat{\Gamma}(E)$, which can be used to diagonalise the old basis:

$$(E - \hat{\mathscr{H}}_{\mathrm{e}} - \hat{\Lambda}(E))|i'\rangle = 0. \tag{8.28}$$

This may be contrasted with the corresponding equation for phonon interactions only:

$$(E_0 - \mathscr{H}_{\mathrm{e}} - \Lambda(E_0))|\bar{i}\rangle = 0. \tag{8.29}$$

The difference between the eigenvalues E and E_0 of the last two equations is solely due to $\mathscr{H}_s$. The Hamiltonian $\mathscr{H}_{\text{eff}}$, whose eigenvalues are $E-E_0$, must therefore represent the effective electronic interaction associated with $\mathscr{H}_s$, modified from its bare form ($\mathscr{H}_s$) by phonon dressing effects. We aim to construct $\mathscr{H}_{\text{eff}}$ and thence the reduction factors.

We expand $\hat{\Sigma}(E)$ in powers of $\mathscr{H}_s$, and consider here only the first-order terms $\hat{\Sigma}_1(E)$ leading to reduction factors linear in $\mathscr{H}_s$. Upon Taylor expansion and use of the perturbation series we obtain the desired equation:

$$(E-E_0-\mathscr{H}_{\text{eff}})|\bar{i}'\rangle=0,$$
$$\mathscr{H}_{\text{eff}}\equiv[1+\Lambda_1(\hat{1})]/[1+\Lambda_1(\mathscr{H}_s)], \tag{8.30}$$

where $\Lambda_1(T)$ corresponds to the series of self-energy diagrams discussed above which are linear in an external interaction T. If the symmetry μ of $\hat{\mathscr{H}}_s$ were the identity irrep, as for the unit operator $\hat{1}$, $\mathscr{H}_{\text{eff}}$ would be a diagonal operator. This corresponds to the requirement that $K(0)=1$; a symmetric interaction does not mix the phonon-stabilised eigenstates, and experiences no Ham reduction. The reduction factors for other symmetries are obtained by comparing the reduced matrix elements of $\mathscr{H}_{\text{eff}}$ with various O^{μ} (operator symmetries in $\mathscr{H}_s$).

8.4.4 Applications of the diagram approach to reduction factors

Higher symmetries

Again in the D_4 example just mentioned, it has been noted in §8.4.2 that the higher, SO(2), symmetry gives the sum rule $K(B_1)=K(B_2)$. We indicate now the manner in which such rules can be proved from the diagram analysis.

The higher symmetries discussed in the literature for a wide variety of Jahn–Teller systems generally correspond to pathological, though sometimes approximately valid, special cases in which various parameters of the theory are equated or related. In particular, the frequencies of a set of lattice modes of different symmetry under the true point group are often equated so that at least formally the set of equi-energetic lattice modes may be regarded as the components of a larger irrep of some higher symmetry. An appropriate restriction on the coupling constants of the different modes is also necessary.

For example, in our standard example of a square molecule, we equate both the frequencies and the couplings of the B_1 and B_2 modes so that they become members of an SO(2) doublet with irrep label 2. In another example, a triplet state in octahedral symmetry may show Jahn–Teller effects on coupling to doublet and to triplet modes; if the doublet and triplet modes are of the same frequency and have coupling strengths in a given numerical relation, they may be treated as a quintet of O(5), and the problem analysed from the viewpoint of this higher symmetry (O'Brien 1969, Judd 1974, 1983a,b, 1987a, Pooler and O'Brien 1977, Payne and Stedman 1983c).

In the case of a doublet interacting with doublet modes in octahedral symmetry, the restriction to linear coupling suffices to introduce an SO(2) symmetry.

In all such cases, we may regard the field theoretic diagrams as possessing an invariance under the higher symmetry group, and all electronic and phonon labels may be relabelled by the irreps of a higher group (for which we use double lines, as in §3.5):

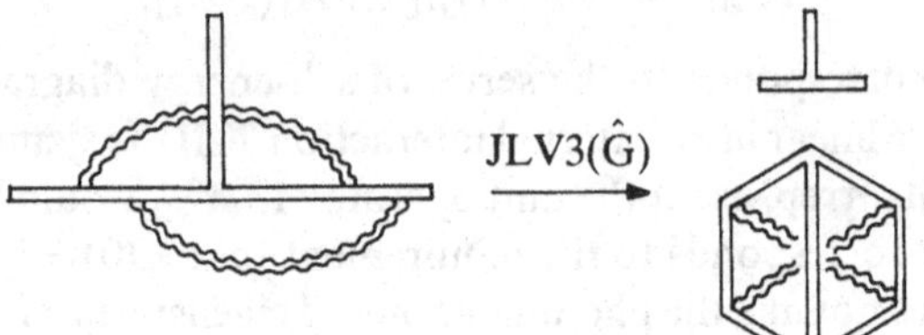

JLV3($\hat{G}$)

The reduction factor analysis of the previous and present subsection then show trivially that the possible variety of reduction factors is limited by the number of irreps of the higher group appearing in the Kronecker square of the (higher group) irrep, Λ say, corresponding to the Jahn–Teller active electronic level. Since such an irrep may reduce to several irreps $\{\mu(\Lambda)\}$ of the point group, all reduction factors for these irreps are equal. The formal reason why the reduction factors $K(B_1)$ and $K(B_2)$ were equal in SO(2) symmetry for our D_4 example of §8.4.2 is that $2(SO(2))\downarrow(B_1+B_2)(D_4)$.

Supersymmetry in Jahn–Teller systems

An unsolved problem is the extension (of this analysis of higher symmetry restrictions on reduction factors) to the supergroup case (Jarvis and Stedman 1984). In this, supersymmetry is the higher symmetry. This mixes the identity of electron (fermion, *à la* Abrikosov) and phonon (boson) operators and interactions, and also demands a numerical relation between ion–phonon interaction constants and those for lattice anharmonicity. We then find restrictions of a rather different sort on the reduction factors. We explain this a little more fully in view of its novelty.

The Hamiltonian $\mathscr{H}_1 = a^\dagger a + b^\dagger b$ for a single electronic state and a single lattice mode is almost trivially supersymmetric. If $|B,F\rangle$ is its eigenstate corresponding to B bosons and F fermions, then the states $|B,1\rangle$ and $|B+1,0\rangle$ are degenerate. The generators of supersymmetry are $S_1 \equiv b^\dagger f$ and its hermitian conjugate. Fermionic anticommutation relations guarantee that $S_1^2=0$, and that $\mathscr{H}_1 = Q_1^2$, where $Q_1 \equiv S_1 + S_1^\dagger$. The property of supersymmetry ($[\mathscr{H}_1, S_1]=0$) is then an inevitable consequence, as may be seen from the Jacobi identity

$$[\{A,B\},C]+[\{B,C\},A]+[\{C,A\},B]=0$$

for the case $A=B=S_1, C=S_1^\dagger$. Other inevitable consequences are the vanishing of the energy of the ground state $|00\rangle$, and the double degeneracy of each excited level (as detailed above).

A less trivial illustration is afforded by including an interaction term. However, the moment one includes a linear coupling with phonons, such as $a^\dagger a\phi$ where the position operator $\phi \equiv (b+b^\dagger)$, the action of the generator S_1 will inevitably throw up terms which are trilinear in boson operators, i.e. cubic anharmonicity in the lattice modes. If the possibility of such additional terms is accepted, the same algebraic structure can be preserved with the new definition

$$S_2 \equiv a^\dagger \beta,\ \beta \equiv b-\alpha\phi^2, \tag{8.31}$$

for the supremely economical but sufficient reason that $S_2^2=0$; the supersymmetry algebra is then guaranteed (Blockley and Stedman 1985). We obtain a corresponding supersymmetric Hamiltonian

$$\mathscr{H}_2 = a^\dagger a[\beta, \beta^\dagger] + \beta^\dagger\beta = a^\dagger a + b^\dagger b - 4\alpha a^\dagger a\phi + 2\alpha\phi - \alpha\phi^3 + \alpha^2\phi^4. \tag{8.32}$$

We note the occurrence of quartic as well as cubic anharmonicity in the final expression.

It is a simple exercise in perturbation theory to show that the ground state $|00\rangle$ is unshifted, and that the excited states remain degenerate in pairs under the combined effects of these various interactions.

These examples may be extended to $i=1,\ldots,n$ mutually degenerate electronic and phonon states. (Since for ionic levels and in the Abrikosov technique of §7.4, the zero of electronic energy is arbitrary, the requirement that electronic and phonon energies should coincide is of no physical consequence.) We simply choose a nilpotent generator of the form $S \equiv \mathbf{a}\cdot\boldsymbol{\beta}^\dagger$, where the vectors denote a set of operators, $\mathbf{a} \equiv (a, a_2, \ldots, a_n)$ etc., and $\boldsymbol{\beta} \equiv \exp(G[\boldsymbol{\phi}])\mathbf{b}\exp(-G[\boldsymbol{\phi}])$, where G is an arbitrary function of the lattice mode position operators $\boldsymbol{\phi}$. We naturally require the same number of electronic as phonon states for an adequate expression of supersymmetry. If, in addition, the set of electronic and lattice operators have the same symmetry under the point group, and the function $G(\boldsymbol{\phi})$ is chosen to be a group invariant, $\boldsymbol{\beta}$ transforms as $\mathbf{b}$ and supersymmetry becomes compatible with the point group symmetry through the invariance of the scalar product definition of S.

We return to the example of the square molecule of §8.2.4 in the case of SO(2) invariance (or equivalently, from a formal viewpoint, the case of the doublet state in octahedral symmetry interacting with a doublet phonon level of the same symmetry). These lend themselves to a supersymmetric extension. The choice $G_1 \equiv -\frac{1}{3}\alpha[\phi_1^3 - 3\phi_1\phi_2^2]$ induces the supersymmetric Hamiltonian whose linear coupling terms are of the standard form, say for the octahedral system:

$$\mathscr{H}_{\text{lin}} = 4\alpha[(a_1^\dagger a_1 - a_2^\dagger a_2)\phi_1 - (a_1^\dagger a_2 + a_2^\dagger a_1)\phi_2] \tag{8.33}$$

and whose anharmonic terms have the form

$$\mathscr{H}_{\text{anh}} = -3G_1 + \alpha^2(\phi_1^2 + \phi_2^2)^2. \tag{8.34}$$

The appearance of the standard linear coupling as well as anharmonic terms of the correct invariant form is not an accident, but a consequence of the preservation of group invariance.

These examples indicate the intimate relation between linear ion–lattice coupling and lattice anharmonicity characteristic of supersymmetry. The same constant α appears in all interactions, fixing their relative strengths. There are no new relations resulting from this extra symmetry among our old reduction factors. Rather, the old reduction factors, covering the effects of phonon dressing on the 'purely electronic' matrix elements, will be constrained through supersymmetry to those factors enshrining the effects of ion–phonon interaction on 'purely lattice' interactions, such as the effect of external stress on the system.

A naive expectation is that supersymmetry would treat together the two vertices of degree three, the linear ion–phonon coupling and cubic anharmonicity:

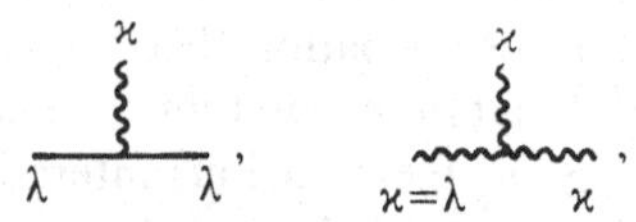

the choice of ion or phonon label corresponding to the choice of a component in the irrep of the supergroup covering both ion and phonon operators. This can at best be only part of the picture, however, since the requirements of supersymmetry inevitably involve a term of second order in the interaction constant and with quartic anharmonicity. The analogues of our group theoretic diagrams in the case of the supergroup must, therefore, be less closely linked to our field theoretic technique. They could, however, be quite closely related to the superfield formalism, and one could hope for an extension of the JLVn theorems in this context. This would require not only the suitable application of a Racah-like algebra for supergroups but also a canonical analysis of the Hamiltonians in terms of superfields, following Salam and Strathdee (1975) and Cooper and Freedman (1983).

Mixed mode symmetries

In case of the square molecule discussed in §8.4.2, or the case of triplet states in octahedral symmetry, there is more than one possible choice for the phonon symmetry $\kappa = \{\kappa_i\}$ even within the restriction of considering only

Jahn–Teller active modes. (This restriction is not accurate in calculating reduction factors, as the example of symmetric modes discussed in §8.4.6 makes clear.)

Each mode symmetry makes its presence felt in a reduction factor calculation in a variety of ways. Diagrams involving just one mode symmetry make their particular, independent, contributions. An obvious question is the relation of the independently evaluated $K(\mu)$ to the final value of the reduction factor when both modes are included. In different contexts, workers in this area tend to assume an additive or multiplicative relation.

The diagram analysis makes several things clear. First, at second order in the phonon interaction, the diagrams produce additive contributions to the scaling factors $d(\mu)$, simply because of the additive form of the perturbative approach to self-energy calculations. However, this does not necessarily mean additive contributions to the $K_{rr'}(\mu)$ since the relation between scaling and reduction factor is nontrivial. To the extent that terms higher than second order in the expansion of the denominator in $K_{rr'}(\mu)$ may be neglected, we have the rule

$$K_{rr'}^{\{\kappa_i\}}(\mu)=1+\Sigma_i[K_{rr'}^{\kappa_i}(\mu)-1] \tag{8.35}$$

from the binomial expansion of equation (8.26). Once we go to fourth order, however, the diagrams involving phonons of both symmetries may be regarded as contributing *interference* terms in the summation even for the scaling factors. There is no *a priori* justification for the multiplicative relation $K_{rr'}^{\{\kappa_i\}}(\mu)=\Pi_i[K_{rr'}^{\kappa_i}(\mu)]$ used by some authors, at least in general. The application of this to the example of the square molecule will be deferred to §8.4.6.

Vanishing, or restricted classes of, diagram contributions

Sometimes certain diagram topologies never contribute, or if at all only in a very restricted way to reduction factors. For example, in the case of a double level in octahedral symmetry interacting with a doublet of phonons, the second-order diagram

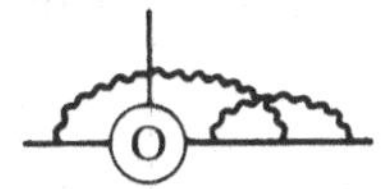

does not contribute to any reduction factor.

The reason for this is that the associated 6j symbol vanishes:

$$\begin{Bmatrix}2&2&2\\2&2&2\end{Bmatrix}\leftrightarrow$$

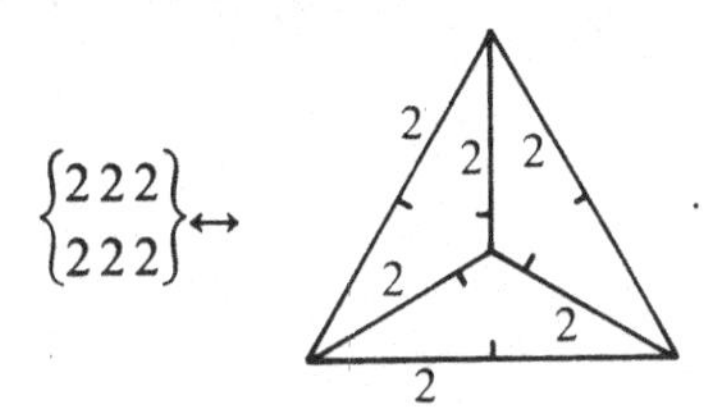

Other zeros of the octahedral 6j symbols similarly simplify the problem for quartet electronic states (Payne and Stedman 1983c).

More generally, classes of diagrams may be identified with particular types of contributions. Again in the case of doublets in octahedral symmetry, each diagram contributes in just one of the following ways:

(a) The diagram contributes to $d(0)$ and $d(\tilde{0})$ only.

In this case (i) the diagram must be reflection-symmetric, and (ii) all insertions of $\mathscr{H}_s$ into the diagram also contribute to $d(0)$ or $d(\tilde{0})$, respectively.

(b) The diagram contributes to $d(2)$ only.

In this case: (i) there must be no more group-theoretic 3jm vertices in the group theory diagram than there were interaction vertices in the parent field theoretic diagram, (ii) $\mathscr{H}_s$ has a unique insertion point in the diagram, (iii) this unique point has the property that there are an even number of ion–phonon vertices on each side.

(c) The diagram contributes to *no* $d(\mu)$.

These rules may be justified by a straightforward analysis of the topology, coupled with the JLVn theorems and the particular interchange phases of the 3jm vertices in this system (Payne and Stedman 1983b).

8.4.5 6j expansion of reduction factors

Introduction

We revert to a diagram analysis of the symmetry constraints on the reduction factor. It is of considerable significance that the symmetry content of contributions to the various $d(\mu)$ are enshrined in nj symbols for the point group in question.

A somewhat different way of applying the JLVn theorems is useful. Consider the (possibly) multiphonon lattice state P which is in existence at the time of the action of the external interaction $\mathrm{H_s}$ corresponding to the propagators threading past the $\mathrm{H_s}$ vertex. We couple these propagators, each of some symmetry κ to some irrep π, diagram by diagram, using the unitarity of point group Clebsch–Gordan symbols. We then apply JLV3 to everything in sight:

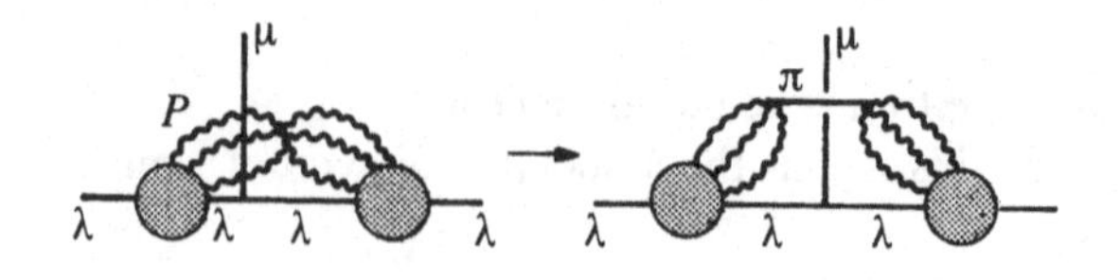

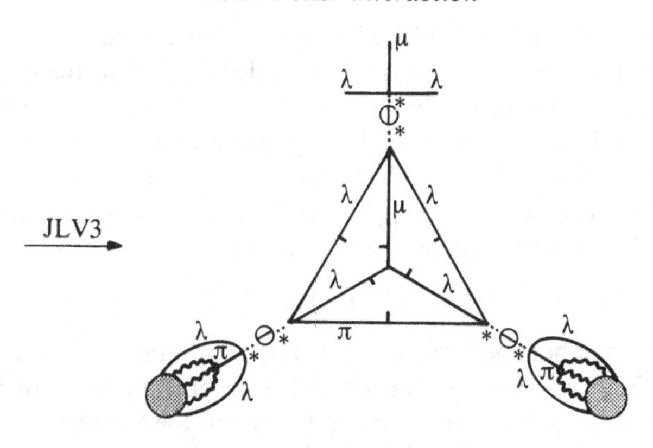

This reduces each $d(\mu)$ to a product of the usual energy factors and reduced matrix elements of the primary interactions. However, we now have essentially split up the monster nj symbol of our previous factorisation into: (a) two closed networks, enshrining the complexities of the formation of the multiphonon state for each amplitude and including the basis transformation nodes that project the multiphonon state onto the symmetry π, together with (b) a single 6j symbol which alone defines the μ-dependence. As a result we may formally expand the scaling factors $d(\mu)$ and hence the reduction factors $K(\mu)$ in terms of 6j symbols times coefficients b^{π} (the material summarised in (a)) which embody the physical coupling strengths, etc.

General reduction factor expansion

At this point we reinsert on the above diagram some extra labels and phases (see Payne and Stedman 1983a) which we have ignored till now, namely the repetition labels associated with a general point group 6j symbol and the various 3j phases arising from careful application of the JLVn theorems:

$$K_{rr'}(\mu)=\Sigma_{\pi ss'} b^{\pi ss'} c_{\mu rr',\pi ss'} \tag{8.36}$$

$$c_{\mu rr',ss'} \equiv \hat{\lambda}\{\lambda\}\{\pi\lambda\lambda s'\}\{\mu\lambda\lambda r'\}\begin{Bmatrix}\mu^* & \lambda & \lambda \\ \pi & \lambda & \lambda\end{Bmatrix}_{rss'r'} \leftrightarrow$$

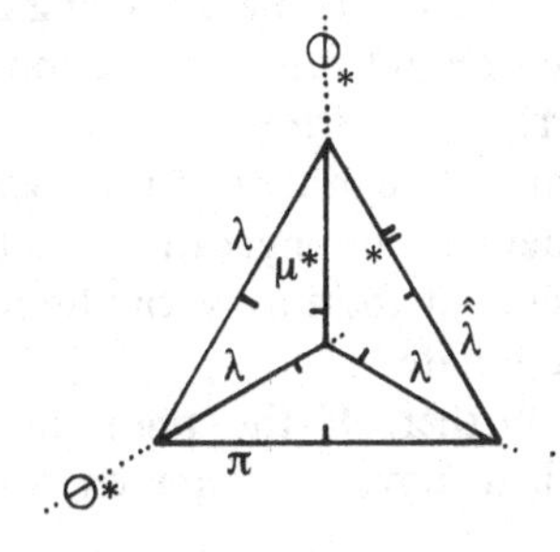

The labels r, r' on the reduction factor express the possibility of nondiagonal terms in these labels when applying JLV3 to the original H_s interaction vertex. The coefficient $b^{\pi ss'}$ appearing in the expansion of $K_{rr'}(\mu)$ are related to those appearing in the expansion of $d^{rr'}(\mu)$ by an equation similar to that linking $d^{rr'}(\mu)$ and $K_{rr'}(\mu)$. With the phase choice specified above for the constants $c_{\mu rr',\pi ss'}$, $c_{\mu rr',000}$ and $c_{000,\pi ss'}$ are each equal to unity, and the coefficients $b^{\pi ss'}$ satisfy the conditions

$$b^{\pi ss'} = (b^{\pi s's})^*, \ \Sigma_{\pi s} b^{\pi ss} = 1. \tag{8.37}$$

Since the 6j symbol obeys an orthogonality condition (§3.3.3), the $c_{\mu rr',\pi ss'}$ are (apart from a trivial normalisation) the elements of an orthogonal matrix, and we may invert equation (8.36) to find the expansion constants $b^{\pi ss'}$ in terms of the reduction factors $K_{rr'}(\mu)$:

$$b^{\pi ss'} = \Sigma_{\mu rr'} |\mu| \, |\pi|^{-1} K_{rr'}(\mu) c_{\mu rr',\pi ss'}. \tag{8.38}$$

The repetition labels s,s' belong to possible repetitions of π in the Kronecker product $[\lambda \times \lambda]$, in exactly the same way that r,r' denote repetitions of μ in this product. The dimension of the spaces $\{\mu rr'\}$ and $\{\pi ss'\}$ are then both $|\lambda|^2$, and there are as many constants $b^{\pi ss'}$ as reduction factors.

Constraints on reduction factors from phonon symmetry

As an example, we may consider the square molecule again. We argued on physical grounds in §8.4.2 that for coupling to $B_1(2)$ modes only, $K(B_1) = 1$. The relevant 6j symbol is

$$\begin{Bmatrix} 2 & 1 & 1 \\ \pi & 1 & 1 \end{Bmatrix},$$

where π ranges over all coupling symmetries derivable from powers of $\kappa = 2$. Hence we have just the cases $\pi = 0$ (for which the 6j is just a 3j phase, and of modulus unity) and $\pi = 2$, for which again the 6j symbol has modulus unity.

Time-even coupling and sum rules

We have a physical restriction on the formalism that we have not yet exploited: π should correspond to the symmetry of a multiphonon state. For first-order self-energy terms, this means a single phonon state $\kappa \equiv \pi$. Now, in view of the assumption of time-even phonon coupling (§7.3.2), we may then restrict π to the irreps appearing in the appropriately symmetrised part of the Kronecker product. The other $b^{\pi ss'}$ must vanish. Hence there is a nontrivial relationship generated amongst the reduction factors, accurate to the extent that first-order contributions with phonons coupling through a time-even interaction dominate.

It can be generalised further. If the phonon modes, of whatever symmetry, are assumed all to have the same energy (an Einstein lattice model), the energy factors $F(\{\omega_k\})$ will be the same in the diagrams associated with a two-phonon intermediate lattice state π for diagrams of somewhat different topology:

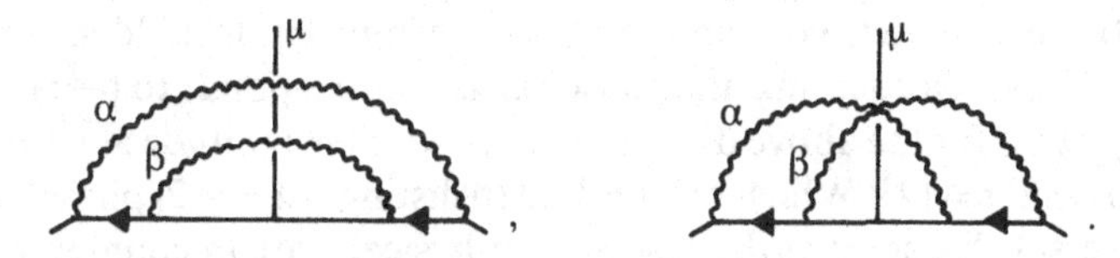

It is an elegant exercise in the group theoretical manipulation of field theoretic diagrams (Payne and Stedman 1983c) to show that the nj symbols from these two diagrams:

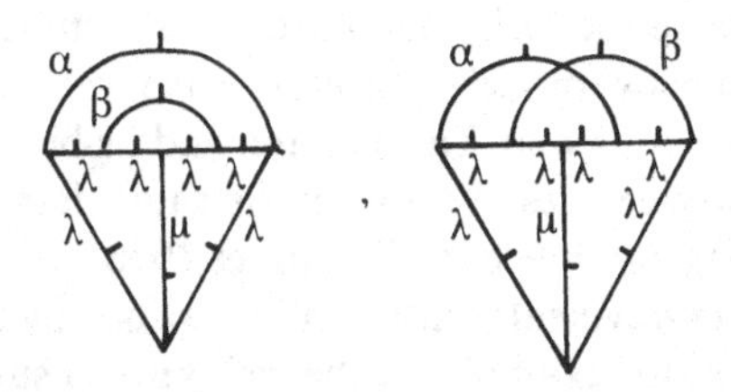

have the same magnitudes and opposite signs, thus contriving to give a cancellation in the final answer when the field theoretic denominators are the same, for the case that π has the forbidden symmetry of the earlier case.

The proof of this follows directly from an application of JLV4 and then JLV3 to the above diagrams:

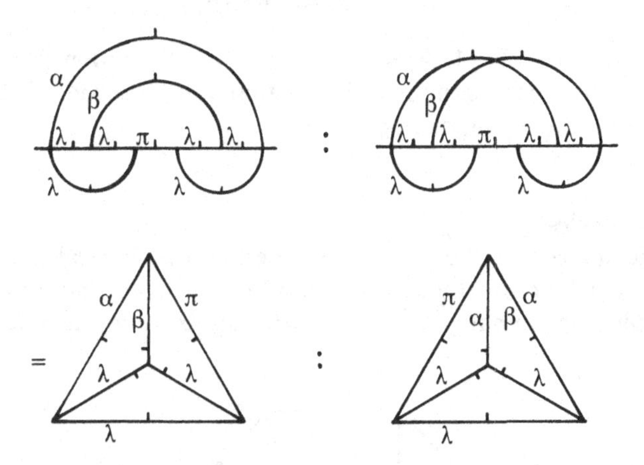

They differ, therefore, by the phase product $\{\alpha\lambda\lambda a\}\{\beta\lambda\lambda b\}\{\pi\lambda\lambda s'\}\{\lambda\}$, where α, β are the irreps describing the two phonons. The first two phase factors cancel, since each of α and β must be in the appropriately symmetrised part of the Kronecker square of λ, i.e. the symmetric square if λ is a true irrep (so that the associated double time-reversal phase $\tau=1$), and the antisymmetric square if λ is a spin irrep ($\tau=-1$, §3.6.4). If, by our hypothesis, the coupling symmetry π is in the opposite part of the symmetrised product, corresponding to a symmetry forbidden for one-phonon time-even coupling, the phase $\{\pi\lambda\lambda s'\}$ corresponds to $(-\tau)$. Hence the *nj* symbols for the above diagrams are of equal magnitude and opposite sign provided $\tau=\{\lambda\}$. While we have had to distinguish the 2j phase and the time-reversal character of the associated physical state (a counterexample to the condition $\tau=\{\lambda\}$ is given in Comment (d) of Problem 3.6, also Comment (f) of Problem 3.12), in practice the condition fails only in very low symmetries and one-dimensional irreps of no interest in Jahn–Teller physics. The number of sum rules so derived between reduction factors is the number of irreps in the forbidden part of the symmetrised square of λ.

Such single-frequency models therefore give the sum rule as being satisfied at fourth order as well. This feature, and its breakdown in the case when the phonon frequencies are not all the same, has been discussed by many authors (Leung and Kleiner 1974, Bates 1978), although the physical origin in terms of time-reversal restrictions is clarified by the above analysis. This analysis applies, for example, to the well known sum rule $q=\frac{1}{2}(1+p)$ for doublet states interacting with doublet phonons in octahedral symmetry (Payne and Stedman 1983b).

We now justify the corresponding result for our earlier example of a square molecule with D_4 symmetry (§8.4.3). The corresponding 6j symbols and *nj* phases may be obtained from Butler (1981) to give the sum rule:

$$1+K(A_2)=K(B_1)+K(B_2). \tag{8.39}$$

(In Butler notation, $E=1$, $A_2=\tilde{0}$, $B_1=2$, $B_2=\tilde{2}$.) Equation (8.39) holds therefore to first order in phonon interaction, and to second order if we equate the B_1 and B_2 mode frequencies.

Symmetric modes

Symmetric modes ($\kappa=0$) of course do not create the classical Jahn–Teller instability. On their own, they do not contribute to any reduction factor, for then the phonon coupling symmetry $\pi=0$ (take $\alpha=0$ in the diagram):

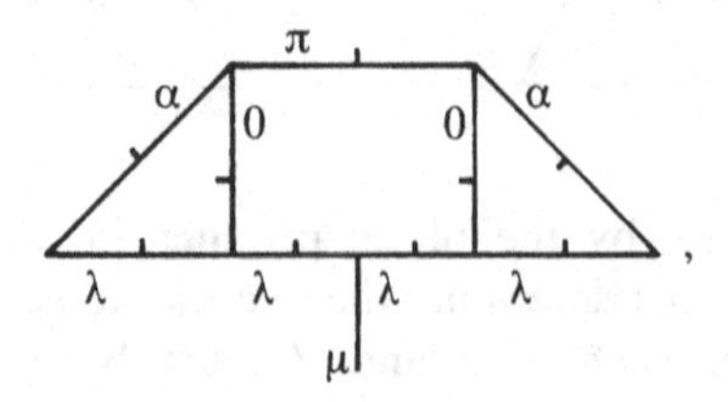

and, if all $b^{\pi ss'}$ are zero for $\pi \neq 0$, $K_{rr'}(\mu)=1$ (from equations 8.36–8.38). However, symmetric couplings may contribute to a reduction factor through interference with another, asymmetric and Jahn–Teller active, phonon mode $\alpha \neq 0$, for then the coupling symmetry π is not necessarily the identity irrep. In practice, it is possible to perform a canonical transformation to transform away the symmetric linear couplings, but the nonlinear symmetric couplings will contribute to reduction factors (Payne and Stedman 1983b).

8.5 Exchange

8.5.1 Dirac–Heisenberg–van Vleck–Serber interaction

The matrix element of an orbital two-particle operator such as the Coulomb interaction between antisymmetrised two-electron states yields terms of the direct and exchange type:

$$\tfrac{1}{2}\langle|\Psi\Phi||1/r_{12}||\Psi'\Phi'|\rangle$$
$$\equiv\langle\Psi_1\Phi_2|1/r_{12}|\Psi_1\Phi_2\rangle-\langle\Psi_1\Phi_2|1/r_{12}|\Phi_1\Psi_2\rangle,$$

where the suffices labelling the electrons will be understood in later equations. The spin–orbital functions used here may be decomposed into their spinors (whose overlaps factorise out) and the remaining direct or exchange orbital matrix element (of the form $\langle\psi\phi|1/r|\psi\phi\rangle$, $\langle\psi\phi|1/r|\phi\psi\rangle$, respectively). However, the manner in which the spinors overlap differs, on account of the electron label permutation, between the direct and exchange term.

Since the early days of quantum mechanics, it has been found useful (see, for example, Martin 1967) to write the exchange term as the matrix element of an *exchange Hamiltonian* between product spin–orbital states. This Hamiltonian is the Coulomb interaction of the direct term, together with (a) a minus sign from the antisymmetrisation sign, (b) a permutation operator $\mathbf{P}^l$ for the electron labels in the orbital parts of the wavefunction (this reverses the order of ψ, ϕ as above), and (c) a permutation operator $\mathbf{P}^\sigma$ for electron labels in the product spinors.

The last of these adaptations may be written as the spin operator:

$$\mathbf{P}^\sigma \equiv 2\mathbf{s}_1\cdot\mathbf{s}_2+\tfrac{1}{2}. \tag{8.40}$$

Hence, the energy of an antisymmetrised state of two electrons with antiparallel spin is greater than that of two electrons with parallel spin; the Pauli exclusion (or Fermi hole) keeps electrons of like spin comparatively far apart, and so reduces their repulsive Coulombic interaction energy. Exchange tends to favour ferromagnetic states.

The normal proof of this form of the spinor permutation operator is merely to note that the right side reproduces the required matrix elements, such as $\langle\uparrow\uparrow||\uparrow\uparrow\rangle=1$, $\langle\uparrow\downarrow||\downarrow\uparrow\rangle=1$, $\langle\uparrow\downarrow||\uparrow\downarrow\rangle=0$, etc.

However, group theory is certainly latent in such a symmetrical expression. The two terms in $\mathbf{P}^\sigma$ are each rotationally invariant, since the permutation operator is a rotational invariant (Dirac 1983). Since $\mathbf{s}_i$ is an irreducible tensor operator, its matrix elements are proportional to 3jm symbols for which the angular momenta are $\frac{1}{2}$, $\frac{1}{2}$ and 1. The term $2\mathbf{s}_i\cdot\mathbf{s}_j$ then is proportional to the coupling tree

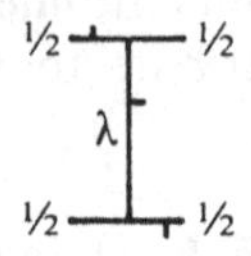

with $\lambda=1$, while the permutation operator $\mathbf{P}^\sigma$ is proportional to the term with $\lambda=0$. The final term in the above expression, $\frac{1}{2}$, is proportional to the left side of the identity

which is thus a diagram counterpart to the expansion of $\mathbf{P}^\sigma$.

The importance of this tree diagram may be verified either by applying JLV4 or 3jm pseudounitarity to any invariant node of degree 4 in spin labels, such as the permutation operator matrix element itself:

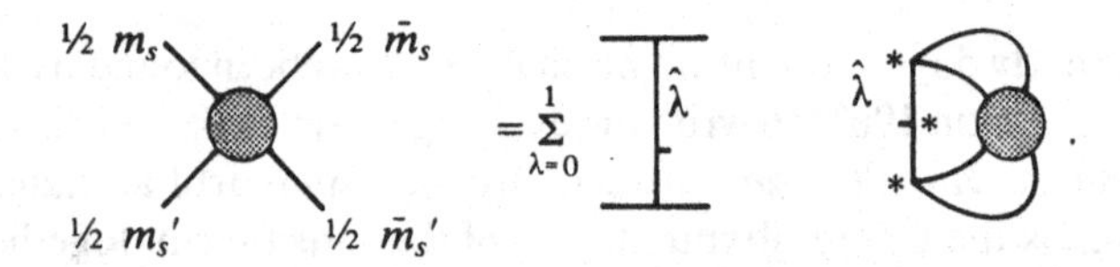

Only these two terms may appear, since $\frac{1}{2}\times\frac{1}{2}=1+0$. The theory of integrity bases for spinor (projective) irreps of SO(3) (Riddell and Stedman 1984) (see §8.6.1) shows that the result may nevertheless be rewritten in a variety of forms, corresponding to the alternative forms of such angular momentum coupling trees as in §1. The only feature of the spin representation of the permutation operator not dictated by these general symmetry arguments is the value of the numerical coefficients, which may be obtained readily from two trivial choices of spin projection.

Spin permutation diagrams are discussed, for example, by Wilson and Gerratt (1979).

8.5.2 Virtual phonon exchange

An extended exercise involving much of the material presented in earlier chapters is the application to assessing the effects of time-reversal symmetry for virtual phonon exchange between Kramers (odd-electron) ions.

Our first task is to generalise the Abrikosov projection method to a pair of ions. When this is done, the surviving terms of interest in pair interaction may be written using matrix propagators in a product basis for the pair states (McKenzie and Stedman 1979):

$$\mathbf{F}(E)=\int_0^\beta \mathrm{d}\tau\ \exp(E\tau)\langle T_\tau\{\mathbf{\Psi}(\tau)\times\mathbf{\Psi}^\dagger(0)\}\rangle, \tag{8.41}$$

where $\mathbf{\Psi}^\dagger$ is a vector of pair level creation operator products:

$$\mathbf{\Psi}^\dagger\equiv(\Psi_{11}^\dagger,\Psi_{12}^\dagger,\ldots,\Psi_{21}^\dagger,\Psi_{22}^\dagger,\ldots),\ \Psi_{mn}^\dagger\equiv a_{im}^\dagger a_{jn}^\dagger,$$

and where i, j label the two ions, and m, n label the various electronic states of each ion. The matrix propagator $\mathbf{F}(E)$ satisfies the Dyson equation, and the expansion of the pair spectrum in powers of the ion–lattice interaction and in terms of the matrix propagators is formally similar to that of the single-ion spectrum. Once again, nonadditive terms do not contribute to the positions of the pair lines, and we may confine our interest to self-energy diagrams formed from the pair propagator.

A typical (fourth-order) self-energy diagram has the form:

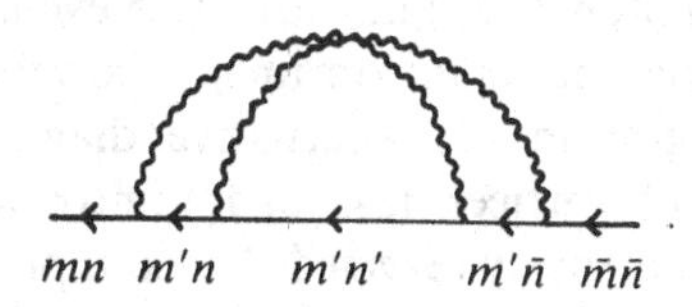

It is convenient for some purposes to separate out the electronic histories of the two ions, and rewrite this term as:

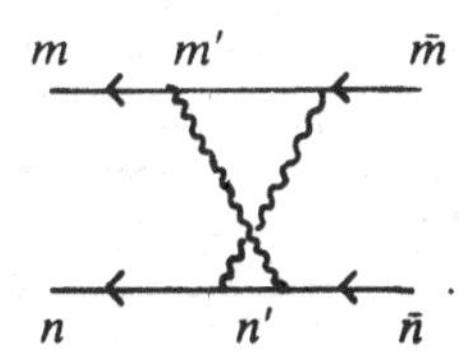

Note, however, that this alternative diagram representation is not deformable in the standard manner, since the horizontal ordering of vertices is of significance.

It is useful and very instructive to assess the consequences of hermiticity and time-reversal symmetry in such diagrams. The time-even character of the ion–phonon coupling, together with the choice of Kramers ions, leads to cancellations amongst numerators of various terms, and changes the character (for example the temperature dependence) of virtual phonon exchange in such systems.

Any contribution to the self-energy may be depicted schematically within the alternative representation in the form

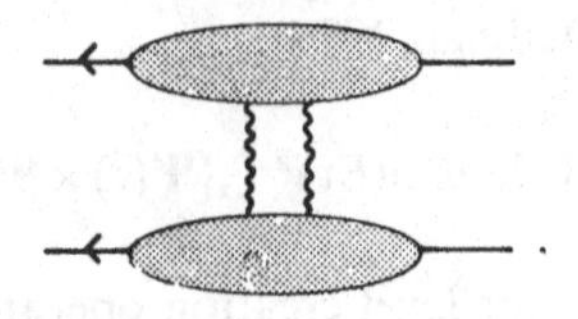

where each bubble may contain an arbitrary number of phonons, all interacting with the same ion. If hermiticity and time-reversal invariance symmetry are combined (HT), each interaction matrix element satisfies $V_{ijk\pm} = V_{\bar{j}\bar{i}\bar{k}\mp}$. This has the consequence of inducing partial or full equivalences between different diagrams.

Suppose that the diagram is manipulated in the following way. It is reflected in a vertical mirror (exchanging left and right). The electronic arrows are returned to their original direction, using HT symmetry. The time-reversal conjugation of the state labels is of no consequence as far as the energy denominators are concerned; time-conjugate states have the same energy, and phonon couplings are time-even. As a result, contributions from a diagram and its mirror image are very closely related. The relationship is strongest in the alternative diagram formalism if the horizontal ordering of vertices (as read jointly for the two lines of propagation) is simultaneously reversed. An example will be given shortly.

We consider now systems of Kramers ions for which by time-reversal symmetry the electronic states are doubly degenerate; we label the ground doublet states of each ion $\{m \equiv \pm\}$. Pair splittings are associated with particular combinations of the matrix elements

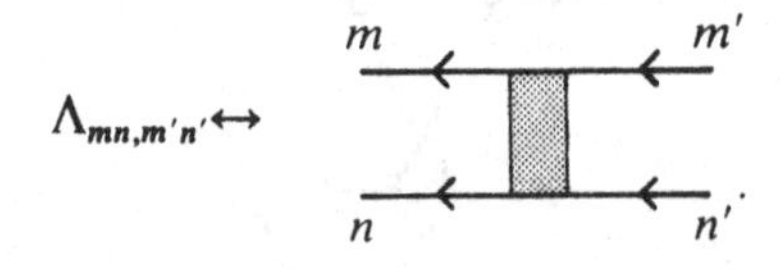

We group the ionic labels as for product states in the form $|P\rangle \equiv |mn\rangle$, where m stands for the doublet state label ($\pm$) of ion 1 and n for that of ion 2; we choose the notation $P \equiv 1$ for $++$, 2 for $+-$, 3 for $-+$ and 4 for $--$ for the four possible product pair states.

The pair states and energies are obtained by diagonalising the matrix $\Lambda_{PP'} \equiv \langle P|H_{VPE}|P'\rangle$. From H and T symmetry, this matrix has the symmetries $\Lambda_{PP'} = \Lambda_{\bar{P'}\bar{P}}$, i.e. $\Lambda_{11} = \Lambda_{44}$, $\Lambda_{22} = \Lambda_{33}$, $\Lambda_{12} = -\Lambda_{34}$, $\Lambda_{13} = -\Lambda_{24}$, from time-reversal invariance. The trace of the matrix does not affect pair splittings. Together these results show that only the combinations S_{11}, S_{12}, S_{13}, S_{14}, S_{23} are independent and affect pair splittings, where we define

$$S_{mn,m'n'} \equiv \tfrac{1}{2}(\Lambda_{mn,m'n'} - \Lambda_{\bar{m}'n,\bar{m}n'}). \tag{8.42}$$

The diagram symmetries discussed above lead to the partial or full cancellation of precisely these combinations. For example, at second order in phonon interaction, we have relations such as

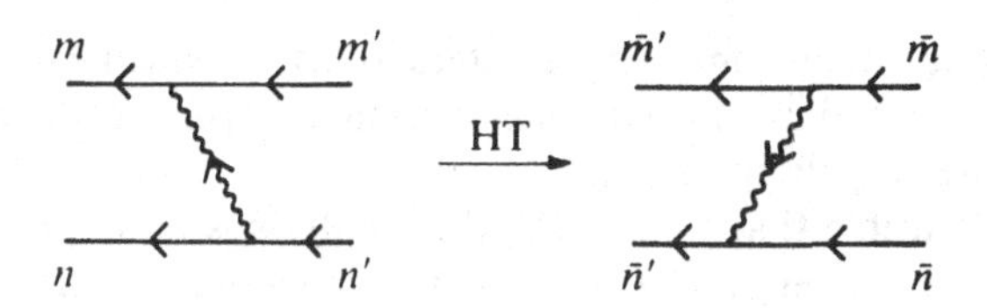

from the HT symmetry discussed above. Hence, second-order interactions with phonons do not contribute to pair splittings in Kramers systems.

When this analysis is extended to fourth order (McKenzie and Stedman 1979) in general partial, rather than complete, cancellations are found, reducing the magnitude of the coupling and qualitatively altering its temperature and energy dependence. Diagrams partnered in the above contributions to the pair splitting parameters $S_{PP'}$ may no longer be exactly opposite in their contributions because they are no longer reflection symmetric in the necessary way; the residual terms are proportional to phonon energies.

Any of several simplifications may still ensure complete cancellation of virtual phonon exchange contributions at fourth order. If only one phonon links the two ions, one may still always find a relative displacement of the single-ion vertex corrections so that under the symmetry operations the contribution exactly cancels that from another diagram, as in the second-order case described above. It is necessary to mirror reverse each interaction on each propagator (and inside each vertex) through hermitian conjugation, and also to time reverse it (these changes are denoted by bars below), so as to obtain cancellation between the diagrams

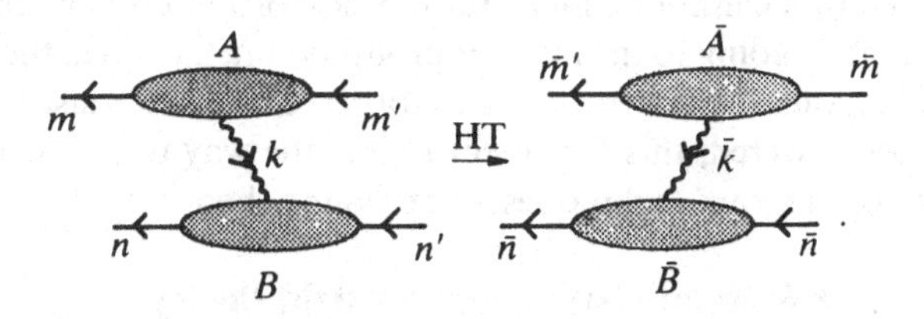

Another situation in which complete cancellation is obtained is when the intermediate electronic states are from the ground manifold, degenerate

with the initial and final states. The following transformations of the field theoretic diagrams illustrate the proof of such cancellation:

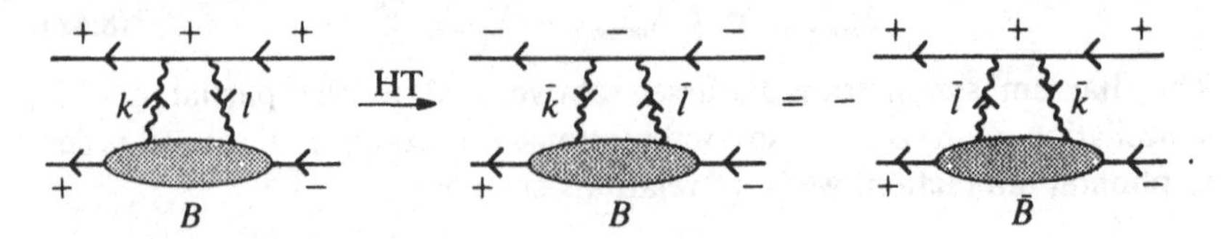

At fourth order, then, only linear interactions in which both phonons couple both ions and in which all intermediate electronic states are excited can contribute to pair splittings.

It seems plausible that underlying this analysis is more group theory; however, it has not yet been developed. One consequence of our diagram approach to group theory may be discerned even in the above field theoretic argument. We should clearly be thinking in terms of the relationship between the individual site symmetry groups and the pair group. The pair splittings correspond to the differences of the eigenvalues of the matrix $\Lambda_{PP'}$, themselves labelled by the pair group irreps. For example, the singlet/triplet classification of the pair eigenstates is related to the irreps $\square\square$, $\begin{smallmatrix}\square\\\square\end{smallmatrix}$ of the group S_2 (§4). If a single phonon linking two ions is inadequate to induce pair splittings, adding any number of phonon interactions, each of which is confined to one ion, will not suffice to induce the splittings either, for the simple reason that pair splittings are necessarily associated with symmetry arguments, and that the application of JLVn within one site symmetry will remove from the corresponding group theoretical diagram all purely local interactions:

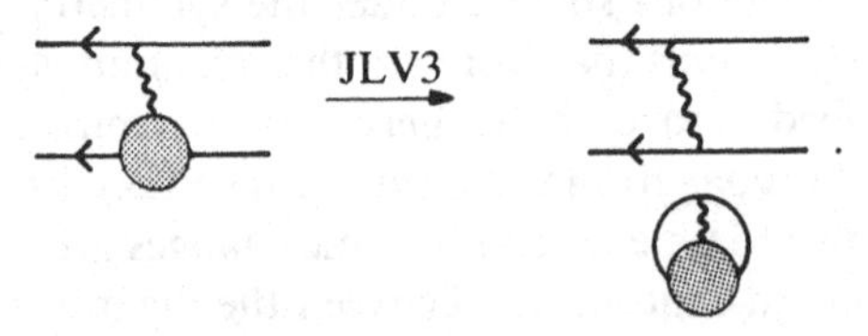

The same reduction cannot be made for the case of a phonon coupling the two ions, except by going to the full pair group symmetry, when the effect of JLVn is innocuous. While this leaves other questions raised by this application unanswered, this argument points the way to a much-needed development of diagram techniques in exchange theory.

8.6 Relativistic quantum field theory

8.6.1 Spinor angular momentum algebra

A similar reduction of invariant diagrams, and in particular angular momentum coupling trees, to that discussed in §1 for vector labels may be performed for spinor labels ($j = \frac{1}{2}$) on the external legs.

One minor change is that the 2jm symbol is nontrivial, since the 2j phase $(-1)^{2j}$ is -1 when $j=\frac{1}{2}$.

A more important change is that the relationships between the various generalised Kronecker deltas associated with the reduction of the angular momentum coupling trees are different, and occur at lower rank.

We shall use an integrity basis theoretic approach to verify this. We note in passing that an isomorphism exists between the invariant spin algebra we are discussing and binor theory, namely the algebra of tensors in an abstract tensor space of (negative) dimension -2 (see Penrose 1971, who quotes a similar diagram relation to the above; also Cvitanovic and Kennedy 1982). The approach of Penrose has been developed by Moussouris (1979) to give a novel and diagrammatic method related to our development for determining general formulae for the Clebsch–Gordan symbols and 3jm symbols of SO(3).

First, we develop the integrity basis theory (e.g. Judd *et al.* 1974) of vectors in SO(3) as confirmation of the identities found in §1. Consider, following Minard *et al.* (1983), the set of polynomials

$$\{P_n(\mathbf{a})\}=\{a_+^p a_0^q a_-^r \mid p+q+r=n, p\geq 0, q\geq 0, r\geq 0\}.$$

This set has a character $\chi_n(\theta)$, where θ is the rotation angle defining the class of SO(3),

$$\chi_n(\theta)=\Sigma_{pqr}\exp[i(p-r)\theta]$$

since $O_\theta(a_m^p)=\exp(ipm\theta)a_m^p$.

We construct from this a generating function for these characters for polynomials of all degrees:

$$F_\theta(A)\equiv\sum_{n=0}^{\infty}\chi_n(\theta)A^n=1/[(1-A)(1-e^{i\theta}A)(1-e^{-i\theta}A)].$$

The number of nth order invariant polynomials is the number of times the identity irrep of SO(3) is contained in the representation with this character. Since $\chi^{(0)}(\theta)=1$, using the character orthogonality theorem this becomes

$$\chi^{(0)}(A)=\frac{1}{\pi}\int_0^{2\pi}\sin^2(\theta/2)F_\theta(A)d\theta.$$

We transform this to a contour integral with $\lambda\equiv e^{i\theta}$:

$$\chi^{(0)}(A)=\frac{-1}{4\pi i}\int(1-\lambda)^2 d\lambda/[\lambda(1-A)(1-\lambda A)(\lambda-A)]=1/(1-A^2).$$

The A^2 in the denominator signals the existence of one member of the integrity basis (a 'denominator invariant') which appears in all its powers in the expansion and (from its form) is itself of second order in the components of $\mathbf{a}$. Clearly this corresponds to the scalar product $\mathbf{a}\cdot\mathbf{a}$.

Repeating this analysis for polynomials formed from the components of two vectors **a**, **b** amounts to the replacement of F_θ by $F_\theta(A,B)=F_\theta(A)F_\theta(B)$, with the result

$$\chi^{(0)}(A,B)=1/[(1-A^2)(1-B^2)(1-AB)],$$

which clearly signals the invariants **a·a**, **a·b**, **b·b**.

Something new happens with three vectors; we obtain

$$\chi^{(0)}(A,B,C) = (1+ABC)/[(1-A^2)(1-B^2)(1-C^2)(1-AB)(1-BC)(1-CA)].$$

There is now a trilinear invariant which is a numerator invariant, i.e. which appears at first order only. This corresponds to the triple scalar product [**abc**]. It appears independently only at first order, since the product of any two such products may be replaced by scalar products (Problem 1.13).

With four vectors, $\chi^{(0)}(A,B,C,D)$ has the expected denominators corresponding to all the possible scalar products, but has a more involved numerator:

$$\chi^{(0)}(A,B,C,D)= \frac{[1+ABC(1-D^2)+ABD(1-C^2)+ACD(1-B^2)+BCD(1-A^2)-A^2B^2C^2D^2]}{(1-A^2)(1-B^2)(1-C^2)(1-D^2)(1-AB)(1-AC)(1-AD)(1-BC)(1-BD)(1-CD)}.$$

The negative terms in the numerator warn us that not all the possible scalar products associated with the denominator invariants are independent. The fifth-order and eighth-order negative terms in the numerator correspond precisely to the relations discussed in Problems 1.11 and 1.14.

We now extend this to spinors, following Riddell and Stedman (1984), noting in passing that our spinors span a projective representation of SO(3), and that strictly one should be talking about the universal covering group spin(3) in the following. For nth order polynomials constructed from components of one Pauli spinor, the character becomes

$$\chi_n(\theta)=\sin[(n+\tfrac{1}{2})\theta]/\sin(\theta/2)$$

and the corresponding generating function is

$$F_\theta(a)=1/[1+a^2-2a\cos(\theta/2)].$$

The contour integration now requires the choice $\lambda=\exp(\tfrac{1}{2}i\theta)$ for the complex variable, and the limits extended from -2π to $+2\pi$, for the integrand to be meromorphic:

$$\chi^{(0)}(a)=(-1)/4\pi i\int\frac{(\lambda^2-1)^2 d\lambda}{\lambda^2(1-a\lambda)(\lambda-a)}=1.$$

No invariants are possible from the unaltered components of one spinor.

However, if we include complex conjugate components in equal numbers, we obtain a generating function

$$F_\theta(A)=(1+A)/[(1-A)(1+A^2-2A\cos\theta)]$$

and hence an invariance component

$$\chi^{(0)}(A) = 1/[1 - A],$$

where A signifies terms bilinear in the spinor components and in their conjugates. Hence, there is one bilinear denominator invariant, which corresponds to the spinor normalisation: if the spinor ψ has components (u,v) the normalisation $u^*u + v^*v$ is an invariant.

If two spinors are so combined with equal numbers of the respective conjugate components we obtain

$$F_\theta(A,B) = F_\theta(A)F_\theta(B),\ \chi^{(0)}(A,B) = 1/[1 - AB].$$

This signals the alternating product $\psi \wedge \phi = u^*{}_1 v_2 - u_2^* v_1$ as another invariant (since $(u^* v^*) \sim (v, -u)$).

A new feature arises when four spinors are so combined; we obtain a numerator $(1 - ABCD)$ in $\chi^{(0)}(A,B,C,D)$ which signals a nontrivial general linear relationship among the six possible alternating products of four spinors. This may be written

$$(\psi_1 \wedge \psi_2)(\psi_3 \wedge \psi_4) + (\psi_1 \wedge \psi_3)(\psi_4 \wedge \psi_2) + (\psi_2 \wedge \psi_3)(\psi_1 \wedge \psi_4) = 0, \quad (8.43)$$

and has any of the diagram forms

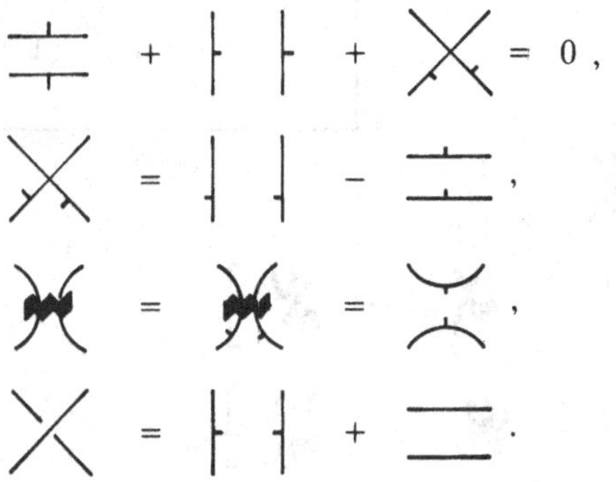

Hence from the third of these relations we have in analogy to Problem 1.14:

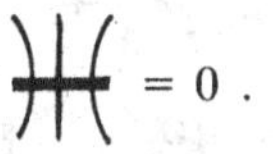

The general proof that alternating and scalar products are the only possible invariants from spinor components is given by Brinkman (1956).

For both vector and spinor labels, one may therefore reduce angular momentum coupling trees to products of Kronecker delta functions (perhaps with an antisymmetric symbol ε_{ijk} for odd-rank vector polynomials) in principle, with due allowance for ambiguities from the above general linear relations. The actual reductions are readily achieved by use of the *aufbau* construction illustrated in §1, and typical results, suitably symmetrised where possible, are reproduced below from Minard *et al.* (1983) and Riddell and Stedman (1984).

$$F_0 = \sqrt{2}\,(\;) = \frac{1}{\sqrt{2}}(\;)$$

$$F_1 = \frac{1}{\sqrt{3}}(\;)$$

$$= \frac{1}{2\sqrt{3}}(2\;\; + \;\;)$$

$F_j \equiv$ ½ ½ $\hat{j}$ ½ ½

$$F_{011} = $$

$$F_{100} = \frac{1}{3\sqrt{3}}$$

$$F_{110} = \frac{1}{3}$$

$$F_{111} = \frac{1}{\sqrt{3}}$$

$F_{jj_1j_2} \equiv$ 1 1 1 * $\hat{j}_1$ $\hat{j}_2$ $\hat{j}$ * 1 1 1

$$F_{102} = \frac{1}{3\sqrt{15}}\left[3\;\; - \;\;\right]$$

$$F_{121} = \frac{1}{3\sqrt{5}}\left[3\;\; - \;\;\right]$$

$$F_{122} = \frac{1}{15\sqrt{3}}\left[9\;\; - 3\left(\;\; + \;\;\right) + \;\;\right]$$

$$F_{211} = \frac{1}{3\sqrt{5}}\left[-3\;\; + 2\;\; + 2\;\;\right]$$

$$F_{221} = \frac{1}{\sqrt{15}}\left[\;\; + 2\;\; - \;\;\right]$$

$$F_{222} = \frac{1}{3\sqrt{5}}\left[2\;\; - \;\; - \;\; + \;\; + \;\; - 2\;\;\right]$$

$$F_{322} = \frac{1}{15\sqrt{7}}\left[15\;\; - 4\;\; - 2\;\; - 2\;\; - \;\;\right]$$

$$F^{1/2}_{000} = \frac{1}{2\sqrt{2}}\left(\ \right)$$

$$F^{1/2}_{110} = \frac{1}{\sqrt{6}}\left(\ \right)$$

J J J J

$$F^{J}_{j_1 j_2 j_3} =$$

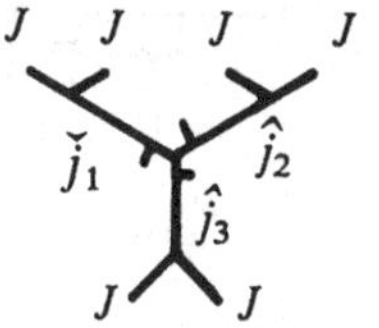

$\check{j}_1$ $\hat{j}_2$ $\hat{j}_3$

J J

$$F^{1/2}_{111} = \frac{1}{\sqrt{3}}\left(\ \right) = \frac{1}{2\sqrt{3}}\left(\ + \ \right)$$

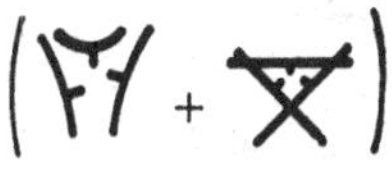

$$= \frac{1}{2\sqrt{3}}\left(2\ + \ + \ + \ \right)$$

$$F^{1}_{111} = \frac{-2}{\sqrt{3}}\left[\ \right]$$

$$F^{1}_{122} = \frac{2}{\sqrt{15}}\left[\ \right]$$

$$F^{1}_{211} = \frac{2}{3\sqrt{5}}\left[3\ + \ \right]$$

$$F^{1}_{222} = \frac{2}{3\sqrt{105}}\left[9\ - 3\left(\ + \ + \ \right) + 2\ \right]$$

J J J J

$$F^{1/2}_{00000} = \frac{1}{2}\left(\ \right)$$

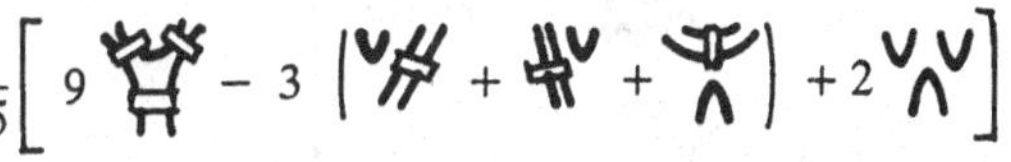

$$F^{J}_{jj_1 j_2 j_3 j_4} \equiv$$

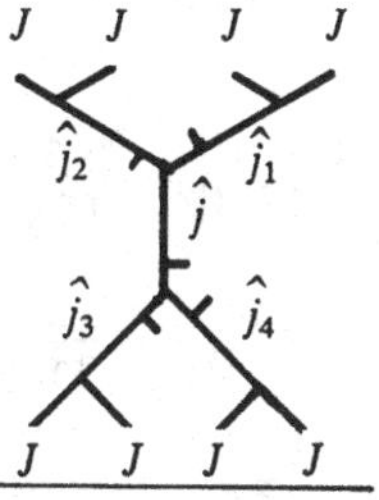

$\hat{j}_2$ $\hat{j}_1$ $\hat{j}$ $\hat{j}_3$ $\hat{j}_4$

J J J J

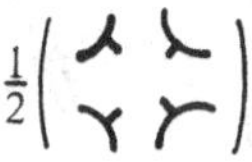

$$F^{1/2}_{01111} = \frac{1}{3}\left(\ \right)$$

$$F^{1/2}_{11001} = \frac{1}{2\sqrt{3}}\left(\ \right)$$

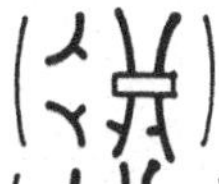

$$F^{1/2}_{11011} = \frac{1}{\sqrt{6}}\left(\ \right) = \frac{1}{2\sqrt{6}}\left(\ + \ \right)$$

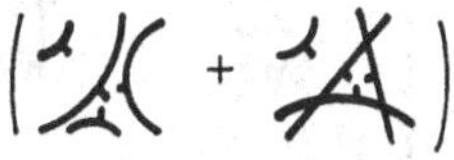

$$F^{1/2}_{11111} = \frac{1}{\sqrt{3}}\left(\ \right) = \frac{1}{4\sqrt{3}}\left(\ + \ + \ + \ \right)$$

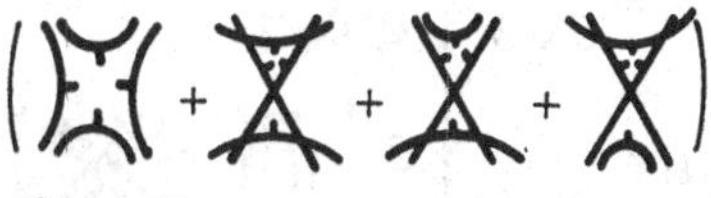

$$= \frac{1}{2\sqrt{3}}\left(\ + \ \right)$$

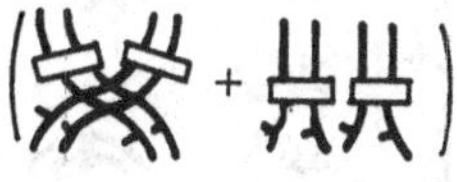

$$F^{1/2}_{21111} = \frac{1}{6\sqrt{5}}\left(3\left[\ + \ \right] - 2\left[\ \right]\right)$$

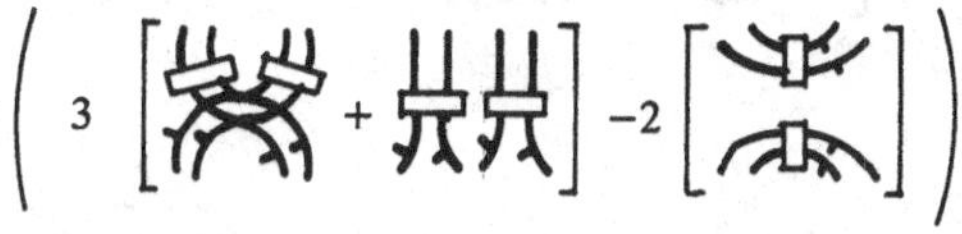

$$F^1_{00000} = \frac{1}{9}$$

$$F^1_{01100} = \frac{1}{3\sqrt{3}}$$

$$F^1_{01111} = \frac{1}{3}$$

$$F^1_{02200} = \frac{1}{9\sqrt{5}} \left[3 \quad - \quad \right]$$

$$F^1_{02222} = \frac{1}{45} \left[9 \quad - 3 \left(\quad + \quad \right) + \quad \right]$$

$$F^1_{10101} = \frac{1}{3\sqrt{3}}$$

$$F^1_{11101} = -\frac{2}{3} \left[\quad \right]$$

$$F^1_{11111} = \frac{1}{2\sqrt{3}} \left[\quad - \quad \right]$$

$$F^1_{11201} = \frac{2}{3\sqrt{15}} \left[3 \quad + \quad \right]$$

$$F^1_{11211} = \frac{-4}{3\sqrt{5}} \left[3 \quad + \quad \right]$$

$$F^1_{11212} = \frac{4}{15\sqrt{3}} \left[9 \quad + \quad 3 \left(\quad + \quad \right) + \quad \right]$$

$$F^1_{12201} = \frac{2}{3\sqrt{5}} \left[\quad \right]$$

$$F^1_{12211} = \frac{-4}{\sqrt{15}} \left[\quad \right]$$

$$F^1_{12212} = \frac{4}{15} \left[3 \quad + \quad \right]$$

$$F^1_{12222} = \frac{4}{5\sqrt{3}} \left[\quad \right]$$

$$F^1_{21111} = \frac{1}{6\sqrt{5}} \left[3 \left(\quad + \quad \right) - 2 \quad \right]$$

$$F^1_{20211} = \frac{2}{3\sqrt{15}} \left[3 \quad + \quad \right]$$

$$F^1_{20202} = \frac{1}{9\sqrt{5}} \left[3 \quad - \quad \right]$$

$$F^1_{21211} = \frac{4}{\sqrt{15}} \left[\quad \right]$$

$$F^1_{21202} = \frac{2}{3\sqrt{5}} \left[\quad \right]$$

$$F^1_{21212} = \frac{4}{3\sqrt{5}} \left[\quad \right]$$

$$F^1_{22211} = \frac{4}{3\sqrt{105}} \left[9 \quad - 3 \left(\quad + \quad \right) + 3 \quad - 2 \quad \right]$$

$$F^1_{22202} = \frac{2}{9\sqrt{35}} \left[9 \quad - 3 \left(\quad + \quad + \quad \right) + 2 \quad \right]$$

$$F^1_{22212} = \frac{4}{3\sqrt{35}} \left[3 \quad - \left(\quad + \quad \right) \right]$$

$$F^1_{22222} = \frac{4}{63\sqrt{5}} \left[27 \quad - 9 \left(\quad + \quad + \quad + \quad \right) - 9 \quad + 6 \left(\quad + \quad \right) + 3 \left(\quad + \quad + \quad \quad \right) - 4 \quad \right]$$

$$F^1_{31212} = \frac{1}{15\sqrt{7}} \left[15 \quad - 9 \quad - 28 \quad + 8 \quad - 12 \left(\quad + \quad \right) \right]$$

$$F^1_{32212} = \frac{2}{15\sqrt{14}} \left[15 \quad - 3 \left(\quad - \quad \right) - 3 \quad - 5 \left(\quad - \quad \right) \right]$$

$$F^1_{32222} = \frac{1}{30\sqrt{7}} \left[15 \left(\quad - \quad \right) - 5 \left(\quad + \quad - \quad - \quad \right) - 24 \quad \right]$$

$$F^1_{42222} = \frac{1}{210} \left[35 \left(\quad + \quad \right) + 40 \left(\quad + \quad + \quad + \quad \right) + 24 \quad - 120 \quad + 26 \quad - 22 \left(\quad + \quad \right) - 25 \left(\quad + \quad + \quad + \quad \right) \right].$$

8.6.2 Quantum electrodynamics

The diagrammatic manipulation of spinorial quantities inevitably raises the question of the feasibility of extending or adapting the JLVn theorems to the symmetry analysis of Feynman diagrams in gauge theories. The paradigms are QED and, more recently, QCD.

While some considerable effort has been channelled into this area, it has been mostly on the formal development or on a general approach to diagram weight (reduced matrix element) calculation suitable for automation within an algebraic or symbolic computing package such as SCHOONSCHIP (van der Bij and Veltman 1984) or REDUCE. Comparatively little has been done in the direction of finding applications of similar simplicity and importance to those discussed earlier in this chapter. But even those that are available hold some interest.

ElBaz and Castel (1972) discuss the applicability of their diagram reduction techniques to SU(3).

Biritz (1975a,b, 1979) has developed a formalism for relativistic field theory, incorporating Lorentz symmetry. Central to this is the identification of a QED vertex with the angular momentum coupling diagram

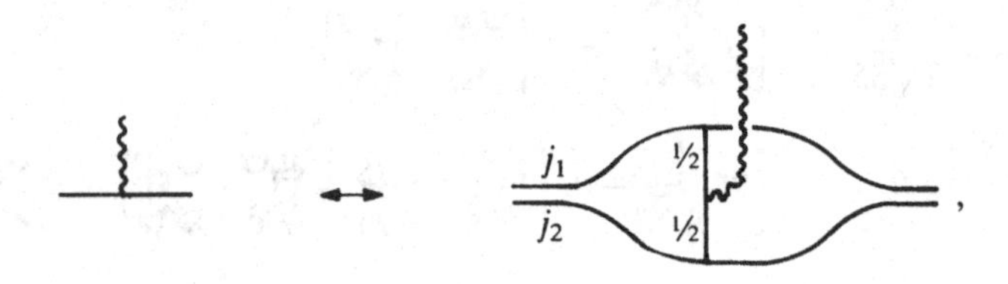

corresponding to the local isomorphism SO(4)$\sim$SO(3)$\times$SO(3), and the connection between the algebra of the Lorentz group SO(3,1) and its compact counterpart SO(4). The vertex is then deformable by JLV3 into the simple vertex times a 9j symbol. In this way, Biritz has investigated magnetic moments and the structure of wave equations for particles of arbitrary spin using the diagram approach to angular momentum algebra.

However, the whole question of the applicability of the results of compact group theory – in order, say, to form analogues of the Racah algebra and the JLVn theorems – in noncompact groups, where unitary irreps are of infinite dimension and the finite-dimensional irreps are nonunitary, needs fuller consideration (see Sharp 1984, for example). However, the gauge groups SU(n) appropriate to the more widely used gauge field theories are compact, as are also many subgroups of the Lorentz group including SO(3); there is no difficulty in applying the JLVn theorems at these levels.

Cvitanovic and collaborators, in a number of important articles (Cvitanovic 1976, 1984, Cvitanovic, Lauwers and Scharbach 1981, Cvitanovic and Kennedy 1982) has given a general development for gauge field theories including QED and QCD, as discussed in §5. Similar analyses are given by Canning (1978) and Butera *et al.* (1980). The more combinatoric aspects associated with the theory of the symmetric group are covered in this context by Mandula (unpublished lecture notes, University of Southampton, 1980) (see §4).

A good place to start is afforded by the work of Kennedy (1982), who has developed in detail the connections between a diagrammatic approach on the lines adopted in this book and the standard manipulations of 4-spinors in QED calculations. This has later been extended by Cvitanovic and Kennedy (1982) to other groups.

For example, much of the labour in QED calculations is centred on the reduction of the trace of a product of 4-matrices, i.e. the Dirac γ-matrices γ^μ and their contraction (on the Lorentz index μ) with four-vectors. The γ matrices give the matrix elements in the defining or fundamental ('electron' for QED, 'quark' for QCD) irrep of the group generators, themselves corresponding to the adjoint ('photon' for QED, 'gluon' for QCD) irrep. The trace of a product of γ-matrices corresponds to a diagram network, closed on itself with respect to the defining irrep labels, to which the JLVn theorems appropriate to the network and to the compact symmetry group under consideration may be applied.

For example, let us take the standard trace rules for the hand reduction of a relatively simple QED expression:

$$\begin{aligned}&\mathrm{tr}(\underset{1}{\gamma^\alpha}\underset{2}{\gamma^\beta}\ldots\underset{2n+1}{\gamma^\omega})=0,\ \mathrm{tr}(\gamma^\mu\gamma^\nu)=4g^{\mu\nu},\\&\mathrm{tr}(\gamma^\mu\gamma^\nu\gamma^\alpha\gamma^\beta)=4[\delta^{\mu\nu}\delta^{\alpha\beta}-\delta^{\mu\alpha}\delta^{\nu\beta}+\delta^{\mu\beta}\delta^{\nu\alpha}]\end{aligned} \tag{8.44}$$

correspond, respectively, to the reductions (Cvitanovic and Kennedy 1982, Kennedy 1982)

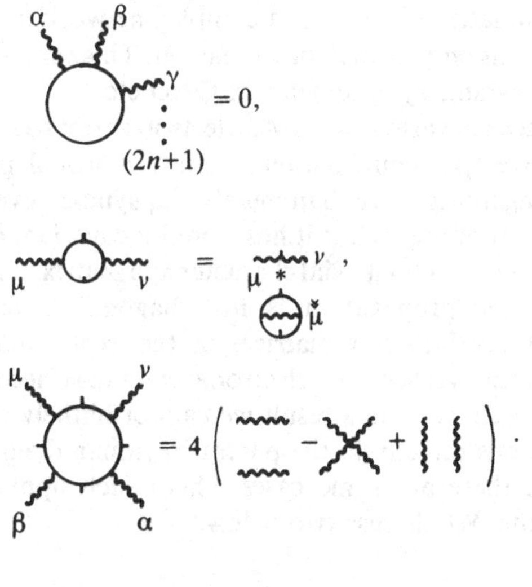

α β γ = 0, (2n+1)

μ ν = μ * ν , μ

μ ν β α = 4 (− +) .

These diagram forms, though merely an elementary illustration of the material of §5 as far as a compact subgroup of SO(3,1) is concerned, make an interesting contrast with the standard development. The anticommutation property of the Dirac matrices corresponds to the antisymmetry (negative 3j phase) of the vertex corresponding to the adjoint irrep. This, with the obvious cyclic symmetry of the trace (also used in §8.3.2 to form circular diagrams), leads to the first result. The second has the form that JLV2 would give; to this extent the noncompact case of SO(3,1) follows the compact SO(4) case. The third, one of an infinite sequence of such relations, might be derived within the SO(3) subgroup for the spatial indices. As such, it corresponds precisely to the reduction of an angular momentum coupling tree given in §8.6.1. Once again, the extension to all labels demands an extension of the analysis at least to SO(4). Conversely, the conventional proofs justify these diagram pinching theorems and hence (using the unitary of Clebsch–Gordan symbols) their application to more complicated networks.

Indeed, many useful extensions for the simplification of higher-order traces, including the Fierz transformation, the Chisholm and the Kahane algorithms, are demonstrated by Kennedy (1982) and by Cvitanovic (1984). They exploit, among many other properties, the relation between diagrams of the following form:

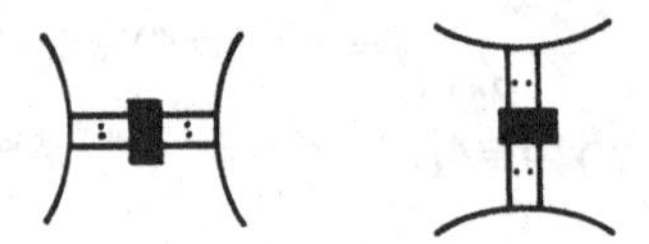

as for the recoupling associated with JLV4. (Boson or adjoint irrep labels have been rewritten in terms of fermion–antifermion, or defining irrep, labels in this figure.)

Such diagrammatic reduction of a complex network may be applied to Feynman diagrams only in an indirect manner. This explains the restricted use of diagram symmetry arguments in QED etc.

This may be seen in various ways. An electron at rest has spin $\frac{1}{2}$, but when moving will have spin contributions from the orbital part. Again, the Feynman propagator is more than merely a 2j symbol, even as regards its group theoretic structure. While it has Lorentz covariant character in the Feynman gauge, it has a nontrivial character as a matrix in the Dirac spinor space, so that the propagation is not diagonal in spin space. The propagators supply further γ matrices in the expression beyond those implicit in the QED vertices. An electronic state may need to be assigned the spin $\frac{3}{2}$, $\frac{5}{2}$ as well as $\frac{1}{2}$. As a result we cannot blithely apply the above diagram reductions directly to the parent Feynman diagrams.

Nevertheless, there are some cases where such application leads to interesting results. We discuss two below.

Chiral anomaly

QED might be reinterpreted, enlarged or modified to accommodate an axial vector (A) coupling. An example is the σ model, or the introduction of a vertex representing interaction of the photon with the spin density in the Dirac sea. (Jauch and Rohrlich 1976, p. 59). In such a theory, there exists a triangle diagram in which an axial current couples to two vector currents:

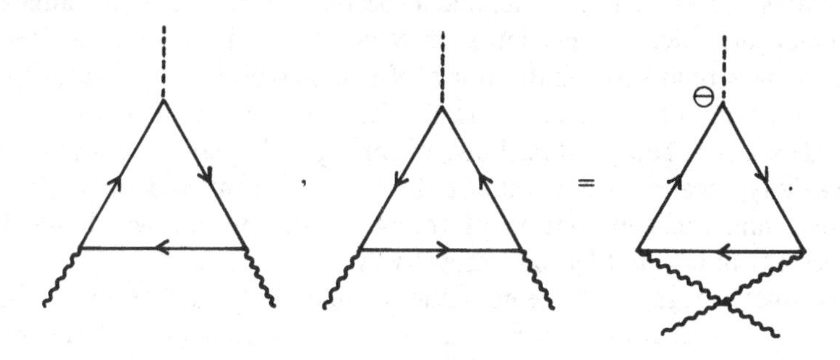

This should be summed against the contributions in which the vector couplings are crossed, and in which the fermion loop is reversed. The nominal cancellation suggested in the AVV case from examining the Feynman integral and from the Ward identities fails on a closer examination. This may be seen as an incompatibility between the various Ward identities (Llewellyn Smith 1979), a lack of invariance of the Feynman integral under a change of origin in the variables of integration (Aitchison 1982), a nonconservation of the axial current even in the zero mass limit, a breakdown of the chirality selection rule for a charged massless Dirac particle interesting with photons (Huang 1982), or a breakdown of the rule that an electron and a positron with the same helicity and opposite chirality should not decay to two photons (Horejsi 1986). In general it represents a breakdown in a classical selection rule when a theory is quantised. Its physical origin is obscure; according to Jackiw (1985, 1986), its co-discoverer, 'the most we can say is that chiral symmetry is broken on introducing a coupling between fermions and a gauge field above a negative energy Dirac sea.'

The interpretation of the formal significance of the anomaly in any theory depends on the theory. QED contains no axial couplings to gauge particles, and the anomaly is not directly embarrassing. Some controversial suggestions have been made that the QED anomaly leads directly to physical effects even within QED which are observable in principle, and that, for example, in the presence of electric and magnetic fields the vacuum generates spin waves in the virtual electrons (Widom *et al.* 1985, Stedman and Bilger 1987, Dolgov, Khriplovich and Zakharov 1987, 1988). Originally, the anomaly was used in an exended theory of the mutual interactions of pions, electrons and photons to deduce the pion decay rate.

However, the nonabelian gauge groups give a more serious species of anomalies, involving gauged currents and the corresponding particle couplings, whose physical import cannot be overlooked. The ultimate theory of matter should be renormalisable and not contain anomalies of this sort. The search has therefore been undertaken for theories in which some extra symmetry principle ensures that the above combination of diagrams cancels. This has dictated a special interest in certain gauge field theories for which a cancellation may be obtained. Superstring theories based on a relatively small choice of symmetry groups have attracted enormous attention recently partly for this reason. The electroweak (SU(2) × U(1)) theory of Glashow, Weinberg and Salam avoids anomalies essentially because of the equal numbers of leptons and quarks in this theory, and the cancellation of triangle diagrams for which the loop represents quark and lepton, respectively.

Symmetry arguments are germane to such analyses. The SU(2) × U(1) anomaly cancellation can be traced to the cancellation of the group theoretical factors associated with the vertices. A unified theory such as SU(5) does not quite remove the mystery, since here contributions from the irreps $\overline{5}$ and 10 cancel. The higher group SO(10) contains both these irreps in the irrep 16, and has a more natural explanation for the freedom from anomalies and so its suitability for being chosen in a grand unified theory (Ross 1986).

The requirement of anomaly cancellation through the choice of symmetry group in any theory corresponds to a condition on a certain trace over the group generators. The group theoretic content of the sum of the triangle diagrams reduces to a sum over group theoretic diagrams of the same topology. The vertices in the group theoretic diagrams correspond to the couplings of left- and right-handed particles, respectively, for the two senses of direction of the arrow on the Feynman triangular loop. We label the associated generators M, N, respectively. On summation in the Feynman diagram these terms are associated with factors $(1-\gamma_5)$, $(1+\gamma_5)$ (the projection operators for left- and right-handed particles), respectively. It is the term proportional to γ_5 that is the troublesome one, and which for an anomaly-free theory should cancel when summed over all types of fermion in the loop (all generations and colours of quarks, for example). This gives the condition:

$$\mathrm{Tr}(M^a\{M^b,M^c\})-\mathrm{Tr}(N^a\{N^b,N^c\})=0 \qquad (8.45)$$

(Leader and Predazzi 1982). This may be checked out for the four generators in the SU(2) × U(1) electroweak theory, as a more abstract vindication of the above numerical calculation. Becher, Bohm and Joos (1984) give a simpler if more restrictive condition of the form

$$\Sigma\mathrm{Tr}(\{X^{\alpha a}, X^{\alpha a}\}X^{\alpha c}]=0, \qquad (8.46)$$

where (§5.5) X^α denotes the generator coupling the adjoint to the defining irrep. Becher *et al.* (1984) also list the Lie groups satisfying this constraint: $SU(2) \sim SO(3)$, $SO(N)$, $N = 5,7,8,9,10,\ldots$, $Sp(2N)$, $N \geq 3$, $G(2)$, $F(4)$, $E(6)$, $E(7)$, $E(8)$.

As an exercise in the analysis of §5, this condition may be given another form more related to our development and more obviously basis-independent. For a compact Lie group we may apply JLV3 to write equation (8.46) as

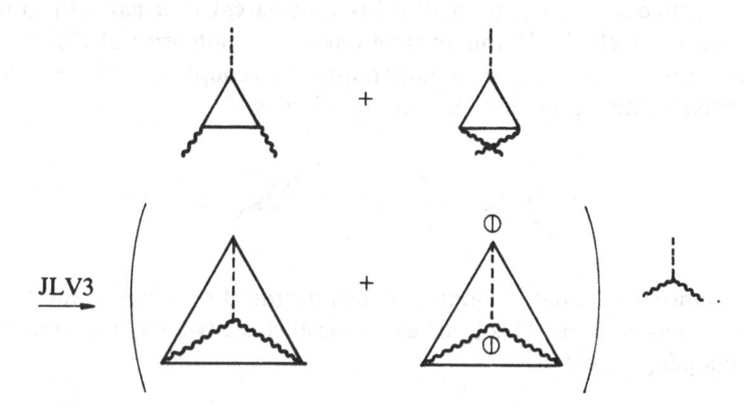

The essential requirement of equation (8.46) is now re-expressed as a condition on the 6j symbols of the group.

Compton scattering

For linearly polarised light, the standard formula (Klein and Nishina 1929) for Compton scattering has the form

$$\frac{d\sigma}{d\Omega} = \left[\frac{e^2}{2mc^2}\right]^2 \left[\frac{\omega'}{\omega}\right]^2 \left[\left[\frac{\omega}{\omega'} + \frac{\omega'}{\omega}\right] - 2 + 4|\mathbf{e}\cdot\mathbf{e}'|^2\right], \tag{8.47}$$

where the incoming and outgoing light beams have frequency and polarisation $(\omega, \mathbf{e})$ and $(\omega', \mathbf{e}')$, respectively. Many standard works (Aitchison 1972, Jackson 1975, Jauch and Rohrlich 1976, Scadron 1979) on QED quote, and Feynman (1961) proves as a beginners' exercise, the following extension of this formula to general polarisation:

$$\frac{d\sigma}{d\Omega} = \left[\frac{e^2}{2mc^2}\right]^2 \left[\frac{\omega'}{\omega}\right]^2 \left[\left[\frac{\omega}{\omega'} + \frac{\omega'}{\omega}\right] - 2 + 4|\mathbf{e}\cdot\mathbf{e}'^*|^2\right]. \tag{8.48}$$

However, equation (8.48) is incorrect. It is ironic that this error, and the correct formula for general polarisation

$$\frac{d\sigma}{d\Omega} = \left[\frac{e^2}{2mc^2}\right]^2 \left[\frac{\omega'}{\omega}\right]^2 \left[\left[\frac{\omega}{\omega'} + \frac{\omega'}{\omega}\right]\left[1 + |\mathbf{e}\cdot\mathbf{e}'^*|^2 - |\mathbf{e}\cdot\mathbf{e}'|^2\right] + 2\left[|\mathbf{e}\cdot\mathbf{e}'^*|^2 + |\mathbf{e}\cdot\mathbf{e}'|^2 - 1\right]\right], \tag{8.49}$$

remained unknown for 22 years (Stedman and Pooke 1982), long after the existence of symbolic manipulation programs such as REDUCE had made the calculation a one-second problem. Feynman's error was to assume that the product of the contractions of the polarisation four-vector and of its complex conjugate reduced to -1. This avoids a long detour in the analysis of the trace of the γ-matrix product appropriate for Compton scattering. However, it can be justified (from equation 8.44) only when the polarisation four-vector is real.

The problem with equation (8.48) is transparent in a naive diagram analysis in which the Feynman diagram for the intensity of Compton scattering is interpreted as an angular momentum coupling diagram. In the Compton scattering process the two amplitudes

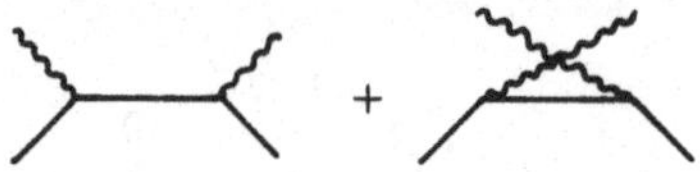

+

are summed; the square modulus is then averaged over initial and final electron polarisations. This leads us to consider the two angular momentum coupling diagrams:

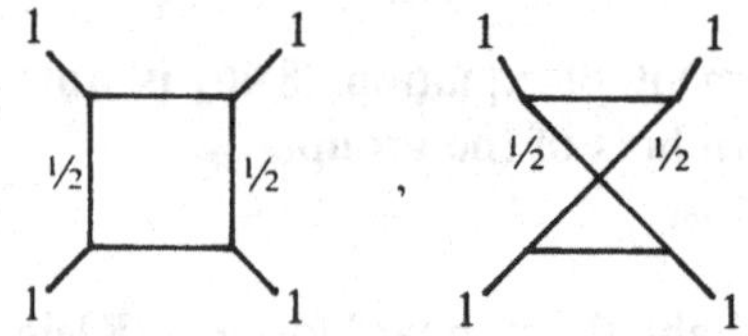

1 1 ½ ½ 1 1 , 1 1 ½ ½ 1 1

in the spirit of §6. If we specialise to the case in which all electronic states have spin $\frac{1}{2}$, we may use equation (8.43) to express the result in the form:

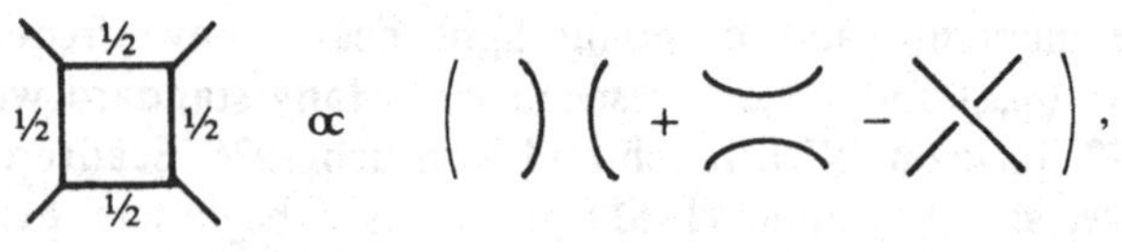

½ ½ ½ ½ $\propto$ ( ) (+ $-$),

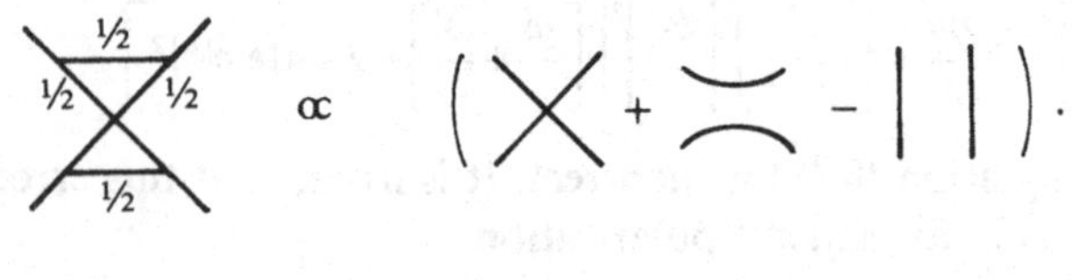

½ ½ ½ ½ $\propto$ (+ $-$ | |).

These diagrams would correspond to differing structures for the intermediate states, and so different energy factors; they could not be added directly. This would lead us to expect that the polarisation vectors should appear in a linear combination of the expressions

$$1+|\mathbf{e}\cdot\mathbf{e}'^{*}|^{2}-|\mathbf{e}\cdot\mathbf{e}'|^{2},\quad |\mathbf{e}\cdot\mathbf{e}'^{*}|^{2}+|\mathbf{e}\cdot\mathbf{e}'|^{2}-1$$

appropriate to each topology (we use the normalisation condition $\mathbf{e}\cdot\mathbf{e}^{*}=1$). This conclusion fails to hold for equation (8.48), but is obeyed by equation (8.49).

Even if we do not assume that intermediate electronic states are restricted to spin $\frac{1}{2}$, we may draw the same conclusion from the diagrams

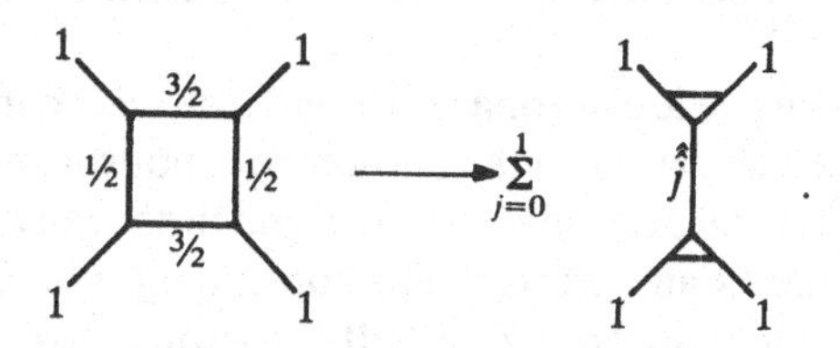

If initial and final electronic states at least have spin $\frac{1}{2}$, their coupling rank in any amplitude, and therefore the coupling rank of the associated photons, is restricted to spin 0 and 1 and necessarily excludes spin 2: $\frac{1}{2}\times\frac{1}{2}=1+0$. (This may be interpreted as the absence for Compton scattering of one characteristic weight in the Placzek (1962) formalism of Raman scattering; see Stedman 1985). Hence the only permissible combinations of photon polarisation vectors are the rank 0 and rank 1 forms:

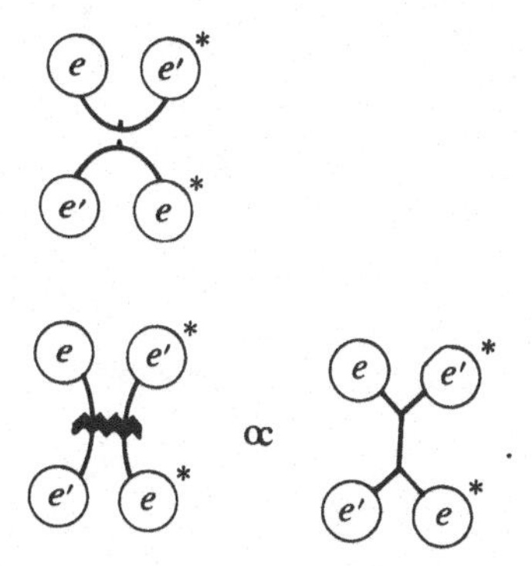

Again, equation (8.49), but not equation (8.48), is consistent with a linear combination of these terms.

Why should such a naive analysis work? First, the initial electronic state is stationary, and so has spin $\frac{1}{2}$. The surprise is that, though the final electronic state is nonstationary, it is safe to assume it has spin $\frac{1}{2}$ for the purposes of the above calculation. Certainly the QED calculation shows that if the initial state were not stationary, the final answer would involve not merely the rotational invariants formed from the polarisation vectors alone, but also those formed from polarisation vectors and momentum vectors, and the simplicity of the above analysis would be shattered. This well illustrates the points that QED diagrams cannot be interpreted directly as angular momentum coupling diagrams in general, and that only stationary electrons can be assumed to have spin $\frac{1}{2}$. The reason why the initial and final photon momentum vectors, and also final state electron momentum vector do not appear in the cross section in this simple case is partly the orthogonality of any photon polarisation and momentum vector, and partly because any scalar product with a photon polarisation vector would introduce a gauge dependence that would have to be cancelled by other such products in the final expression.

We close with this little cautionary tale and historical curiosity. Feynman kindly commented in a letter to the author: 'It is unusual to find an error in a formula that has been around for so long, and in so many books, and done by so many people as an exercise.' The concluding moral is obvious. The new research student in the admittedly fearsome world of theoretical physics should not allow himself or herself to be brainwashed by an apparently overwhelming burden of personal ignorance. One should take heart and check everything.

Appendix
Tables of point group properties

These tables indicate the consequences of group–subgroup branching $O(3)\downarrow G$, where G is a point group. They are included to back up the examples of this book, and of §6 in particular, and to permit their fuller development. They are a subset of the tables in Stedman (1985).

Table A1 compares a variety of popular notation for point group irreps. We standardise on the notation of Butler (1981).

Tables A2 and A3 list the number of occurrences of the identity of each point group in the O(3) irrep J^{π} up to $J=4$. These suffice to determine for any point group the *number* of geometrical factors associated with any physical process, such as those discussed in §§6.3, 6.4.

The remaining tables give details of the forms of group-symmetrised basis states. This permits in principle the determination of the explicit *form* of any geometrical factor. The tables give matrix elements $\langle J^{\pi}c|J^{\pi}\phi\lambda\mu\rangle$ with kets in a cartesian basis (for $J=1$, 2) or spherical tensor basis for $J=3$. ϕ is a parentage label, while λ denotes an irrep of the group G in question and μ is a component label, usually an irrep of a subgroup. Following Butler (1981), and Reid and Butler (1982), we use the group chain as a definition of the coordinate system. In an $O\supset D_3$ branching, the z axis will be along the body diagonal of the cube, while in an $O\supset D_4$ branching, the z axis will be perpendicular to a cube face (Reid and Butler 1982). In this manner data for several groups in a chain may be presented together.

Phase choices are implicit, not merely in the forms of these matrix elements, but also in the choice of group chain (Butler 1981). For example, $\tilde{0}^-$ (D_{2h}) branches to $0(C_s)$ if the intermediate irrep and group is $0(C_{2v})$, but to $1(C_s)$ if via $0^-(C_{2h})$. The reason is that these two C_s groups have differing orientation; one has a σ_y and the other a σ_z reflection plane. The chain C_{4v}–C_{2v}–C_s is calculated here in a basis for which the y axis is the rotation axis. We include the corrections to Butler (1981) discussed by Reid and Butler (1982) and the extensions to C_∞ etc. by Reid (1984).

For $J=1$, we depart from the Butler (1981) phase choice, which is based on the Condon and Shortley (1935) convention, in favour of the Fano–Racah phase convention (based on the addition of a factor i^l to the Condon and Shortley definition of the spherical harmonic Y^l_m. This ensures that at least for $J=0, 1$ the 2jm symbols are trivial and the 3jm symbols are real in a cartesian basis. This alteration will introduce only an overall phase factor which is of no consequence in the determination of any geometrical factor.

For $J=3$, we use the standard spherical (SO(3)–SO(2)) basis for c, and the results are essentially transcribed from Butler (1981). We quote for brevity nontrivial sets of m, f_m values where $f_m \equiv \langle 3^\pi m | 3^\pi \phi \lambda \mu \rangle$. With one exception the $J=3$ kets are omitted from this table for the chain $(S)O_3$–$O_{(h)}$–$T_{(h)}$–$D_{2(h)}$–$C_{2(h)}$ since this chain is included only to cover the groups T, T_h; the exception is to cover the one occurrence of 0(T) in $J^\pi = 3^\pi$.

Parity superscripts may be ignored in rotation groups. Irrep labellings are often different for reflection–rotation groups when branchings from negative-parity states are considered; we add the relevant alterations in square brackets below the original labels. Where the irrep labelling is not specified for rotation–reflection groups, it is as given for the main chain, except that parity labels may be ignored.

The necessity of the '2j phase' for multiplicity lines in group–subgroup branching (Problem 3.6 and §3.6.1) is noted by an asterisk on the appropriate subgroup irrep label. In this case, the product of 2j factors for the chain $O(3) \supset G$ is -1 rather than $+1$.

Examples of the use of these tables are given in §§6.2.2 and 6.3.3.

Table A1.

This gives the relationship between Butler (1981) notation for point group irreps and the notations of Griffith (1962) and Koster *et al.* (1963). A discussion is given on p. 190 of Butler (1981). In general, Butler chooses $0^{(+)}$ to denote the identity irrep, and $\pm n$ denotes a complex conjugate pair. The digit order (0,1,2,3) and conjugation ($\tilde{0}$, etc.) has a basis residing more in logic than in convention, to do with the order at which any irrep appears in the Kronecker powers of the irrep $\frac{1}{2}$. Inversion groups have the same irrep labels as the parent rotation group, with the addition of the usual parity suffices.

G	0	$\tilde{0}$	1	$\tilde{1}$	2	$\tilde{2}$	3	$\tilde{3}$	± 1	± 2
K	A_1	—	T_1	T_2	V	—	U	—	—	—
O,T_d	A_1	A_2	T_1	T_2	E	—	—	—	—	—
	(Γ_1	Γ_2	Γ_4	Γ_5	Γ_3					
T	A_1	—	T	—	—	—	—	—	—	E
	(Γ_1	—	Γ_4	—	—	—	—	—	—	Γ_2,Γ_3)
D_∞,$C_{\infty v}$	A_1	A_2	E_1	—	E_2	—	E_3	—	—	—
D_6,C_{6v},D_{3h}	A_1	A_2	E_1	—	E_2	—	B_1	B_2	—	—
	(Γ_1	Γ_2	Γ_5	—	Γ_6	—	Γ_3	Γ_4)		
D_5,C_{5v}	A_1	A_2	E_1	—	E_2	—	—	—	—	—
D_4,D_{2d},C_{4v}	A_1	A_2	E	—	B_1	B_2	—	—	—	—
	(Γ_1	Γ_2	Γ_5	—	Γ_3	Γ_4)				
D_3,C_{3v}	A_1	A_2	E	—	—	—	—	—	—	—
	(Γ_1	Γ_2	Γ_3)							
C_∞	A_1	—	—	—	—	—	—	—	E_1	E_2
C_6,C_{3h}	A_1	—	—	—	—	—	A_2	—	E_1	E_2
	(Γ_1	—	—	—	—	—	Γ_4	—	Γ_5,Γ_6	Γ_2,Γ_3)
C_5	A_1	—	—	—	—	—	—	—	E_1	E_2
C_4,S_4	A	—	—	—	B	—	—	—	E	—
	(Γ_1	—	—	—	Γ_2	—	—	—	Γ_3,Γ_4)	
C_3	A	—	—	—	—	—	—	—	E	—
	(Γ_1	—	—	—		—	—	—	Γ_2,Γ_3)	
D_2,C_{2v}	A_1	B_1	B_2	B_3	—	—	—	—	—	—
	(Γ_1	Γ_3	Γ_2	Γ_4)						
C_2,C_s	A	—	B	—	—	—	—	—	—	—
	(Γ_1	—	Γ_2)							
C_i	A(Γ_1)									

Table A2.

Numbers of occurrences of the identity irrep 0(G) in the irrep J(SO(3)) for rotation groups, and of the identity $0^+(G_i)$ in the irrep J^+(O(3)) for inversion groups.

	J				
G G_i	0	1	2	3	4
SO(3),O(3)	1	—	—	—	—
K,K_h	1	—	—	—	—
O,O_h	1	—	—	—	1
T,T_h	1	—	—	1	1
D_∞,$D_{\infty h}$	1	—	1	—	1
D_6,D_{6h}	1	—	1	—	1
D_5,D_{5h}	1	—	1	—	1
D_4,D_{4h}	1	—	1	—	2
D_3,D_{3h}	1	—	1	1	2
D_2,D_{2h}	1	—	2	1	3
C_∞,$C_{\infty h}$	1	1	1	1	1
C_6,C_{6h}	1	1	1	1	1
C_5,C_{5i}	1	1	1	1	1
C_4,C_{4h}	1	1	1	1	3
C_3,C_{3i}	1	1	1	3	3
C_2,C_{2h}	1	1	3	3	5
C_1,C_i	1	3	5	7	9

Table A3.

Numbers of occurrences of the identity 0(G) of the rotation–reflection group G in the spin–parity irrep J^π(O(3)). H is the rotation group for which J^+ has the same reduction as for G.

		J^π									
G	(H)	0^+	0^-	1^+	1^-	2^+	2^-	3^+	3^-	4^+	4^-
T_d	(O)	1	—	—	—	—	—	—	1	1	—
$C_{\infty v}$	(D_∞)	1	—	—	1	1	—	—	1	1	—
C_{6v}	(D_6)	1	—	—	1	1	—	—	1	1	—
D_{3h}	(D_6)	1	—	—	—	1	—	—	1	1	1
C_{5v}	(D_5)	1	—	—	1	1	—	—	1	1	—
D_{2d}	(D_4)	1	—	—	—	1	1	—	1	2	1
C_{4v}	(D_4)	1	—	—	1	1	—	—	1	2	1
C_{3v}	(D_3)	1	—	—	1	1	—	1	2	2	1
C_{3h}	(C_6)	1	—	1	—	1	—	1	2	1	2
S_4	(C_4)	1	—	1	—	1	2	1	2	3	2
C_{2v}	(D_2)	1	—	—	1	2	1	1	2	3	2
C_s	(C_2)	1	—	1	2	3	2	3	4	5	4

Table A4.

Basis transformation matrix elements $\langle J^\pi c | J^\pi \phi \lambda \mu \rangle$ for all point groups, for $0 < J \leq 3$. The notation is discussed at the start of this appendix.

				π	Factor	$J^\pi c$			
O_3——	O_h——	D_{4h}——	C_4	$\pm$					
SO_3——	O——	D_4——	C_4	$\pm$					
O_3——	T_d——	D_{2d}——	S_4	$+$					
[O_3——	T_d——	D_{2d}——	S_4]	$-$					
J^π	ϕ	λ	μ			$1^\pi x$	$1^\pi y$	$1^\pi z$	
1^π	1^π [$\tilde{1}$]	1^π	1^π [-1]		$1/\sqrt{2}$ [$-1/\sqrt{2}$]	i	-1	—	
1^π	1^π [$\tilde{1}$]	1^π	-1^π [1]		$1/\sqrt{2}$ [$-1/\sqrt{2}$]	$-$i	-1	—	
1^π	1^π [$\tilde{1}$]	$\tilde{0}^\pi$ [$\tilde{2}$]	$0^{\pi *}$ [2]		1	—	—	i	
						$2^\pi xx$	$2^\pi yy$	$2^\pi zz$	
2^π	2^π	0^π [2]	0^π [2]		$1/\sqrt{30}$	-1	-1	2	
2^π	2^π	2^π [0]	2^π [0]		$1/\sqrt{10}$ [$-1/\sqrt{10}$]	1	-1	—	
						$2^\pi yz$	$2^\pi zx$	$2^\pi xy$	
2^π	$\tilde{1}^\pi$ [1]	1^π	1^π [-1]		$1/\sqrt{20}$ [$-1/\sqrt{20}$]	$-$i	-1	—	
2^π	$\tilde{1}^\pi$ [1]	1^π	-1^π [1]		$1/\sqrt{20}$ [$-1/\sqrt{20}$]	i	-1	—	
2^π	$\tilde{1}^\pi$ [1]	$\tilde{2}^\pi$ [$\tilde{0}$]	$2^{\pi *}$ [0]		$1/\sqrt{10}$	—	—	$-$i	
						m	f_m	m'	$f_{m'}$
3^π	1^π [$\tilde{1}$]	$\tilde{0}^\pi$ [$\tilde{2}$]	$0^{\pi *}$ [2]		1	0	-1	—	—
3^π	1^π [$\tilde{1}$]	1^π	1^π [-1]		$1/\sqrt{8}$ [$-1/\sqrt{8}$]	1	$-\sqrt{3}$	-3	$-\sqrt{5}$

3^π	1^π $[\tilde{1}]$	1^π	-1^π $[1]$	$1/\sqrt{8}$ $[-1/\sqrt{8}]$	3	$-\sqrt{5}$	-1	$-\sqrt{3}$
3^π	$\tilde{1}^\pi$ $[1]$	1^π	1^π $[-1]$	$1/\sqrt{8}$ $[-1/\sqrt{8}$	1	$-\sqrt{5}$	-3	$\sqrt{3}$
3^π	$\tilde{1}^\pi$ $[1]$	1^π	-1^π $[1]$	$1/\sqrt{8}$ $[-1/\sqrt{8}]$	3	$\sqrt{3}$	-1	$-\sqrt{5}$
3^π	$\tilde{1}^\pi$ $[1]$	$\tilde{2}^\pi$ $[\tilde{0}]$	$2^{\pi *}$ $[0]$	$1/\sqrt{2}$	± 2	-1		
3^π	$\tilde{0}^\pi$ $[0]$	2^π $[0]$	2^π $[0]$	$1/\sqrt{2}$	± 2	∓ 1		

				π	Factor	$J^\pi c$			
O_3—	$D_{\infty h}$—	D_{6h}—	C_{6h}	$\pm$					
SO_3—	D_∞—	D_6—	C_6	$\pm$					
O_3—	$D_{\infty h}$—	D_{3h}—	C_{3h}	$+$					
$[O_3$—	$D_{\infty h}$—	D_{3h}—	$C_{3h}]$	$-$					
J^π	ϕ	λ	μ			$1^\pi x$	$1^\pi y$	$1^\pi z$	
1^π	$\tilde{0}^\pi$	$\tilde{0}^\pi$ $[\tilde{3}]$	$0^{\pi *}$ $[3]$		1 $[-1]$	—	—	i	
1^π	1^π	1^π $[2]$	1^π $[-2]$		$1/\sqrt{2}$ $[-1/\sqrt{2}]$	i	-1	—	
1^π	1^π	1^π $[2]$	-1^π $[2]$		$1/\sqrt{2}$ $[-1/\sqrt{2}]$	$-i$	-1	—	
						$2^\pi xx$	$2^\pi yy$	$2^\pi zz$	$2^\pi xy$
2^π	0^π	0^π $[3]$	0^π $[3]$		$1/\sqrt{30}$	1	1	-2	—
2^π	2^π	2^π $[1]$	2^π $[-1]$		$1/\sqrt{20}$ $[-1/\sqrt{20}]$	1	-1	—	i
2^π	2^π	2^π $[1]$	-2^π $[1]$		$1/\sqrt{20}$ $[-1/\sqrt{20}]$	1	-1	—	$-i$

					$2^\pi yz$	$2^\pi zx$
2^π	1^π [2]	1^π	1^π [−2]	$1/\sqrt{20}$ [$-1/\sqrt{20}$]	i	1
2^π	1^π [2]	1^π	-1^π [2]	$1/\sqrt{20}$ [$-1/\sqrt{20}$]	−i	1

					m	f_m
3^π	$\tilde{0}^\pi$	$\tilde{0}^\pi$ [$\tilde{3}$]	$0^{\pi *}$ [3]	1 [−1]	0	1
3^π	1^π	1^π [2]	1^π [−2]	1 [−1]	1	−1
3^π	1^π	1^π [2]	-1^π [2]	1 [−1]	−1	−1
3^π	2^π	2^π [1]	2^π [−1]	1 [−1]	2	1
3^π	2^π	2^π [1]	-2^π [1]	1 [−1]	−2	−1
3^π	3^π	3^π [0]	3^π [0]	$1/\sqrt{2}$	±3	1
3^π	3^π	$\tilde{3}^\pi$ [$\tilde{0}$]	$3^{\pi *}$ [0]	$1/\sqrt{2}$	±3	±1

				π	Factor	$J^\pi c$		
O_3—	K_h—	D_{5h}—	C_{5h}	±				
SO_3—	K—	D_5—	C_5	±				
O_3—	K_h—	C_{5v}—	C_5	+				
[O_3—	K_h—	C_{5v}—	C_5]	−				
J^π	ϕ	λ	μ			$1^\pi x$	$1^\pi y$	$1^\pi z$
1^π	1^π	$\tilde{0}^\pi$ [0]	$0^{\pi *}$		1	—	—	i
1^π	1^π	1^π	1^π		$1/\sqrt{2}$	i	−1	—
1^π	1^π	1^π	-1^π		$1/\sqrt{2}$	−i	−1	—

					$2^\pi xx$	$2^\pi yy$	$2^\pi zz$	$2^\pi xy$
2^π	2^π	0^π [$\tilde{0}$]	0^π	$1/\sqrt{30}$	-1	-1	2	—
2^π	2^π	2^π	2^π	$1/\sqrt{20}$	1	-1	—	i
2^π	2^π	2^π	-2^π	$1/\sqrt{20}$	1	-1	—	$-$i
					$2^\pi yz$	$2^\pi zx$		
2^π	2^π	1^π	1^π	$1/\sqrt{20}$	i	-1		
2^π	2^π	1^π	-1^π	$1\surd\sqrt{20}$	i	1		
					m	f_m	m'	f_m
3^π	$\tilde{1}^\pi$	$\tilde{0}^\pi$ [0]	$0^{\pi *}$	1	0	-1	—	—
3^π	3^π	1^π	1^π	1	1	1	—	—
3^π	3^π	1^π	-1^π	1	-1	1	—	—
3^π	3^π	2^π	2^π	$1/\sqrt{5}$	2	$-\sqrt{2}$	-3	$\sqrt{3}$
3^π	3^π	2^π	-2^π	$1/\sqrt{5}$	3	$\sqrt{3}$	-2	$\sqrt{2}$
3^π	$\tilde{1}^\pi$	2^π	2^π	$1/\sqrt{5}$	2	$-\sqrt{3}$	-3	$-\sqrt{2}$
3^π	$\tilde{1}^\pi$	2^π	-2^π	$1/\sqrt{5}$	3	$-\sqrt{2}$	-2	$\sqrt{3}$

O_3 —	O_h —	D_{4h} —	D_{2h} —	C_{2h}	$\pm$				
SO_3 —	O —	D_4 —	D_2 —	C_2	$\pm$				
J^π	ϕ	λ	μ	ν	π	Factor	$J^\pi c$		
							$1^\pi x$	$1^\pi y$	$1^\pi z$
1^π	1^π	$\tilde{0}^\pi$	$\tilde{0}^\pi$	$0^{\pi *}$		1	—	—	i
1^π	1^π	1^π	1^π	1^π		1	—	-1	—
1^π	1^π	1^π	$\tilde{1}^\pi$	$1^{\pi *}$		1	-1	—	—
							$2^\pi xx$	$2^\pi yy$	$2^\pi zz$
2^π	2^π	0^π	0^π	0^π		$1/\sqrt{30}$	-1	-1	2
2^π	2^π	2^π	0^π	0^π		$1/\sqrt{10}$	-1	1	—
							$2^\pi yz$	$2^\pi zx$	$2^\pi xy$
2^π	$\tilde{1}^\pi$	1^π	1^π	1^π		$1/\sqrt{10}$	—	-1	—
2^π	$\tilde{1}^\pi$	1^π	$\tilde{1}^\pi$	$1^{\pi *}$		$1/\sqrt{10}$	1	—	—
2^π	$\tilde{1}^\pi$	$\tilde{2}^\pi$	$\tilde{0}^\pi$	$0^{\pi *}$		$1/\sqrt{10}$	—	—	$-$i

						m	f_m	m'	$f_{m'}$
3^π	1^π	$\tilde{0}^\pi$	$\tilde{0}^\pi$	$0^{\pi *}$	1	0	-1	—	—
3^π	1^π	1^π	1^π	1^π	1/4	± 1	$-\sqrt{3}$	± 3	$-\sqrt{5}$
3^π	1^π	1^π	$\tilde{1}^\pi$	$1^{\pi *}$	1/4	± 1	$\pm\sqrt{3}$	± 3	$\mp\sqrt{5}$
3^π	$\tilde{1}^\pi$	1^π	1^π	1^π	1/4	± 1	$-\sqrt{5}$	± 3	$\sqrt{3}$
3^π	$\tilde{1}^\pi$	1^π	$\tilde{1}^\pi$	$1^{\pi *}$	1/4	± 1	$\pm\sqrt{5}$	± 3	$\pm\sqrt{3}$
3^π	$\tilde{1}^\pi$	$\tilde{2}^\pi$	$\tilde{0}^\pi$	$0^{\pi *}$	$1/\sqrt{2}$	± 2	-1	—	—
3^π	$\tilde{0}^\pi$	2^π	0^π	0^π	$1/\sqrt{2}$	± 2	± 1	—	—

					π Factor $J^\pi c$			
O_3——	O_h——	C_{4v}—— (${C_{ny},\sigma_z}$ basis)	C_{2v}——	C_s	+ [−]			
J^π	ϕ	λ	μ	ν		$1^\pi x$	$1^\pi y$	$1^\pi z$
1^π	1^π	1	1	0	1	—	—	i
			[$\tilde{1}$]	[1]	[−1]			
1^π	1^π	$\tilde{0}$	$\tilde{0}$	1	1	—	1	—
		[0]	[0]	[0]	[−1]			
1^π	1^π	1	$\tilde{1}$	1	1	−i	—	—
			[1]	[0]				
						$2^\pi xx$	$2^\pi yy$	$2^\pi zz$
2^π	2^π	0	0	0	$1/\sqrt{30}$	−1	2	−1
		[$\tilde{0}$]	[$\tilde{0}$]	[1]				
2^π	2^π	2	0	0	$1/\sqrt{10}$	−1	—	1
		[$\tilde{2}$]	[$\tilde{0}$]	[1]	[$-1/\sqrt{10}$]			
						$2^\pi yz$	$2^\pi zx$	$2^\pi xy$
2^π	$\tilde{1}^\pi$	$\tilde{2}$	$\tilde{0}$	1	$1/\sqrt{10}$	—	−1	—
		[2]	[0]	[0]				
2^π	$\tilde{1}^\pi$	1	$\tilde{1}$	1	$1/\sqrt{10}$	i	—	—
			[1]	[0]				
2^π	$\tilde{1}^\pi$	1	1	0	$1/\sqrt{10}$	—	—	−i
			[$\tilde{1}$]	[1]	$-1/\sqrt{10}$			

						m	f_m	m'	$f_{m'}$
3^π	1^π	1	1	0	1	0	-1	—	—
			[$\tilde{1}$]	[1]	[-1]				
3^π	1^π	$\tilde{0}$	$\tilde{0}$	1	1/4	±1	$-\sqrt{3}$	±3	$-\sqrt{5}$
		[0]	[0]	[0]					
3^π	1^π	1	$\tilde{1}$	1	1/4	±1	$\pm\sqrt{3}$	±3	$\mp\sqrt{5}$
			[1]	[0]					
3^π	$\tilde{1}^\pi$	$\tilde{2}$	$\tilde{0}$	1	1/4	±1	$-\sqrt{5}$	±3	$\sqrt{3}$
		[2]	[0]	[0]					
3^π	$\tilde{1}^\pi$	1	$\tilde{1}$	1	1/4	±1	$\mp\sqrt{5}$	±3	$\mp\sqrt{3}$
			[1]	[0]					
3^π	$\tilde{1}^\pi$	1	1	0	$1/\sqrt{2}$	±2	-1		
			[$\tilde{1}$]	[1]	[$-1/\sqrt{2}$]				
3^π	$\tilde{0}^\pi$	2	0	0	$1/\sqrt{2}$	±2	∓1		
		[$\tilde{2}$]	[$\tilde{0}$]	[1]	[$-1/\sqrt{2}$]				

					π Factor	$J^\pi c$			
O_3	O_h	T_h	D_{2h}	C_{2h} $\pm$					
SO_3	O	T	D_2	C_2 $\pm$					
J^π	ϕ	λ	μ	ν		$1^\pi x$	$1^\pi y$	$1^\pi z$	
1^π	1^π	1^π	$\tilde{0}^\pi$	$0^{\pi *}$	1	—	—	$-i$	
1^π	1^π	1^π	1^π	1^π	1	—	-1	—	
1^π	1^π	1^π	$\tilde{1}^\pi$	$1^{\pi *}$	1	$-i$	—	—	
						$2^\pi xx$		$2^\pi yy$	$2^\pi zz$
2^π	2^π	2^π	0^π	0^π	$1/\sqrt{60}$	$-1-\sqrt{3}i$		$-1+\sqrt{3}i$	2
2^π	2^π	-2^π	0^π	0^π	$1/\sqrt{60}$	$-1+\sqrt{3}i$		$-1-\sqrt{3}i$	2
						$2^\pi yz$	$2^\pi zx$	$2^\pi xy$	
2^π	$\tilde{1}^\pi$	$1^{\pi *}$	$1^{\pi *}$	$1^{\pi *}$	$1/\sqrt{10}$	—	$-i$	—	
2^π	$\tilde{1}^\pi$	$1^{\pi *}$	$\tilde{1}^{\pi *}$	1^π	$1/\sqrt{10}$	-1	—	—	
2^π	$\tilde{1}^\pi$	$1^{\pi *}$	$\tilde{0}^{\pi *}$	0^π	$1/\sqrt{10}$	—	—	1	
						m	f_m	m'	$f_{m'}$
3^π	$\tilde{0}^\pi$	0^π	0^π	0^π	1/3	±3	$\pm\sqrt{2}$	0	$\sqrt{5}$

					π	Factor	$J^\pi c$			
O_3	$D_{\infty h}$	D_{6h}	D_{3d}	C_{3i}	$\pm$					
SO_3	D_{∞}	D_6	D_3	C_3	$\pm$					
O_3	$C_{\infty v}$	C_{6v}	C_{3v}	C_3	$+$					
[O_3	$C_{\infty v}$	C_{6v}	C_{3v}	C_3]	$-$					
J^π	ϕ	λ	μ	ν			$1^\pi x$	$1^\pi y$	$1^\pi z$	
1^π	$\tilde{0}^\pi$ [0]	$\tilde{0}^\pi$ [0]	$\tilde{0}^\pi$ [0]	$0^{\pi *}$		1 [−1]	—	—	i	
1^π	1^π	1^π	1^π	1^π		$1/\sqrt{2}$ [$-1/\sqrt{2}$]	i	−1	—	
1^π	1^π	1^π	1^π	-1^π		$1/\sqrt{2}$	−i	−1	—	
							$2^\pi xx$	$2^\pi yy$	$2^\pi zz$	$2^\pi xy$
2^π	0^π [$\tilde{0}$]	0^π [$\tilde{0}$]	0^π [$\tilde{0}$]	0^π		$1/\sqrt{30}$	1	1	−2	—
2^π	2^π	2^π	$1^{\pi *}$	$-1^{\pi *}$		$1/\sqrt{20}$ [$-1/\sqrt{20}$]	i	−i	—	−1
π	2^π	2^π	$1^{\pi *}$	$1^{\pi *}$		$1/\sqrt{20}$	i	−i	—	1
							$2^\pi yz$	$2^\pi zx$		
2^π	1^π	1^π	1^π	1^π		$1/\sqrt{20}$ [$-1/\sqrt{20}$]	i	1		
2^π	1^π	1^π	1^π	-1^π		$1/\sqrt{20}$	−i	1		
							m	f_m		
3^π	$\tilde{0}^\pi$ [0]	$\tilde{0}^\pi$ [0]	$\tilde{0}^\pi$ [0]	$0^{\pi *}$		1 [−1]	0	1		
3^π	1^π	1^π	1^π	1^π		1 [−1]	1	−1		
3^π	1^π	1^π	1^π	-1^π		1	−1	−1		
3^π	2^π	2^π	$1^{\pi *}$	$-1^{\pi *}$		1 [−1]	2	i		
3^π	2^π	2^π	$1^{\pi *}$	$1^{\pi *}$		1	−2	−i		
3^π	3^π	3^π [$\tilde{3}$]	$0^{\pi *}$ [$\tilde{0}$]	$0^{\pi *}$		$1/\sqrt{2}$ [$-1\sqrt{2}$]	±3	i		
3^π	3^π	$\tilde{3}^\pi$ [3]	$\tilde{0}^{\pi *}$ [0]	0^π		$1/\sqrt{2}$ [$-1/\sqrt{2}$]	±3	±i		

References

The prevalence of Feynman diagrams, and the plague of their corruptions, would be a poor monument for a subversive man.

Maddox (1988)

Abragam, A.A. and Bleaney, B. 1970 *Electron Paramagnetic Resonance of Transition Metal Ions* (Oxford: Clarendon)

Abrikosov, A.A. 1965 *Physics* **2** 5–20

Abrikosov, A.A., Gor'kov, L.P. and Dzyaloshinskii, I.Y. 1963 *Methods of Quantum Field Theory in Statistical Physics*, 2nd edn. (Oxford: Pergamon)

Adams, B.G., Paldus, J. and Čížek, J. 1977 *Int. J. Quantum Chem.* **11** 849–67

Agrawala, V.K. and Belinfante, J.G. 1968 *Ann. Phys.* **49** 130–70

Aitchison, I.J.R. 1972 *Relativistic Quantum Mechanics* (London: Macmillan), p. 162

Aitchison, I.J.R. 1982 *An Informal Introduction to Gauge Field Theories* (Cambridge University Press), pp. 148–55

Aitken, A.C. 1958 *Determinants and Matrices* (Edinburgh: Oliver & Boyd)

Andrews, D.L. 1980 in *Laser Advances and Applications*, ed. B.S. Wherett (Chichester: Wiley), pp. 135-8

Arima, A. 1983 in *Symmetries in Nuclear Structure*, eds. K. Abrahams, K. Allaart and A.E.L. Dieperink (New York: Plenum), pp. 93–117

Axe, J.D. 1964 *Phys. Rev. A* **136** 42-5

Barnes, S.E. 1985 *Phys. Rev. Lett.* **55** 2192–5

Barnes, S.E. 1986 *Phys. Rev. B* **33** 3209–46

Barrie, R. and Rystephanick, R.G. 1966 *Can. J. Phys.* **44** 109–38

Barron, L.D. 1982 *Molecular Light Scattering and Optical Activity* (Cambridge University Press)

Barron, L.D. 1986 *Chem. Phys. Lett.* **123** 423–7

Barron, L.D. 1987 *BioSystems* **20** 7–14

Barron, L.D. and Buckingham, A.D. 1972 *Molec. Phys.* **23** 145–70

Barron, L.D. and Buckingham, A.D. 1975 *Ann. Rev. Phys. Chem.* **26** 381–96

Bates, C.A. 1978 *Phys. Reports* **35** 187–304
Becher, P., Bohm, M. and Joos, H. 1984 *Gauge Theories of Strong and Electroweak Interaction* (New York: Wiley), p. 142
Bell, J.S. 1959 *Nuclear Phys.* **12** 117–24
Bennett, C.L. 1987 *Phys. Rev. A* **35** 2409–28
Bickerstaff, R.P. 1984 *J. Math. Phys.* **25** 2808–14
Bickerstaff, R.P. 1985 *Lett. Math. Phys.* **10** 1–6
Bickerstaff, R.P. and Damhus, T. 1985 *Int. J. Quantum Chem.* **27** 381–91
Biggs, N.L., Lloyd, E.K. and Wilson, R.J. 1976 *Graph Theory 1736–1936* (Oxford: Clarendon), ch. 1
Billy, N. and Lhuillier, C. 1982 *Ann. Physique Fr.* **4** 309–51
Billy, N. and Lhuillier, C. 1983 *Ann. Physique Fr.* **8** 231–71
Biritz, H. 1975a *Phys. Rev. D* **12** 2254–65
Biritz, H. 1975b *Nuovo Cim. B* **25** 449–78
Biritz, H. 1979 *Int. J. Theor. Phys.* **18** 601–88
Bishton, S.S. and Newman, D.J. 1968 *J. Phys. Chem. Solids* **29** 1245–53
Black, R.J. and Stedman, G.E. 1982 *Int. J. Quantum Chem.* **21** 823–31
Blockley, C.A. and Stedman, G.E. 1985 *Eur. J. Phys.* **6** 218–24
Bloembergen, N. 1982 *Rev. Mod. Phys.* **54** 685
Boerner, H. 1963 *Representations of Groups* (Amsterdam: North-Holland)
Bolotin, A., Levinson, Y. and Tomalchev, V.V. 1964 *Leit. Fiz. Rinkings* **1** 33
Borcherds, P.H. and Alfrey, G.F. 1975 *J. Phys. C* **8** 2655–66
Briat, B. 1981 *Molec. Phys.* **42** 347–70
Briggs, J.S. 1971 *Rev. Mod. Phys.* **43** 189–230
Brink, D.M. and Satchler, G.M. 1968 *Angular Momentum*, 2nd edn (London: Oxford University Press)
Brinkman, H.C. 1956 *Applications of Spinor Invariants in Atomic Physics* (Amsterdam: North-Holland)
Brueckner, K.A. 1955 *Phys. Rev.* **100** 36–45
Buckingham, A.D., Graham, C. and Raab, R.E. 1971 *Chem. Phys. Lett* **8** 622–4
Buckingham, A.D. and Raab, R.E. 1975 *Proc. R. Soc. Lond. A* **345** 365–78
Buckingham, A.D. and Shatwell, R.A. 1978 *Chem. Phys.* **35** 353–4
Buckingham, A.D. and Shatwell, R.A. 1980 *Phys. Rev. Lett.* **45** 21–3
Burgos, E. and Bonadeo, H. 1987 *Molec. Phys.* **60** 1415–24
Butera, P., Cicuta, G.M. and Enriotti, M. 1980 *Phys. Rev. D* **21** 972–8
Butler, P.H. 1975 *Phil. Trans. R. Soc. A* **277** 545–85
Butler, P.H. 1981 *Point Group Applications, Methods and Tables* (New York: Plenum)
Callen, H.B. 1961 in *Fluctuation, Relaxation and Resonance in Magnetic Systems*, ed. D. ter Haar (Edinburgh: Oliver & Boyd)
Canning, G.P. 1978 *Phys. Rev. D* **18** 395–410
Cheng-tian Feng and Judd, B.R. 1982 *J. Phys. A: Math., Gen.* **15** 2273–84
Churcher, C.D. 1982 *Molec. Phys.* **46** 621–8
Churcher, C.D. and Stedman, G.E. 1981a *J. Phys. C* **14** 2237–64
Churcher, C.D. and Stedman, G.E., 1981b *J. Phys. C* **14** 5707–23
Churcher, C.D. and Stedman, G.E. 1982 *J. Phys. C* **15** 5507–20
Churcher, C.D. and Stedman, G.E. 1983 *J. Phys. B* **16** 1303–38
Cohen-Tannoudji, C. and Reynaud, S. 1977 *J. Phys B* **10** 345–63, 365–83

Coleman, A.J. 1968 *Adv. Quantum Chem.* **4** 83–108
Condon, E.U. and Shortley, G.H. 1935. *The Theory of Atomic Spectra* (Cambridge University Press)
Conzett, H., Goldstein, G.R. and Moravcsik, M.J. 1985 *Phys. Rev. Lett.* **54** 604–7
Cooper, F. and Freedman, B. 1983 *Ann. Phys.* **146** 262–88
Cornwell, J.F. 1984 *Group Theory in Physics*, vols. 1 & 2 (London: Academic)
Craig, D.P. and Thirunamachandran, T. 1983 *Adv. Quantum Chem.* **16** 97–160
Craig, D.P. and Thirunamachandran, T. 1987 *Proc. R. Soc. Lond. A* **410** 337–51
Craigie, N.S., Hidaka, K., Jacob, M. and Renard, F.M. 1983 *Phys. Reports* **99** 104–236
Creutz, M. 1983 *Quarks, Gluons and Lattices* (Cambridge University Press)
Crichton, J.H. and Yu, S. 1973 *Ann. Phys.* **75**, 77–102
Cvitanovic, P. 1976 *Phys. Rev. D* **14** 1536–53
Cvitanovic, P. 1983 *Field Theory* (Copenhagen: NORDITA)
Cvitanovic, P. 1984 *Group Theory Part I* (Copenhagen: NORDITA)
Cvitanovic, P. and Kennedy, A.D. 1982 *Phys. Scripta* **26** 5–14
Cvitanovic, P., Lautrup, B. and Pearson, R.B. 1978 *Phys. Rev. D* **18** 1939–40
Cvitanovic, P., Lauwers, P.G. and Scharbach, P.N. 1981 *Nucl. Phys. B* **186** 165–86
Damhus, T., Harnung, S.E. and Schaffer, C.E. 1984 *Theor. Chim. Acta* **65** 317–447
Danos, M. 1971 *Ann. Phys. N.Y.* **63** 319–34
de Figueiredo, I.M.B. and Raab, R.E. 1980 *Proc. R. Soc. Lond. A* **369** 501–16
de Figueiredo, I.M.B. and Raab, R.E. 1981 *Proc. R. Soc. Lond. A* **375** 425–41
Derome, R. and Sharp, W.T. 1965 *J. Math. Phys.* **6** 1584–90
Dirac, P.A.M. 1983 *Principles of Quantum Mechanics*, 4th edn., revised (Oxford: Clarendon)
Dolgov, A.D., Khriplovich, I.B. and Zakharov, V.I. 1987 *Pisma Zh. Eksp. Teor. Fiz.* **45** 511; 1988 *Nucl. Phys., B* **309** 591–6
Doniach, S. and Sondheimer, E.H. 1974 *Green's Functions for Solid State Physicists* (Reading, MA: Benjamin), p. 232
Downer, M.C. and Bivas, A. 1984 *Phys. Rev. B* **28** 3677–96
ElBaz, E. 1981 *Nuovo Cim. A* **63** 257–320
ElBaz, E. 1985 *J. Math. Phys.* **26** 728–31
ElBaz, E. 1986 *J. Math. Phys.* **27** 53–60
ElBaz, E. and Castel, B. 1971 *Am. J. Phys.* **39** 868–76
ElBaz, E. and Castel, B. 1972 *Graphical Methods of Spin Algebras in Atomic, Nuclear and Particle Physics* (New York: Dekker)
ElBaz, E., Massot, J.N. and Lafoucriere, J. 1966 *Nucl. Phys.* **12** 189–208
ElBaz E. and Nahabetian, R. 1977 *J. Phys. A* **10** 1063–77
Elliott, J.P. and Dawber, P.G. 1979 *Symmetry in Physics*, vols 1, 2 (London: Macmillan)
Englman, R. 1972 *The Jahn–Teller Effect in Molecules and Crystals* (New York: Wiley)
Fano, U. and Racah, G. 1959 *Irreducible Tensorial Sets* (New York: Academic)
Fetter, A.L. and Walecka, J.D. 1971 *Quantum Theory of Many-particle Systems* (New York: Wiley)
Feynman, R.P. 1961 *Theory of Fundamental Processes* (New York: Benjamin), p. 162
Feynman, R.P. 1985 *QED; the strange theory of light and matter* (Princeton University Press), pp. 3ff

Fletcher, J.R. and Pooler, D.R. 1982 *J. Phys. C* **15** 2695–707
Friedberg, R. and Hartmann, S.R. 1988 *J. Phys. B* **21** 683–712
Fujimoto, Y., Morikawa, M. and Sasuki, M. 1986 *Phys. Rev. D* **33** 590–3
Gaudin, M. 1960 *Nucl. Phys.* **15** 89–91
Gauthier, N. and Walker, M.B. 1976 *Can. J. Phys.* **54** 9–25
Gilmore, R. 1974 *Lie Groups, Lie Algebras and Some of their Applications* (New York: Wiley)
Goldstein, G.R. and Moravcsik, M.J. 1984a *Phys. Rev. D* **30** 55–62
Goldstein, G.R. and Moravcsik, M.J. 1984b *Phys. Rev. Lett.* **53** 1885–92
Graham, C. 1980 *Proc. R. Soc. Lond. A* **369** 517–35
Grange, P. and Richert, J. 1976 *Z. Phys. A* **277** 249–59
Grewe, N. and Keiter, H. 1981 *Phys. Rev. B* **24** 4420–44
Griffith, J.S. 1962 *The Irreducible Tensor Method for Molecular Symmetry Groups* (Englewood Cliffs, N.J: Prentice-Hall)
Gruber, B. and O'Raifeartaigh, L. 1964 *J. Math. Phys.* **5** 1796–804
Haase, R.W. and Butler, P.H. 1984 *J. Phys. A* **17** 47–74
Haase, R.W. and Butler, P.H. 1986 *J. Phys. A* **19** 1065–82
Ham, F.S. 1972 in *Electron Paramagnetic Resonance*, ed. S. Geschwind (New York: Plenum)
Hamer, C.J., Irving, A.C. and Preece, T. 1986 *Nucl. Phys. B* **270** 553–74
Hamermesh, M. 1962 *Group Theory and its Applications* (Reading, MA.: Addison-Wesley)
Harary, F. 1969 *Graph Theory* (Reading, MA: Academic)
Heine, V. 1960 *Group Theory and Quantum Mechanics* (Oxford: Pergamon)
Horejsi, J. 1986 *J. Phys. G* **12** L7–10
Huang, K. 1982 *Quarks, Leptons and Gauge Fields* (Singapore: World Scientific), p. 235
Huang, K-N. 1979 *Rev. Mod. Phys.* **51** 215–36
Huang, K-N. and Starace, A.F. 1978 *Phys. Rev. A* **18** 354–79
Hugenholtz, N.M. 1965 *Rep. Progr. Phys.* **28** 201–47
Jackiw, R. 1985 *Making Sense out of Anomalous Gauge Theories* (Preprint CTP#1300, MIT)
Jackiw, R. 1986 *Helv. Phys. Acta* **59** 835–43
Jackson, J.D. 1975 *Classical Electrodynamics*, 2nd edn (New York: Wiley), p. 682
Jarvis, P.D. and Stedman, G.E. 1984 *J. Phys. A* **17** 757–76
Jauch, J.M. and Rohrlich, R.F. 1976 *The Theory of Photons and Electrons* (New York: Springer)
Jerphagnon, J., Chelma, D. and Bonneville, R. 1978 *Adv. Phys.* **27** 609–50
Jucys, A.P. and Bandzaitas, A.A. 1964 *The Theory of Angular Momentum in Quantum Mechanics* (Vilnius: Institute for Physics and Mathematics of the Academy of Science of Lithuanian SSR) (in Lithuanian)
Jucys, A.P., Levinson, I.B. and Vanagas, V.V. 1960 *Mathematical Apparatus of the Theory of Angular Momentum* (Vilnius: Institute for Physics and Mathematics of the Academy of Science of Lithuanian SSR) (in Lithuanian); see also the English translation, Yutsis, A.P. *et al.* (1962)
Judd, B.R. 1962 *Phys. Rev.* **127** 750–61
Judd, B.R. 1967 *Second Quantization and Atomic Spectroscopy* (Baltimore: Johns Hopkins University)
Judd, B.R. 1974 *Can. J. Phys.* **52** 999–1044

Judd, B.R. 1983a in *The Dynamical Jahn–Teller Effect for Localised Systems*, eds. Yu E. Perlin and M. Wagner (Amsterdam: North-Holland)
Judd, B.R. 1983b *Found. Phys.* **13** 51–9
Judd, B.R. 1985 *Reports Progr. Phys.* **48** 907–53
Judd, B.R. 1987a *J. Phys. C* **14** 375–84
Judd, B.R. 1987b *J. Phys. C* **20** L903–6
Judd, B.R., Lister, G.M.S. and Suskin, M.A. 1986 *J. Phys. B: Atom. Molec.* **19** 1107–14
Judd, B.R., Miller, W., Patera, J. and Winternitz, P. 1974 *J. Math. Phys.* **15** 1787–99
Judd, B.R. and Pooler, D.R. 1982 *J. Phys. C* **15** 591–8
Kawabata, A. 1971 *J. Phys. Soc. Japan* **30** 68–85
Kennedy, A.D. 1982 *Phys. Rev. D* **26** 1936–55
Kibler, M. and Elbaz, E. 1979 *Int. J. Quantum Chem.* **16** 1161–94
Killingbeck, J. 1970 *Reports Progr. Phys.* **33** 533–644
Kittel, C. 1963 *Quantum Theory of Solids* (New York: Wiley), p. 157
Klein, O. and Nishina, Y. 1929 *Z. Phys.* **52** 853–68
Koster, G.F., Dimmock, J.O., Wheeler, R.G. and Statz, H. 1963 *Properties of the 32 Point Groups* (Cambridge, MA: MIT)
Koutecky, J., Paldus, J. and Čižek, J. 1985 *J. Chem. Phys.* **83** 1722–35
Kubo, R. 1966 *Reports Progr. Phys.* **29** 255–84
Kuo, T.T.S., Shurpin, J., Tam, K.C., Osnes, E. and Ellis, P.J. 1981 *Ann. Phys.* **132** 237–76
Kuramoto, Y. 1983 *Z. Phys. B* **53** 37–52
Kwok, P.C. and Schultz, T.D. 1969 *J. Phys. C* **2** 1196–205
Lagrange, J-L. 1965 *Mécanique Analytique* (Paris: Blanchard)
Lamb, W.E., Schlicher, R.R. and Scully, M.O. 1987 *Phys. Rev. A* **36** 2763–72
Langhoff, P.W., Epstein, S.T. and Karplus, M. 1984 *Rev. Mod. Phys.* **44** 602–44
Lawson, R.D. and Macfarlane, M.H. 1965 *Nuclear Phys.* **66** 80–96
Lax, M. 1974 *Symmetry Principles in Solid State and Molecular Physics* (New York: Wiley)
Leader, E. and Predazzi, E. 1982 *An Introduction to Gauge Theories and the New Physics* (Cambridge University Press)
Leech, J.W. and Newman, D.J. 1969 *How to Use Groups* (London: Methuen)
Leggett, A.J. 1982 *Physica B* **109–10** 1393–403
Leung, C.H. and Kleiner, W. 1974 *Phys. Rev. B* **10** 4434–46
Lindgren, I. and Morrison, J. 1982 *Atomic Many-Body Theory*, Springer series in chemical sciences #13 (Berlin: Springer-Verlag)
Livanova, A. 1980 *Landau: A Great Physicist and Teacher*, trans. J.B. Sykes (Oxford: Pergamon)
Llewellyn Smith, C.H. 1979 *Topics in QCD*. 1979 Boulder Institute Lectures (New York: Plenum)
Lulek, T. 1975 *Acta Phys. Polonica A* **48** 513–27
Luypaert, R. and van Craen, J. 1977 *J. Phys. B* **10** 3627–36
Luypaert, R. and van Craen, J. 1979 *J. Phys. B* **12** 2613–23
McClain, W.M. 1971 *J. Chem. Phys.* **55** 2789–96
McClain, W.M. and Harris, R.A. 1977 *Excited States*, vol. 3, ed. E.C. Lim (New York: Academic), pp. 1–56
McClain, W.M. and Harris, R.A. 1983 *J. Chem. Phys.* **79** 3689–93

McKenzie, B.J. and Stedman, G.E. 1976 *J. Phys. A* **9** 187–95
McKenzie, B.J. and Stedman, G.E. 1978 *J. Phys. C* **11** 589–603
McKenzie, B.J. and Stedman, G.E. 1979 *J. Phys. C* **12** 5061–75
Maddox, J. 1988 *Nature* **331** 653
Manson, N.B., Newman, D.J. and Wong, K.Y. 1977 *J. Phys. C* **10** 4619–29
Martin, D.H. 1967 *Magnetism in Solids* (London: Iliffe)
Mason, S.T. and Starace, A.F. 1982 *Rev. Mod. Phys.* **54** 389–405
Mattuck, R.D. 1967 *A Guide to Feynman Diagrams in the Many-Body Problem* (New York: McGraw-Hill)
Mattuck, R.D. and Theumann, A. 1971 *Adv. Phys.* **20** 721–45
Merzbacher, E. 1961 *Quantum Mechanics* (New York: Wiley), ch. 21
Messiah, A. 1962 *Quantum Mechanics*, vol. 2 (Amsterdam: North-Holland)
Minard, R.A., Stedman, G.E. and McLellan, A.G. 1983 *J. Chem. Phys.* **78** 5016
Moussouris, J.P. 1979 in *Advances in Twistor Theory*, eds. L.P. Hughston and R.S. Ward, Research Notes in Mathematics # 37 (London: Pitman)
Mukhopadhyay, A. 1984 *Int. J. Quantum Chem.* **25** 965–1002
Mukhopadhyay, A. and Pickup, B.T. 1984 *Int. J. Quantum Chem.* **26** 125–43
Nencka-Ficek, H. and Lulek, T. 1982 *Physica* **114A** 66–7
Newman, D.J. 1971 *Adv. Phys.* **20** 197–256
Newman, D.J. and Balasubramanian, G. 1975 *J. Phys. C* **8** 37–44
Newman, D.J. and Wallis, J. 1976 *J. Phys. A* **9** 2021–33
Ng, B. and Newman, D.J. 1986 *J. Phys. C* **19** L585–8
Nienhuis, G. 1984 *J. Phys. B* **17** 587–604
O'Brien, M.C.M. 1969 *Phys. Rev.* **187** 407–18
Ofelt, G. 1962 *J. Chem. Phys.* **37** 511–20
Ojima, I. 1981 *Ann. Phys.* **137** 1–32
Paldus, J., Adams, B.G. and Čížek, J. 1977 *Int. J. Quantum Chem.* **11** 813–48
Paldus, J. and Čížek, J. 1975 *Adv. Quantum Chem.*, ed. P-O, Lowdin, vol. 9 (New York: Academic), p. 107
Parikh, J.C. 1978 *Group Symmetries in Nuclear Structure* (New York: Plenum)
Payne, S.H. and Stedman, G.E. 1977 *J. Phys. C* **10** 1549–59
Payne, S.H. and Stedman, G.E. 1983a *J. Phys. C* **16** 2679–2704
Payne, S.H. and Stedman, G.E. 1983b *J. Phys. C* **16** 2705–2723
Payne, S.H. and Stedman, G.E. 1983c *J. Phys. C* **16** 2725–48
Penrose, R. 1971 in *Combinatorial Mathematics and its Applications*, ed. J.A. Welsh (New York: Academic), p. 234
Pickup, B.T. and Mukhopadhyay, A. 1984 *Int. J. Quant. Chem.* **26** 101–23
Piepho, S. and Schatz, P.N. 1983 *Group Theory in Spectroscopy with Applications to MCD* (New York: Wiley)
Placzek, G. 1962 *The Rayleigh and Raman Scattering.* UCRL Trans 526(1) (Livermore: University of California)
Pooler, D.R. and O'Brien, M.C.M. 1977 *J. Phys. C.* **10** 3769–97
Prakash, J.S. 1980 *J. Phys. A* **13** 3347–55
Psaltakis, G.C. and Cottam, M.G. 1980 *J. Phys. C* **13** 6009–23
Psaltakis, G.C. and Cottam, M.G. 1981 *Phys. Status Solidi B* **103** 709–16
Raghavacharyulu, I.V.V. 1973 *J. Phys. C* **6** L455–7
Reid, M.F. 1984 *J. Phys. A* **17** 1755–9
Reid, M.F. 1988 *J. Phys. Chem. Solids* **49** 185–9

Reid, M.F. and Butler, P.H. 1982 *J. Phys. A* **15** 2327–35
Reid, M.F. and Richardson, F.S. 1983 *J. Chem. Phys.* **79** 5735–42
Reid, M.F. and Richardson, F.S. 1984 *J. Phys. Chem.* **88** 3579–86
Riddell, A.G. and Stedman, G.E. 1984 *Phys. Rev. A* **30** 1727–33
Rosensteel, G., Ihrig, E. and Trainor, L.E.H. 1975 *Proc. R. Soc. Lond. A* **344** 387–401
Ross, G.G. 1986 *Grand Unified Theories* (New York: Benjamin/Cummings)
Ross, H.J., Sherborne, B.S. and Stedman, G.E. 1989 *J. Phys. B* **22** 459–73
Rotenberg, M., Bivins, R., Metropolis, N. and Wooten, J.K. 1959 *The 3j and 6j symbols* (Cambridge, MA: Technology Press)
Rudzikas, Z. and Kaniauskas, J. 1984 *Quasispin and Isospin in Atomic Theory* (Vilnius: Mokslas)
Running, T. 1927 *Graphical Mathematics* (New York: Wiley)
Sakurai, J.J. 1967 *Advanced Quantum Mechanics* (Reading, MA: Addison-Wesley)
Salam, A. 1988 in *Superstrings – a theory of everything?* eds. P.C.W. Davies & J. Brown (Cambridge University Press) p. 179
Salam, A. and Strathdee, J. 1975 *Phys. Rev. D* **11** 1521–35
Sandars, P.G.H. 1969a in *Lectures in Theoretical Physics, Brandeis University Summer Institute*, Vol. 1, eds. M. Cretien and E. Lipworth (New York: Gordon and Breach), pp. 171–216
Sandars, P.G.H. 1969b *Adv. Chem. Phys.* **14** 365–419
Sandars, P.G.H. 1977 *J. Phys. B* **10** 2983–95
Saunders, R.W. and Young, W. 1980 *J. Phys. C* **13** 103–20
Scadron, M.D. 1979 *Advanced Quantum Mechanics and its Applications* (New York: Springer), p. 218
Schrieffer, J.R. 1964 *Theory of Superconductivity* (New York: Benjamin)
Schrieffer, J.R. 1973 *Science* **180** 1243–8
Schultz, T.D. 1964 *Quantum Field Theory and the Many Body Problem* (London: Gordon & Breach)
Schweber, S.S. 1986 *Rev. Mod. Phys.* **58** 449–508
Sharp, W.T. 1984 *Racah Algebra and the Contractions of Groups* (Department of Mathematics, University of Toronto)
Shen, Y.R. 1984 *Principles of Non-linear Optics* (New York: Wiley)
Shore, B.W. and Menzel, D.H. 1968 *Principles of Atomic Spectra* (New York: Wiley)
Silverman, M.P. 1986 *J. Opt. Soc. Am.* *A3* 830–7
Silverstone, H.J. and Holloway, T.T. 1971 *Phys. Rev. A* **4** 2191–9
Skinner, R. and Weil, J.A. 1978 *J. Mag. Res.* 223–41
Soliverez, C.E. 1967 *AM. J. Phys.* **35** 624–7
Stedman, G.E. 1968 *Contemp. Phys.* **9** 49–69
Stedman, G.E. 1971a *J. Phys. C* **4** 1022–35
Stedman, G.E. 1971b *Am. J. Phys.* **39** 205–14
Stedman, G.E. 1972 *J. Phys. C* **5** 121–33
Stedman, G.E. 1974 *J. Phys. A* **7** L48–51
Stedman, G.E. 1975 *J. Phys. A* **8** 1021–35
Stedman, G.E. 1976a *J. Phys. A* **9** 1999–2019
Stedman, G.E. 1976b *J. Phys. C* **9** 535–51
Stedman, G.E. 1978 *J. Phys. C.* **11** 1017–30
Stedman, G.E. 1983a *Am. J. Phys.* **51** 750–5
Stedman, G.E. 1983b *Eur. J. Phys.* **4** 156–61
Stedman, G.E. 1985 *Adv. Phys.* **34** 513–87

Stedman, G.E. 1987 *J. Phys. A* **20** 2629–43
Stedman, G.E. and Bilger, H.R. 1987 *Phys. Lett. A* **122** 289–91
Stedman, G.E. and Butler, P.H. 1980 *J. Phys. A* **13** 3125–40
Stedman, G.E. and Cade, N.A. 1973 *J. Phys. C* **6** 474–8
Stedman, G.E. and Kaiser, A.B. 1987 *J. Phys. C* **20** 3943–51
Stedman, G.E. and Pooke, D.M. 1982 *Phys. Rev. D* **26** 2172–4
Stevens, K.W.H. and Toombs, G.A. 1965 *Proc. Phys. Soc.* **85** 1301–2
Sturge, M.D. 1967 in *Solid State Physics*, eds. F. Seitz, D. Turnbull and H. Ehrenreich, vol. 20 (New York: Academic), p. 91.
Suzuki, K. and Okamoto, R. 1983 *Progr. Th. Phys. Japan* **70** 439–51
Sztucki, J. and Strek, W. 1986 *Phys. Rev. B* **34** 3120–6
Taylor, P.L. 1970 *A Quantum Approach to the Solid State* (Englewoods, NJ: Prentice-Hall), ch. 7
Tinkham, M. 1964 *Group Theory and Quantum Mechanics* (New York: McGraw-Hill)
Tomalchev, V.V. 1969 *Adv. Chem. Phys.* **14** 421–520
Valentini, A. 1988 *Phys. Rev. Lett.* **61** 1903–5
van der Bij, J. and Veltman, M. 1984 *Nucl. Phys. B* **231** 205–34
van Zanten, A.J. and de Vries, E. 1973 *J. Math. Phys.* **14** 1423–33
Watanabe, H. 1966 *Operator Methods in Ligand Field Theory* (Englewoods, NJ: Prentice-Hall), pp. 63ff
Weyl, H. 1931 *The Theory of Groups and Quantum Mechanics*, trans. H.P. Robertson (London: Methuen)
Widom, A., Friedman, M.H., Srivastava, Y. and Feinberg, A. 1985 *Phys. Lett.* **108** 377–9
Wigner, E.P. 1959 *Group Theory*, (trans J.J. Griffin) (New York: Academic), pp. 332–48
Wilson, S. 1984 *Electronic Correlation in Molecules* (Oxford: Clarendon)
Wilson, S. and Gerratt, J. 1979 *J. Phys. B* **12** 339–44
Wise, M.B. and Trainor, L.E.H. 1977 *Can. J. Phys.* **55** 994–1004
Wulfman, C.E. 1986 *Chem. Phys. Lett.* **127** 26–32
Wybourne, B.G. 1965 *Spectroscopic Properties of Rare Earths* (New York: Interscience)
Wybourne, B.G. 1970 *Symmetry Principles and Atomic Spectroscopy* (New York: Wiley)
Wybourne, B.G. 1973 *Int. J. Quantum Chem.* **7** 1117–37
Wybourne, B.G. 1974 *Classical Groups for Physicists* (New York: Wiley)
Yang, D. Hsieh-Yen and Wang, Y-L. 1974 *Phys. Rev. B* **10** 4714–23
Yeung, Y.Y. and Newman, D.J. 1986 *J. Chem. Phys.* **86** 6717–21
Yuratich, M.A. and Hanna, D.C. 1976a *J. Phys. B* **9** 729–50
Yuratich, M.A. and Hanna, D.C. 1976b *Optics Commun.* **18** 134
Yuratich, M.A. and Hanna, D.C. 1977 *Molec. Phys.* **33** 671–82
Yutsis, A.P., Levinson, I.B. and Vanagas, V.V. 1962 *Mathematical Apparatus of the Theory of Angular Momentum* (Jerusalem: Israel Program of Scientific Translations); see Jucys *et al.* (1960)
Ziman, J.M. 1960 *Electrons and Phonons* (Oxford University Press)
Ziman, J.M. 1985 in *Information Sources in Physics*, ed. D.F. Shaw (London: Butterworth)

Index of first named authors

General Index